Topics in Applied Physics Volume 55

Topics in Applied Physics Founded by Helmut K. V. Lotsch

The Physics of
Hydrogenated
Amorphous Silicon I

Structure, Preparation, and Devices

Edited by J. D. Joannopoulos and G. Lucovsky

With Contributions by D. E. Carlson
J. D. Joannopoulos   D. Kaplan   J. C. Knights
P. G. LeComber   G. Lucovsky   A. Madan
W. E. Spear   M. J. Thompson

With 209 Figures

Springer-Verlag Berlin Heidelberg GmbH 1984

Professor *John D. Joannopoulos*, PhD

Massachusetts Institute of Technology, Department of Physics
Cambridge, MA 02139 USA

Professor *Gerald Lucovsky*, PhD

North Carolina State University, Department of Physics
Raleigh, NC 27607, USA

ISBN 978-3-662-30858-5   ISBN 978-3-540-38741-1 (eBook)
DOI 10.1007/978-3-540-38741-1

Library of Congress Cataloging in Publication Data. Main entry under title: The Physics of hydrogenated amorphous silicon. (Topics in applied physics; v. 55–56) Contents: 1. Structure, preparation, and devices – 2. Electronic and vibrational properties. 1. Silicon. I. Joannopoulos, J. D. (John D.), 1947–. II. Lucovsky, G. III. Series. QC611.8.S5P49 1983 537.6′22 83-16732

Originally published by Springer-Verlag Berlin Heidelberg New York Tokyo in 1984
Softcover reprint of the hardcover 1st edition 1984

Typesetting: Schwetzinger Verlagsdruckerei, Schwetzingen

2153/3130-543210

Preface

During the past several years there has grown an enormous experimental and theoretical activity associated with amorphous silicon and its alloys. This is based on the exciting possibilities emerging from the doping of hydrogenated amorphous silicon. Experimental and theoretical efforts have been directed at obtaining an understanding of the underlying physics of a variety of interesting and unusual phenomena associated with this material. In addition, major effort has also been expended towards the exploration of the technological consequences of these phenomena. This two part series presents a broad, as well as in-depth, overview of the entire field of amorphous silicon and its alloys. At the present, sufficient progress and understanding exist that such volumes should be useful and timely. Briefly, the present Volume I concentrates on structure, preparation techniques, and device applications. Volume II concentrates on theoretical and experimental investigations of a variety of electronic and vibrational phenomena. Each chapter is written as a critical review with a conscious effort to separate fact and interpretation. In addition, the contributions represent constructive reviews that help define future directions of research whenever possible. The chapters are written at the level of a graduate student which should be helpful to both students and scientists who are not experts in this area. Finally, in an attempt to add an archival flavor to the reviews, many representative results and comprehensive lists of citations are presented.

Cambridge, Raleigh
December 1983

J. D. Joannopoulos · G. Lucovsky

Contents

Contributors

Carlson, David E.
    RCA Laboratories, Princeton, NJ 08540, USA

Joannopoulos, John D.
    Department of Physics, Massachusetts Institute of Technology,
    Cambridge, MA 02139, USA

Kaplan, Daniel
    Laboratoire Central de Recherches, Thomson-CSF, F-91401 Orsay,
    France

Knights, John C.
    Xerox Palo Alto Research Center, 3333 Coyote Hill Road,
    Palo Alto, CA 94304, USA

LeComber, Peter G.
    Carnegie Laboratory of Physics, University of Dundee,
    Dundee, Scotland

Lucovsky, Gerald
    Department of Physics, North Carolina State University,
    Raleigh, NC 27607, USA

Madan, Arun
    Solar Energy Research Institute, 1617 Cole Boulevard,
    Golden, CO 80401, USA

Spear, Walter E.
    Carnegie Laboratory of Physics, University of Dundee,
    Dundee, Scotland

Thompson, Malcolm J.
    Xerox Palo Alto Research Center, 3333 Coyote Hill Road,
    Palo Alto, CA 94304, USA

1. Introduction

John D. Joannopoulos and Gerald Lucovsky

For many years it had been thought to be a basic and fundamental fact that an amorphous semiconductor could not be substitutionally doped. Mott had explained, in a simple and intuitive way, that the absence of periodic steric constraints in a disordered system could allow the normal valency of any impurity atom to be completely satisfied. Thus the usual mechanism for introducing extra electrons or holes into a solid could be tremendously impeded.

Nevertheless, in 1975, W. E. Spear and P. G. LeComber demonstrated a new amorphous silicon material which could be doped. It could be doped both n-type and p-type and its room temperature conductivity could be made to vary over more than ten orders of magnitude. It was discovered that a very important feature of this material was that it contained significant amounts of bonded hydrogen. The role of hydrogen in the atomic structure, the electronic structure, and the doping mechanism remains an outstanding problem to be addressed by both theoretical and experimental investigations. Nonetheless, doping of the material opened up possibilities for the fabrication of thin-film electronic devices, including photovoltaic solar cells. These exciting prospects have led to a recent explosion of experimental and theoretical work on hydrogenated amorphous silicon and other closely related materials. These include investigations of the structural, electronic, vibrational, optical, luminescent, magnetic, transport, and photoconductive properties of these materials. The investigations to date have revealed a wealth of interesting and unusual physical and chemical phenomena.

This volume is part I of a two part series presenting an overview of the physics that underlies a virtual industry associated with the study of amorphous silicon and related alloys. Volume I concentrates on the structure, preparation techniques, and device applications of hydrogenated amorphous silicon. Specifically, Chapter 2 by J. C. Knights presents an in-depth discussion of the structural and chemical characterization of the material. Chapters 3, 4, and 5 (by W. E. Spear and P. G. LeComber, M. J. Thompson, and D. Kaplan) describe, respectively, the three most important techniques for depositing thin films of doped amorphous silicon. These include glow-discharge decomposition of silane in the presence of dopant gases, the reactive sputtering of silicon and the chemical vapor deposition (CVD) of silicon followed by post-deposition hydrogenation. These three chapters include detailed discussions of the fundamental and technological aspects of the deposition processes, as well as discussions of some of the important properties of the resultant films. Finally, Chapters 6 by D. E. Carlson and 7 by A. Madan present a thorough overview of the recent applied work that underlies the fabrication of solar cells and other important electronic devices. These chapters draw heavily on the material discussed in the preceding chapters, and also serve as an introduction to the material that follows in the second volume [1.1].

Volume II focuses on the presentation of recent fundamental theoretical and experimental investigations of electronic, magnetic, transport, vibrational, and localization phenomena associated with hydrogenated amorphous silicon and related alloys. The material is complementary to the material presented in this volume. The most important theme connecting the material in the two volumes is the intimate relationship that exists between the deposition technique and the resulting physical properties. This is a manifestation of the non-equilibrium character of the deposition technologies and the metastable nature of the resultant material.

The information contained in both volumes should prove to be quite useful for both scientists and technologists who are working at the frontiers of this field of disordered materials.

Abbreviations Frequently Used in the Text

a-Si	amorphous silicon
c-Si	crystalline silicon
μc-Si	microcrystalline silicon
a-Si : H	hydrogenated amorphous silicon
a-Si : F	fluorinated amorphous silicon
CVD	chemical vapor deposition
PES	photoemission spectroscopy
XPS	x-ray photoemission spectroscopy
UPS	ultraviolet photoemission spectroscopy
EDC	energy distribution curve
DOS	density of states
CRN	continuous random network
RDF	radial distribution function
SCL	space charge limited
SCR	space charge region
ESR	electron spin resonance
LESR	light induced electron spin resonance
NMR	nuclear magnetic resonance
ODMR	optically detected magnetic resonance

PL	photoluminescence
DLTS	deep level transient spectroscopy
EXAFS	extended x-ray absorption fine structure
PDS	photothermal deflection spectroscopy
SIMS	secondary ion mass spectroscopy
GDOS	glow discharge optical spectroscopy
LEED	low energy electron diffraction
SEM	scanning electron microscopy
TEM	transmission electron microscopy
MIS	metal-insulator-semiconductor
MOS	metal-oxide-semiconductor
MS	metal-semiconductor
FET	field effect transistor
FF	fill factor
TCO	transparent conductive oxide
ITO	indium-tin-oxide
SB	Schottky barrier
TSC	thermally stimulated current

Reference

1.1 J. D. Joannopoulos, G. Lucovsky (eds.): *The Physics of Hydrogenated Amorphous Silicon II,* Topics Appl. Phys., Vol. 56 (Springer, Berlin, Heidelberg, New York, Tokyo 1984)

2. Structural and Chemical Characterization

John C. Knights

With 44 Figures

At an undergraduate physics level, crystalline semiconductor physics is the physics of perfection – infinite regular lattices, Bloch waves and symmetry on all sides. In the real world of current-day semiconductor research, however, most of the problems and activities are associated with understanding and controlling deviations from that perfection – bulk defects, surfaces and interfaces. From a conceptual standpoint the situation for amorphous semiconductors is very similar; we have passed the point of understanding the major properties of a random network of covalently bonded atoms and are primarily concerned with understanding the effects of deviation from that structure. From an empirical and primarily electronic viewpoint, these deviations can be easily identified; they are clearly associated with variations in material preparation technique or with subsequent treatment such as annealing or particle bombardment, i.e., they are extrinsic, and they produce profound effects on such measurables as charge transport, photoconductivity, luminescence, etc. From a structural standpoint, however, both conceptually and practically these deviations are more difficult to identify and categorize. When does a distortion in a random network become a defect? Is there the equivalent of a dislocation in an amorphous solid? In the specific case of hydrogenated amorphous silicon, these problems are compounded by a number of features.

First, the material can only be prepared in thin film form – maximum reported thicknesses are typically in the 10–100 μm range, thus limiting the types and accuracy of measurement probes that can be applied. Second, the material is compositionally unstable above its deposition temperature (~ 200–300 °C) restricting the use of such measurements as atomic diffusion, defect annealing, etc. Finally, material produced by most of the currently used deposition processes typically contains impurities such as oxygen, carbon and nitrogen in concentrations often above their solubility limits in crystalline silicon. This makes the unambiguous identification of structural defects from electronic measurements difficult because of the whole spectrum of impurity-related phenomena that might, as is known in the case of crystalline silicon, be present.

Despite these problems, considerable progress has been made in understanding the nature of the structural defects and in the chemical characterization of hydrogenated amorphous silicon. It is the goal of this chapter to critically review this progress and point to areas where future progress might be made.

2.1 Historical Perspective

Before starting the review, it is helpful to place the work on structural and chemical characterization of a-Si : H in some historical perspective. The current wave of activity in this field dates back to the work of *Chittick* [2.1] and coworkers at the Standard Telecommunications Laboratories in Harlow, England in 1968–1970. During a broad study of materials produced by glow-discharge decomposition of gases, they discovered that when silane or germane were used by themselves (or with a dopant phosphine in the case of germane), amorphous materials with properties quite different from sputtered or evaporated amorphous silicon and germanium were produced. While the work at STL was not continued, *Spear* and *LeComber* at the University of Dundee, noting these interesting differences, started a research program to more fully characterize the materials' electronic properties. In the case of the work with silane, this led to the landmark paper [2.2] of 1975 in which the successful *n* and *p*-type doping of the material was reported. Somewhat prior to this, the group at Harvard headed by *Paul* reported that the addition of hydrogen [2.3] to the sputtering gas during the preparation of amorphous silicon produced material with electronic properties similar to those of (undoped) material produced by glow discharge decomposition of silane. Close on the heels of these reports came confirmation that the materials prepared from glow discharge decomposition of silane contained hydrogen [2.4, 5]. It was at this point, in 1976, that widespread efforts to structurally and chemically characterize the material began, almost six years after the work on electronic properties had commenced. This delay can be attributed in hindsight to a number of factors: the interests and backgrounds of the early workers in the field, the fact that the electronic properties were so fascinating in their own right, and an early report [2.6] of transmission electron microscopy on a-Si : H that indicated essentially featureless amorphous films at the 10–100 Å scale. More than anything else, however, was the failure to look back in the literature where chemists ranging from *Ogier* [2.7] (1880) and *Emeleus* and *Stewart* [2.8] (1945) to *Jolly* and coworkers [2.9] (1950–65) had published evidence of glow-discharge production of silicon subhydrides that would have compelled an early study of the hydrogen issue. This "johnnie-come-lately" position of structural and chemical characterization combined with the difficulties mentioned earlier has led to a separation from the electronic characterization that exists to this day. There have been only limited attempts to correlate specific electronic and structural (or chemical) properties and those working on electronic properties have been able to proceed with empirical material optimization with scant, if any, regard for the developments in structural and chemical characterization.

It is my opinion that this era of separation is coming to an end for a number of reasons. First, the empirical optimization of electronic properties has become increasingly more difficult as device structures for, e.g., solar cells become more complex, thus a better understanding of the growth process and its effects via structure and chemistry on the electronic properties has become imperative. Second, as more device-related work is done, it is clear that reproducibility of results becomes a much more important issue and the precise control of dopant profiles and the content of impurities such as oxygen and nitrogen is being recognized as an essential element in achieving such reproducibility. Finally, the "onion" that is the complete understanding of amorphous hydrogenated silicon has had several layers peeled relatively easily; the next layer that contains, for example, a much more sophisticated understanding of defect chemistry will resist anything but a unified attack.

2.2 The Dimensions of the Problem

To assist the reader who is unfamiliar with the literature on a-Si:H, this section is designed to give a broad overview of the conclusions drawn to date, indicating where controversy still exists and where there is reasonable consensus. Subsequent sections are intended to provide a more detailed discussion of these same areas thus permitting the reader to judge for himself the relative merits of the different positions.

2.2.1 Atomic-Scale Structure

It is generally accepted that amorphous Si:H is indeed amorphous – evidence for this comes from transmission electron diffraction, x-ray and neutron diffraction, EXAFS, ^{29}Si NMR and Raman scattering. This latter measurement has also been used to identify the onset of microcrystallinity which is brought about either by doping fluorinated films prepared by plasma decomposition of SiF_4/H_2 mixes or by decomposing silane under conditions of near equilibrium.

The amorphous nature of the material is characterized by a well-preserved first-neighbor separation, i.e., no change in bond length from crystalline silicon, but a broadened distribution of second-neighbor separations due to variations $\sim \pm 5\%$ of the bond angle.

2.2.2 Hydrogen Content

The most uncontroversial conclusion is that regardless of preparation technique, material that exhibits "useful" electronic properties contains hydrogen in the ~ 2–16 atomic percent range. It is not clear that hydrogen in these quantities is a priori necessary to obtain useful properties, only that experimentally these are the contents found. It is generally accepted that removing all or part of the incorporated hydrogen from material that is electronically "useful" creates defects which degrade the useful properties. Materials with hydrogen contents ranging from < 1 to ~ 60 at.% have been reported that typically contain high concentrations of electronically active defects. It is still a matter of contention whether the amorphous silicon-hydrogen system should be treated as a simple binary or as a multiphase system; drawing a demarcation line around the region of useful electronic properties and treating inside and outside that line as two different materials is a commonly used but questionable approach.

2.2.3 Other Impurities

In contrast to hydrogen content measurement on which much effort has been expended, few groups have reported contents of impurities other than those deliberately added such as phosphorus or boron. Those groups that have studied other impurities find oxygen, nitrogen, carbon and inert gases such as argon where these have been used in sputtering. The impurity contents reported are significantly (orders of magnitude) higher than in crystalline silicon.

This is illustrated in Fig. 2.1 by a SIMS (secondary ion mass) spectroscopy profile through a plasma-deposited layer showing oxygen contents ~ 0.3 at.%, and carbon and nitrogen in the 10^{18}–10^{19} range. The comparison with the single crystal substrate is clear. Reports on sputtered material put oxygen contents as high as 10 at.% and argon contents in the 0–2 at.% range. The origin of these impurities is a subject of current investigation; "dirty" vacuum systems, impure targets and post deposition oxidation have all been implicated by various groups but no systematic study has yet been published.

2.2.4 Structural Inhomogeneity

There is now considerable evidence that structural inhomogeneity, specifically voids or microcracks, exists in much of the material that is not electronically "useful", independent of the deposition technique used in preparation. Figure 2.2 shows a transmission electron micrograph of magnetron-sputtered material illustrating a "microcrack" network – similar results will be presented later for diode sputtered and plasma-deposited material. Whether this inhomogeneity is universal and just difficult to detect in useful material is a subject of contention, as are its precise origins. Although there are strong correlations between the presence of structural inhomogeneity and high levels of electrically active defects, no direct connections between the electrical and structural defects have been made nor has any strong case been made for any inhomogeneity in electrical properties although a number of unusual

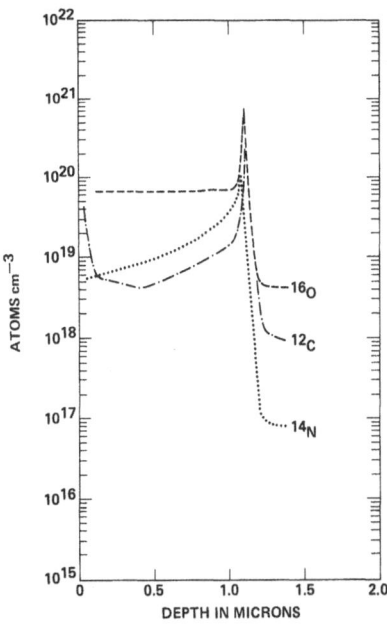

Fig. 2.1. Concentration profiles obtained by SIMS of oxygen, carbon and nitrogen in an ~ 1 µm thick plasma-deposited a-Si : H film on a crystalline silicon substrate. The density of silicon atoms is ~ 5 × 10^{22} cm^{-3}; the levels shown in the substrate represent the combined instrument background and actual concentrations

Fig. 2.2. Transmission electron micrograph of dc magnetron sputtered a-Si : H. The micrograph width is equivalent to approximately 2000 Å [2.40]

observations concerning electrical behavior are consistent with charge transport in an inhomogeneous medium.

2.2.5 Compositional Inhomogeneity

A combination of hydrogen evolution measurements (where the sample is essentially decomposed by heating to high temperatures) and proton magnetic resonance spectra have been interpreted as indicating an inhomogeneous hydrogen distribution commensurate with the structural inhomogeneity. This interpretation, specifically the inhomogeneity deduction from NMR, is not without controversy although semantic differences may be partly to blame. The NMR evidence is remarkable for the fact that it crosses the demarcation line between useful and not-useful material.

NMR measurements have also been used to suggest that another impurity, boron, can also be inhomogeneously distributed and a combination of transmission electron microscopy, annealing and NMR seem to indicate that fluorine, when present, is also inhomogeneously distributed. No evidence exists one way or the other for any other impurities.

2.2.6 Surfaces and Interfaces

Following in the footsteps of crystalline silicon, attention to the bulk properties has preceded consideration of surfaces and interfaces. From a structural and chemical standpoint, the results to date are concentrated in the area of oxidation and metal – a-Si : H contacts with ellipsometry and interference enhanced Raman spectroscopy being major tools. None of the conclusions drawn so far are particularly controversial and indicate broadly that a-Si : H is comparable to crystalline silicon in its chemical reactivity at ambient temperatures.

2.3 Atomic-Scale Structure

As outlined in the above section, there is now no controversy over the fact that a-Si : H is indeed amorphous. This has been established by the use of the standard techniques, namely, diffraction studies using electrons, x-rays and neutrons as well as by experiments aimed primarily at elucidating different phenomena: Raman spectroscopy, EXAFS (Extended X-ray Absorption Fine Structure) and nuclear magnetic resonance of the ^{29}Si isotope.

2.3.1 Diffraction Techniques

Table 2.1 lists the diffraction techniques (of which EXAFS is a member) and the problems and benefits of applying them to a-Si : H. The first studies of

Table 2.1. Diffraction techniques applied to a-Si:H

Technique	Sample thickness	Substrate requirements	Information expected	Limitations
X-ray	10 μm to >1 mm	Low Z, e.g. Be, Al	RDF of Si, Si–Si bond length, inhomogeneities from small-angle scattering	Insensitive to hydrogen
Electron (transmission)	0.01–0.15 μm	Self-supported or carbon film	RDF of Si, Si–Si bond length, inhomogeneities from small-angle scattering	Insensitive to hydrogen
Neutron	10 μm to >1 mm	Low cross section, e.g. V, Al	Potentially (with deuteration) partial RDFs Si–Si, Si–H, inhomogeneities from small-angle scattering	Self-scattering gives data analysis problem
EXAFS[a]	≈ 200 μm	Low Z	Si–Si distance, high Z (As, Ga, Ge) impurity environment	Si K edge is in a difficult energy range, insensitive to hydrogen

[a] EXAFS, extended x-ray absorption fine-structure spectroscopy

x-ray and electron diffraction on a-Si:H films were made by *Mosseri* et al. [2.10] and *Barna* et al. [2.6]. Subsequent detailed electron diffraction measurements were performed by *Graczyk* [2.11] and neutron diffraction measurements by *Postol* et al. [2.12]. As indicated in Table 2.1, the sample requirements for the different techniques are quite different. Electron microscopy can be performed on ≤ 100 Å thick films weighing $\leq 10^{-7}$ g (for 1 mm²) whereas, for example, the experiments of Postol et al. used ~ 3.5 g.

In order to obtain the requisite sample masses for the x-ray and neutron experiments, both groups relied on sample preparation techniques which produced high defect density material. Mosseri et al. scraped material off the walls of a glow-discharge system that were at room temperature during deposition. This material is known to contain ~ 35 at.% hydrogen and high electron spin densities. Postol et al. reported that in their material (produced by high rate magnetron sputtering) there was intense small-angle scattering, a feature previously associated with high defect densities, even though the substrates were held at high temperature (200–250 °C). In contrast, Barna et al. and Graczyk, using electron diffraction, were able to examine glow-discharge material deposited under essentially optimum conditions as well as, in the latter case, high defect density samples. The reduced density function $G(r)$ is related to the radial distribution function (RDF) $J(r)$ by the expression

$$J(r) = rG(r) + R\pi r^2 P_0 \, ,$$

where P_0 is the average atomic density. This function is shown for three different glow-discharge samples as determined by Graczyk in Fig. 2.3 a–c. The dotted lines in these figures are the values calculated for a continuous random network. Sample (b) corresponds to low defect-density material whereas (a) and (c) both correspond to high defect densities. Graczyk and Barna et al. both concluded that the low defect density material was structurally well described by the continuous random network structure originally proposed by *Polk* [3.13] modified to include 20% of a staggered configuration of tetrahedral units to account for some additional structure observed at $r = 4.55$ Å. On annealing to $\sim 580\,°C$, Graczyk observed crystallization as illustrated in Fig. 2.4, which demonstrates the radical difference in diffraction patterns between amorphous and crystalline. Both Graczyk and Mosseri

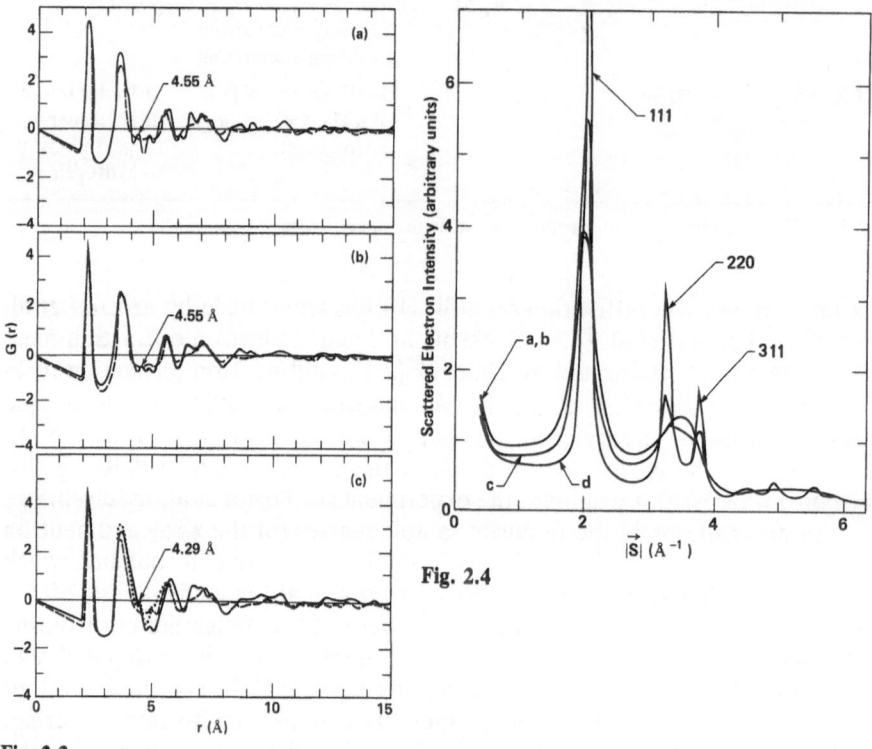

Fig. 2.3

Fig. 2.4

Fig. 2.3 a–c. Reduced density function versus *r* from transmission electron diffraction for a-Si : H plasma-deposited onto cleaved NaCl; (a) plasma pressue $(P) = 0.05$ torr, substrate temperature $(T_s) = 25\,°C$, (b) $p = 0.3$ torr, $T_s = 250\,°C$, (c) $p = 0.3$ torr, $T_s = 25\,°C$. (——) experimental data; (– – –) calculated function for a continuous random network [2.11]

Fig. 2.4. Experimental electron intensity profiles at room temperature from an a-Si : H sample (Fig. 2.3 b) as a function of increasing temperature up to crystallization. Traces shown are for (a) $T = 30$, (b) 560, (c) 580, and (d) 582 °C [2.11]

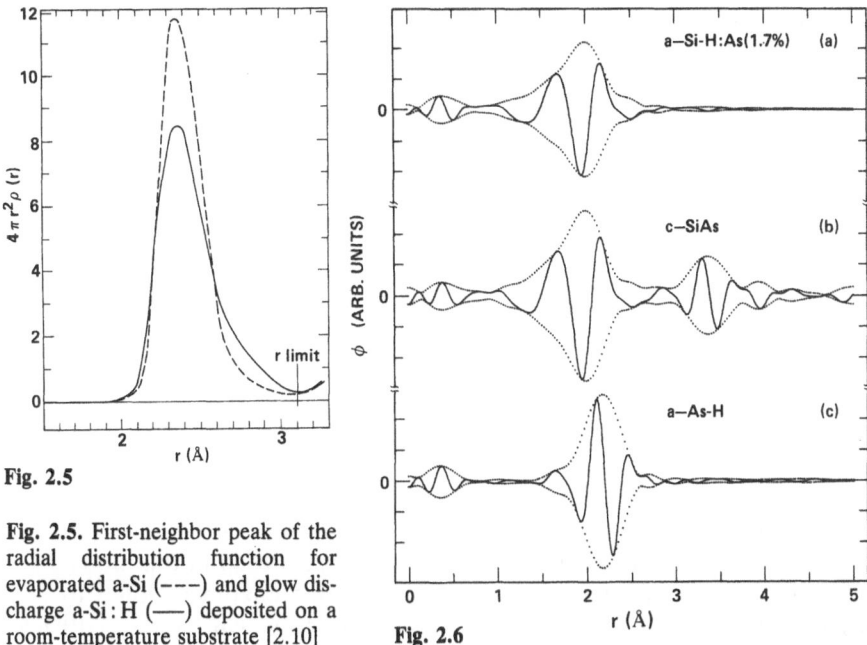

Fig. 2.5

Fig. 2.5. First-neighbor peak of the radial distribution function for evaporated a-Si (– – –) and glow discharge a-Si:H (——) deposited on a room-temperature substrate [2.10]

Fig. 2.6

Fig. 2.6a–c. The real part (——) and the magnitude of the Fourier transform $\varphi(r)$ of the EXAFS on the As K-shell absorption in (**a**) 1.7 at. % As in amorphous Si–H, (**b**) c-SiAs, and (**c**) amorphous As–H. The vertical units are arbitrary, but identical for each set of data. The data were all transformed using a square window with k between 3.75 and 13.85 Å^{-1}, broadened by a Gaussian $\sigma_\omega = 0.7$ Å^{-1} [2.14]

et al. concluded that the material deposited on low temperature substrates showed significant deviations from the random network configuration. For the thicker of the two samples in Fig. 2.3c, Graczyk noted that there was an additional peak at 4.29 Å that he proposed might be due to densely packed polymeric (SiH$_2$) chains. He suggested that this sample might, therefore, consist of a mixture of polymeric and random network regions. Mosseri et al. concluded that the hydrogen in their low temperature material was associated with asymmetry in the first Si–Si peak when compared to that measured for evaporated unhydrogenated material; this is illustrated in Fig. 2.5. They attributed this effect to inhomogeneity suggesting that as a function of increasing hydrogen content the material possesses increasing amounts of large (seven or greater) membered rings eventually reaching the linear (SiH$_2$) chain configuration, although they stop short of suggesting the two-component inhomogeneity proposed by Graczyk.

EXAFS measurements by *Knights* et al. [3.14] have further demonstrated the lack of short-range order in the specific case of the environment of arsenic dopant atoms. EXAFS is a particularly sensitive tool for determining fluctuations in interatomic spacings. The observed modulation of the x-ray absorption cross section (in this case above the arsenic K absorption edge) is due to the constructive and destructive interference of portions of a spherical electron wave reflected from the surrounding atoms. Because amplitudes as opposed to intensities are being summed and because the total detected signal is a sum over a large number of absorbing arsenic atoms, any random fluctuations in the interatomic spacing will significantly reduce the magnitude of the oscillation. The results of this study are shown after Fourier transformation into real space in Fig. 2.6. The fact that in the crystalline SiAs four shells of atoms can be discerned while only the nearest-neighbor shell can be observed in the case of both a-Si:As:H and a-As:H is clear evidence for random fluctuations in next-nearest neighbor spacings. Specifically it is found that (i) the mean nearest neighbor spacing (As–Si) is identical within experimental error to that in the crystalline silicon arsenide, 2.38 ± 0.01 Å (ii) the width of the nearest-neighbor peak is the same as in the crystal, i.e., the distribution in nearest-neighbor spacings is ≤ 0.02 Å, and (iii) by modeling, a distribution in next-nearest neighbor distances of at least 0.2 Å greater than that of the first is required to explain the suppression of the corresponding EXAFS peak. It should be added that the samples used in this study were prepared by plasma decomposition of silane/arsine mixtures diluted 10% in argon and deposited onto 230 °C substrates at deposition rates ~ 5–10 Å/s, conditions known to produce material on the borderline between "useful" and "not useful". Finally, all the above techniques, as indicated in Table 2.1, are insensitive to hydrogen by virtue of its low electron density. However, in both the diffraction studies and in the EXAFS the apparent reduction in the normal coordination of the silicon (arsenic) due to the presence of bonded hydrogen was detected and found to agree approximately with the reduction expected from independent hydrogen content measurements.

2.3.2 ^{29}Si NMR

As discussed later in this chapter, nuclear magnetic resonance has proved to be a very effective tool in extracting information about the spatial distribution of hydrogen in a-Si:H. Several groups also recognized that measurements of ^{29}Si NMR using a special technique known as cross-polarization magic-angle sample spinning (CPMASS) [2.15] had the potential for producing unique information about the SiH$_x$ ($x = 1.3$) environments. Unfortunately the information was not forthcoming for a reason that is directly related to the amorphous nature of the material and as such is worthy of discussion.

The aim of the experiment is to observe the resonance line of ^{29}Si under conditions where all line broadening mechanisms (see later discussion of proton NMR) such as spin-spin interactions are removed. In principle, the only residual perturbation of the position of the resonance from the Zeeman energy is the chemical shift term due to variations in the orbital electron screening of the nucleus from the applied field. This screening can arise either from different bond charges, e.g., those associated with SiH, SiH_2, SiH_3 or from electron density changes due to bond angle distortion in the amorphous silicon matrix. In the former case, discrete lines corresponding to the discrete chemical shifts would be expected and are observed as essentially delta function peaks in many hydrocarbon solids, e.g., for ^{13}C in CH_2, CH_3, etc. In the case of bond angle distortion, depending on the average magnitude and distribution, a broad featureless spectrum should be seen.

Experimentally, a combination of static and pulsed magnetic fields is applied that performs the following functions: (i) magnetization of the protons, (ii) transfer of magnetization from the protons to ^{29}Si, and (iii) decoupling of the protons from the ^{29}Si. Simultaneously the sample is spun at rates ~ 3000 Hz at the "magic angle" which causes the anisotropic components of the chemical shift tensors to average to zero. The free induction decay of the ^{29}Si is measured and after averaging to obtain acceptable signal to noise, the signal is Fourier transformed to give the resonance lineshape.

Both plasma-deposited and sputtered material have been studied, the former by *Reimer* et al. [2.16], the latter by *Lamotte* et al. [2.17] and *Jeffrey* et al. [2.18]. All the groups found, in contrast to the hydrocarbon case, very broad lines with only Lamotte et al. observing any structure. These latter authors concluded that the two shoulders they observed on the upfield (negative frequency shifted) side of the main peak may be due to SiOH bonds associated with oxidation of the sample, while the bulk of the line is due to Si, SiH and SiH_2 whose lines are broadened by the distribution of screening fields arising from the amorphous network. Jeffrey et al. found that the center of their line is shifted ~ 73 ppm (of the reference frequency of silicon in tetramethylsilane) upfield, while Reimer et al. observed for plasma-deposited material shifts of 50 and 42 ppm for samples both with and without polysilane $[(SiH_2)_n]$ vibrational modes present. The difference between sputtered and plasma-deposited material is real and must reflect some change in the bulk susceptibility. Both groups concluded that the line broadening cannot be explained by a superposition of discrete lines such as SiH, SiH_2 and that broadening due to a distribution of screening charge must be invoked. Jeffrey et al., using a model proposed by *Guttman* et al. [2.19] for charge redistribution in an a-Si network, calculated the expected distribution of chemical shifts and found that this calculation is in good agreement with their experimental linewidth.

The overall conclusion from the ^{29}Si resonance experiments is clear; the ^{29}Si nuclei, whether bound to silicon or to one or two hydrogens, experience a random screening field from the orbital electrons that is consistent with a fully-disordered solid.

2.3.3 Raman Scattering and Microcrystallinity

Raman spectra from a-Si with and without hydrogen have been studied by a number of authors [2.20, 21]. In the energy region corresponding to the crystalline LO and TO phonons, a-Si exhibits a broad featureless peak centered at ~ 480 cm^{-1}. In contrast, crystalline silicon exhibits a sharp peak at 522 cm^{-1} corresponding to the zone center LO and TO transitions. It is generally accepted that the broad peak in amorphous silicon arises from a relaxation of the crystalline selection rules that depend on a well-defined wave vector k. All vibrational modes then contribute to the observed signal and the amorphous spectrum reflects the vibrational density of states. Clearly if one starts with an infinite crystal and progressively reduces the size, at some point there must be a breakdown in the selection rules. Several groups have studied this theoretically and it has been observed experimentally in graphite [2.22] and boron nitride [2.23]. The transition is gradual and involves both a broadening and a shift, the shift arising from the existence of significant dispersion in the phonon frequencies.

Recently several groups have shown that microcrystalline Si:H and Si:F:H, otherwise described as amorphous with increased intermediate range order, can be produced by a variety of techniques. These include chemical transport of silicon in a hydrogen plasma [2.24], plasma deposition from SiH$_4$ under conditions of near equilibrium [2.25], arsenic doping of

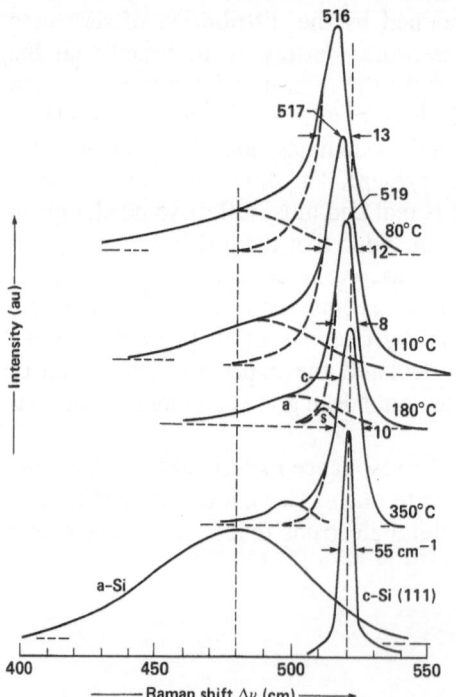

Fig. 2.7. Raman spectrum of (5145 Å excitation) polycrystalline silicon deposited by hydrogen plasma transport of silicon onto different temperature substrates. Spectra run under the same conditions on single crystal [111] silicon and plasma-deposited a-Si:H are shown for comparison [2.28]

material deposited from SiF_4/H_2 plasmas [2.26] and phosphorus doping under high rf power conditions in a SiH_4 plasma [2.27]. The interest from the perspective of a structural understanding of a-Si:H in the transition from amorphous to microcrystalline. This is illustrated in Fig. 2.7. As the crystallite size is reduced, the Raman peak at 522 cm^{-1} broadens and shifts to a lower wave number as would be expected from the phonon dispersion curve for silicon [2.28]. At the same time an amorphous background builds up which is attributed to disordered grain boundary regions. Finally, when the crystallite size, determined by electron microscopy, reaches approximately 30 Å, the crystalline peak disappears altogether. *Veprek* et al. [2.24] have shown that this trend is associated with an increase in the lattice spacing as the crystallite size is reduced leading to an abrupt transition to the amorphous state. These authors suggest that as the crystallite size decreases, the strain energy builds up until the crystalline form is unstable and transforms into the amorphous form.

The conclusion from this work is clear and supports the other results in this section: amorphous Si:H is truly amorphous and the ideas discussed in the late '60s and early '70s concerning microcrystalline units bonded together as a structural model for tetrahedral amorphous semiconductors is not appropriate for this material.

2.4 Structural Inhomogeneity

Before discussing the specific results concerning structural inhomogeneity in a-Si:H, it is helpful to place this facet of the review in the context of the much larger body of work on thin films in general.

Thin films deposited by any technique undergo very similar processes at an atomic level during deposition. Species, be they atoms or molecular fragments, are adsorbed onto the substrate. Depending on the surface mobility and desorption rates, these species migrate until they encounter a sufficiently attractive potential to prevent further migration. This potential might arise from surface defects in the substrate or a fluctuation in the density of the migrating species sufficient to cause a local condensation. This process results in the nucleation of the film at discrete points. Subsequent migrating species attach at these nuclei causing growth in three dimensions until neighboring clusters have dimensions approaching half the internuclear spacing. Up to this point it is immaterial from a process viewpoint whether the clusters are crystalline or amorphous; however, as the clusters actually come into contact, the results can be divided into three classes based on whether the material is crystalline, amorphous or vitreous with the substrate temperature close to T_g, the glass transition temperature. In the first class, unless the growth has been epitaxial on a single crystal substrate, crystalline grain

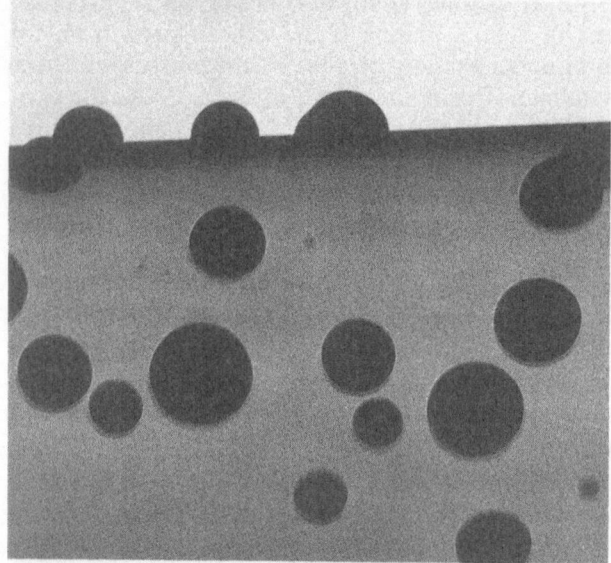

Fig. 2.8

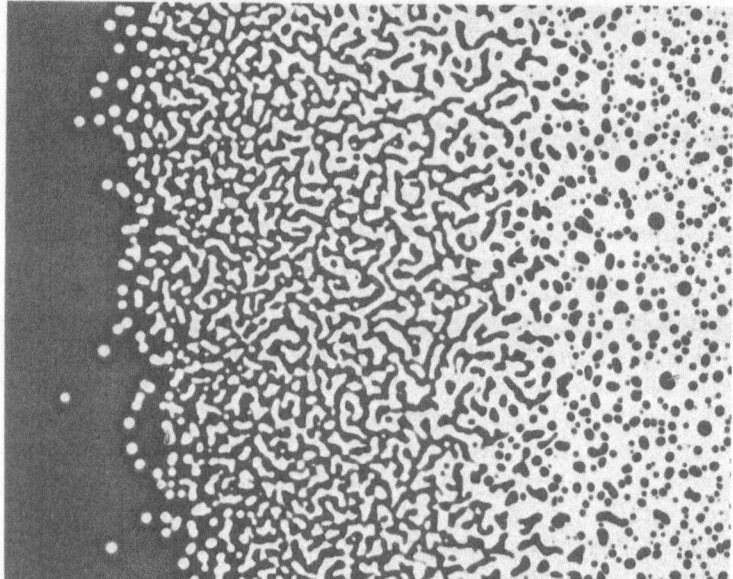

Fig. 2.9

Fig. 2.8. Transmission electron micrograph of a discontinuous film of selenium evaporated onto a carbon support film at $\sim 30\,°C$. The support film was subsequently broken to permit the sideways view. The diameter of the hemispheres at their base is $\sim 1000–5000\,\text{Å}$

Fig. 2.9. Transmission electron micrograph of a film of selenium evaporated onto an (invisible) carbon support film. A mask was used to reduce the deposition rate over half the area shown to permit the temporal development of the microstructure to be spatially resolved

boundaries are formed. Impurities that do not incorporate into the grains themselves due to energetic considerations are typically segregated to these boundary regions and a host of effects due to these "contaminated" grain boundaries are observed in many physical properties such as magnetic permeability, electrical conductivity and corrosion resistance.

In the case of a vitreous material, when the substrate temperature is higher than T_g, the clusters behave like liquid droplets as illustrated in Figs. 2.8, 9. These figures show the behavior of selenium evaporated onto a substrate at $\sim 40\,°C$ demonstrating both the effects of surface tension in controlling cluster profiles and the flowing together of the clusters upon contact. Similar results have been observed for antimony [2.29] and arsenic triselenide [2.30]. It is clear that the concept of grain boundaries is not applicable to this class of material. As a consequence there is no "memory" of the initial nucleation sites, a fact easily illustrated by comparing the top surface of a selenium coated copier drum with the uncoated aluminum blank – the latter is considerably rougher than the former which is smooth and truly glassy in appearance.

The final class, that to which hydrogenated amorphous silicon belongs, is amorphous either with no glass transition temperature or deposited at a substrate temperature well below T_g. The behavior on cluster-contact in these materials typically leads to an amorphous equivalent of grain boundaries; the materials may in fact be better described as "polyamorphous". The extent to which these boundaries are observable depends on a number of material and deposition technique parameters but in most materials they can be observed in transmission electron microscopy as periodic fluctuations in electron density which have been variously characterized as supernetworks, microvoid networks and island structures. They have been observed in evaporated and sputtered amorphous silicon and germanium [2.31, 32] (without hydrogen) and are particularly evident in sputtered amorphous rare-earth transition metal compounds such as gadolinium-cobalt [2.33] and terbium-iron [2.34]. In a similar fashion to their polycrystalline counterparts, these materials' physical properties can be strongly affected by the existence of these boundaries, very marked changes in resistance to oxidation and/or chemical attack by etches being observed in the case of amorphous silicon and germanium [2.35, 36] and profound changes in magnetic properties in the case of the rare earth-transition metal compounds [2.37].

2.4.1 Experimental Results

Electron microscopy has been both the most widely employed and the most successful technique in addressing the issue of structural inhomogeneity. The first report by *Barna* et al. [2.6] of transmission microscopy on plasma-deposited material indicated that in contrast to evaporated a-Si, there was no evidence of voids or microstructure on the scale of 10–1000 Å. Later work by

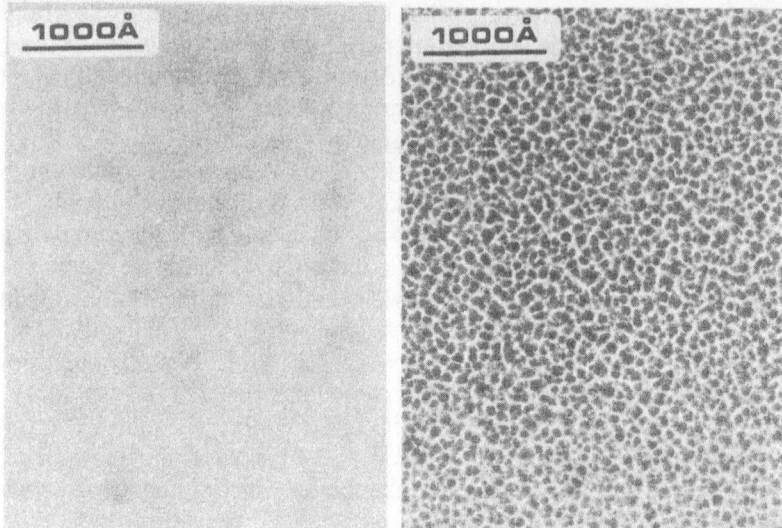

Fig. 2.10

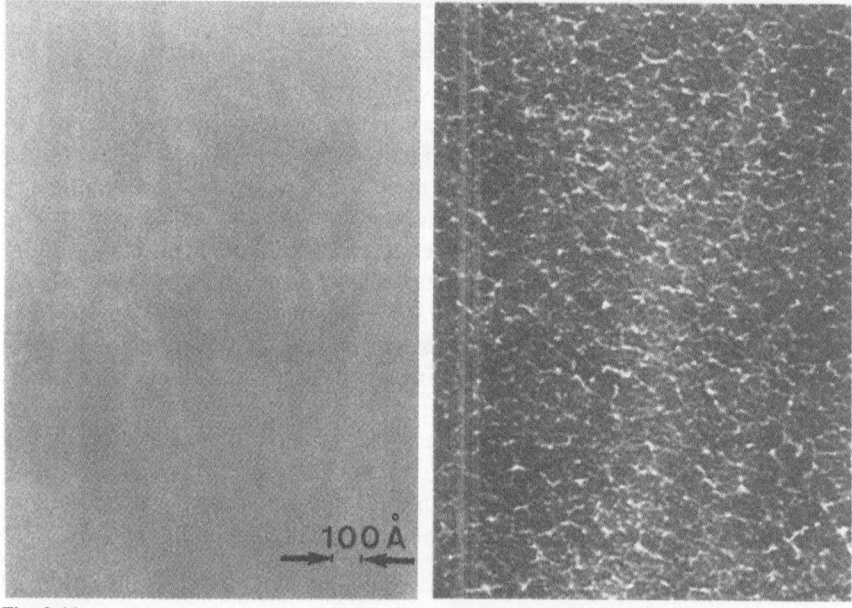

Fig. 2.11

Fig. 2.10. Transmission electron micrographs of high (*left*) and low (*right*) electronic quality plasma-deposited a-Si : H films showing the respective absence and presence of structural inhomogeneity [2.38]

Fig. 2.11. Transmission electron micrographs of high (*left*) and low (*right*) electronic quality diode sputtered a-Si : H films showing the respective absence and presence of structural inhomogeneity. The left-hand micrograph was provided by A. Chenevas-Paule of C.E.A. Grenoble; the right is from [2.39]

Fig. 2.12. Fracture surface perpendicular to the film surface of a diode sputtered a-Si:H film (from A. Chenevas-Paule, C.E.A. Grenoble; with permission)

a number of groups [2.38–40] indicated that while this finding is true in some regions of the deposition parameter space, it is not universally true in either plasma deposited or sputtered material. Figure 2.10 shows the two extremes of behavior observed in plasma deposited material, while Fig. 2.11 shows corresponding micrographs for sputtered material [2.41]. It is clear from these and other published micrographs that periodic density fluctuations on the scale ~ 100 Å are a major feature of thin ($< \sim 1000$ Å) a-Si:H films deposited under certain conditions. Studies of fracture surfaces of thick films using scanning electron microscopy show that the observation of supernetworks of density fluctuations in transmission microscopy is *always* associated with a columnar growth morphology whereas films featureless in transmission microscopy have correspondingly featureless fracture surfaces; these results are illustrated in Figs. 2.12, 13. These figures also illustrate the phenomena, noted first by *Ross* and *Messier* [2.42], of a change in scale in columnarity as the sample thickness increases. The apparent "facet" width increases as thickness increases, reaching as will be shown later, values ≥ 1 μm and there is an associated increase in the roughness of the top surface of the film resulting in a cauliflower appearance in many cases.

The observations of inhomogeneity in transmission microscopy were closely followed by reports [2.12, 43, 44] of intense small angle x-ray and neutron scattering in both plasma-deposited and sputtered material. These results were obtained on very thick (≥ 20 μm) layers and both confirmed the electron microscopy results and added some new information. The confirma-

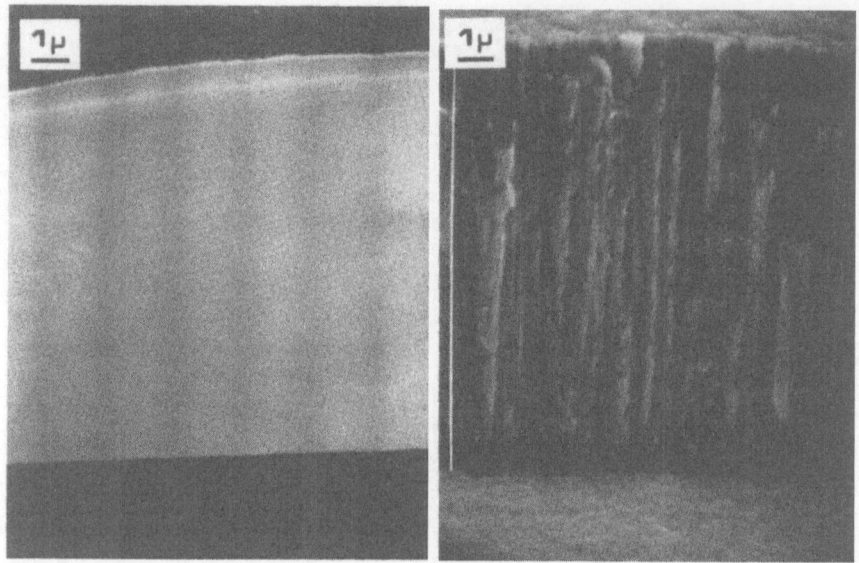

Fig. 2.13. Fracture surfaces of high (left) and low (right) electronic quality plasma-deposited a-Si : H films [2.38]

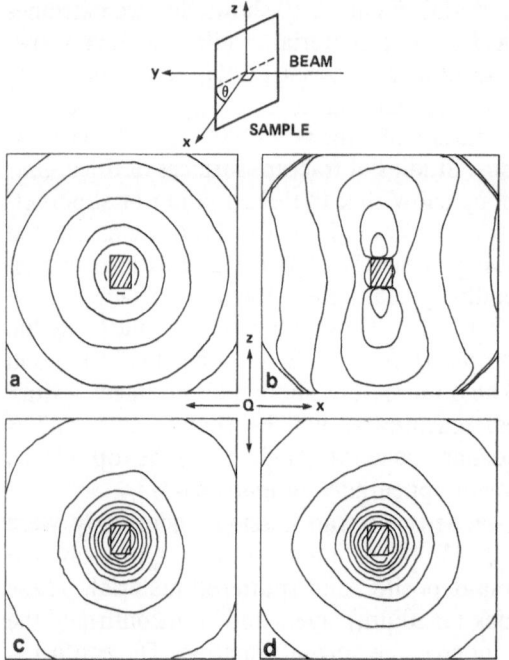

Fig. 2.14 a–d. Neutron scattering intensity contours for two plasma-deposited a-Si : H samples with the sample surface normal (**a, c**) and at 60° (**b, d**) to the incident beam [2.44]

tion is illustrated in Fig. 2.14 a, b where scattered neutron intensity contours
from a film normal to the beam and tilted at 60° show strong and anisotropic
scattering that can be well modeled by treating the material as clustered
cylinders with a mean diameter ~ 60 Å, similar to the island dimension seen
in TEM. In films with both low densities of electrically active defects (free
spins) and featureless TEM micrographs, small angle scattering, if present,
was below the background level of detection; this was found to be true both
of plasma-deposited and sputtered material. The new information is illustrated in Fig. 2.14 c, d where a pair of scattered neutron intensity contour
maps from a film with a high density of free electron spins shows *isotropic*
scattering. Surprisingly, thin (≤ 1000 Å) films prepared under the same conditions show *no evidence* of density fluctuations in transmission microscopy.

The origin of the supernetwork of low density material in the initial stages
of film formation was addressed in the introduction. The experimental evidence seems to support the "polyamorphous" model over at least a part of
the deposition parameter space. Specifically there is evidence that surface
structure such as grain boundaries produced in an aluminum foil substrate
[2.45] can influence the shape and location of the "islands" and, as shown in
Fig. 2.15, factors relating to the deposition environment, in this case the
"inert" gas used as a diluent of silane during plasma deposition [2.46] can
influence the island dimensions, i.e., the nucleation center density. The origin of the columnar growth subsequent to the initial nucleation and the origin
of the density fluctuations causing isotropic scattering are not so easily an-

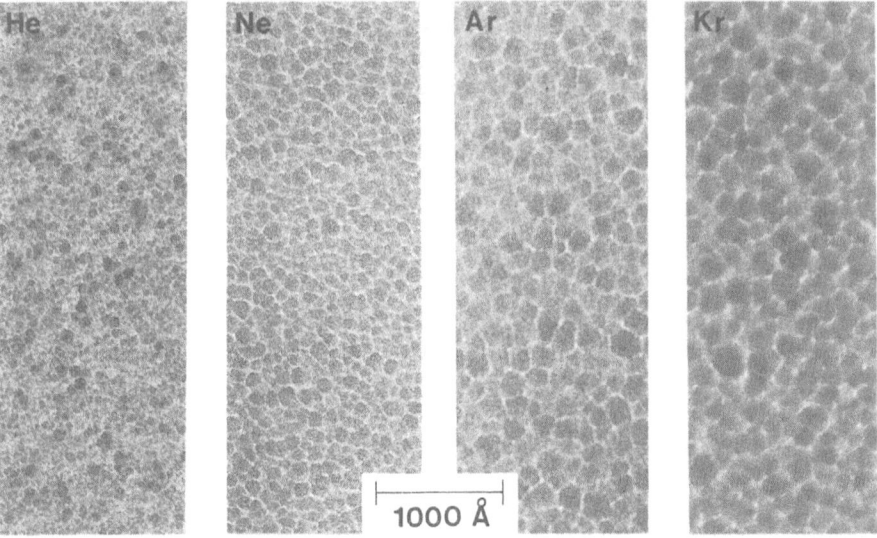

Fig. 2.15. Transmission electron micrographs of a-Si : H films prepared from different SiH₄-inert
gas mixtures [2.46]

swered but electron microscopy has provided some clues on both subjects. Taking the latter issue first, *Chenevas-Paule* [2.47] and coworkers have observed an interesting class of defects in a somewhat "nonstandard" material (rehydrogenated sputtered amorphous silicon), material deposited in the absence of hydrogen and then hydrogenated by exposure to a hydrogen plasma. Figure 2.16 shows this material after removal from a silicon substrate with a hydrofluoric acid etch. Stereo microscopy reveals that this structure is not, as might appear, incipient "grain" boundaries but microscopic tubes of diameter ~ 10 Å running at an angle to the film surface. The authors suggest that the hydrogen bonds preferentially to disclinations [2.48] in the amorphous material and this hydrogenated material is selectively removed by etching during the substrate removal step. Structures of this type are not visible in TEM prior to etching but could cause isotropic small angle scattering. Whether these are part of distorted grain boundaries or, as suggested by *Sadoc* and *Mosseri* [2.48], a completely separate class of defects, is a question for further study.

Chenevas-Paule et al. further observed the coexistence of columnar and approximately isotropic microstructure in the same sample; Fig. 2.17 shows a scanning electron micrograph of a fracture surface showing a fine "crack" pattern inside the larger column structure. This type of sample would give rise to anisotropic scattering at small angles but more isotropic at larger angles.

In addition, *Schiff* et al. [2.49] have reported a "dopant-precipitated" morphology that occurs over a certain range of doping concentration when B_2H_6 is added to SiH_4/Ar mixtures. The fracture surfaces of thick films show an almost globular appearance. As will be discussed later, boron doping, even in the absence of directly observed microstructure, produces effects in hydrogen evolution indicative of interconnected void networks. The fact that, at least in SEM, this microstructure is not anisotropic, suggests that it probably will give rise to isotropic neutron scattering.

The origin of the columnar structure in most thin film deposition processes has been connected to some form of shadowing of the intercolumnar regions from a directional stream of atoms or molecules. Initial roughness in the substrate or self-shadowing due to fluctuations in the growing film's topography have been variously invoked as initiators of the process. Once a surface topography develops, it is perpetuated by the fact that sites in the "valleys" subtend smaller collection angles for impinging atoms than sites on the "hills". The direction of the impinging stream relative to the substrate surface and the energy of the arriving species and the surface mobility are important factors in determining the details of the morphology. Two points are specifically relevant to this discussion. The first is that in the sputter deposition of metal films it has been shown that enhanced ion bombardment of the growing film achieved by negatively biasing the film relative to the plasma effectively suppresses the columnar morphology [2.50] at the expense of incorporating large amounts (atomic percentages) of the inert gas used in

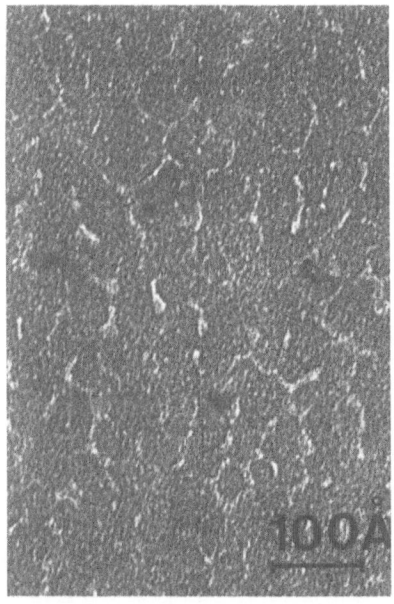

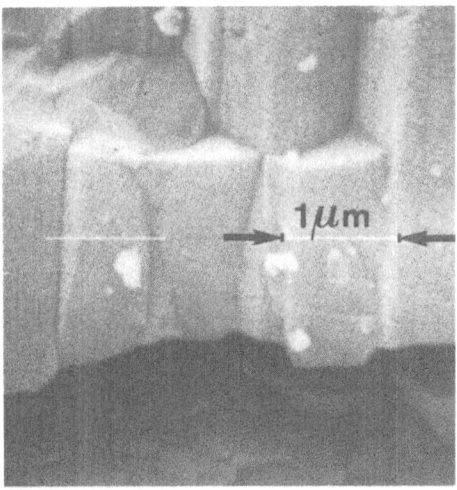

Fig. 2.17

Fig. 2.16

Fig. 2.16. Transmission electron micrograph of a sputtered a-Si : H film removed from a fused silica substrate with hydrofluoric acid. Stereo microscopy shows the light areas to be microscopic tubules ~ 10 Å in diameter (from A. Chenevas-Paule, C.E.A. Grenoble; with permission)

Fig. 2.17. Scanning electron micrograph of the fracture surface of a diode sputtered a-Si : H film showing approximately isotropic substructure within columns (from A. Chenevas-Paule, C.E.A. Grenoble; with permission)

sputtering, usually argon. The model proposed by *Thornton* [2.51] for this, the "shot-peening" model, invokes a billiard-ball knockon process in which arriving high energy ions dislodge already attached atoms and literally fill the intergranular regions with these dislodged atoms, trapping argon in small voids or vacancies. The second is that, as might be intuitively expected from the shadowing argument, the columnar morphology becomes more' pronounced as the angle of incidence of the impinging stream relative to the substrate becomes greater. Both these effects are observed in amorphous silicon deposited both by sputtering and plasma deposition.

Figure 2.18 shows side-by-side scanning electron micrographs of fracture surfaces of two regions of a plasma-deposited film corresponding to positions indicated on the diagram of the deposition configuration. In the position near the edge of the mask, the columnar morphology is clearly more pronounced than in the center and is also not perpendicular to the film surface. The straightforward interpretation is that the flux of SiH_x radicals is sufficiently directional to self-shadow and that the mask perturbs that flux. On the other hand, since the mean free path of the free radicals is ~ 1 mm and since the electric field at the junction of the mask and substrate is at $\sim 45°$ to the

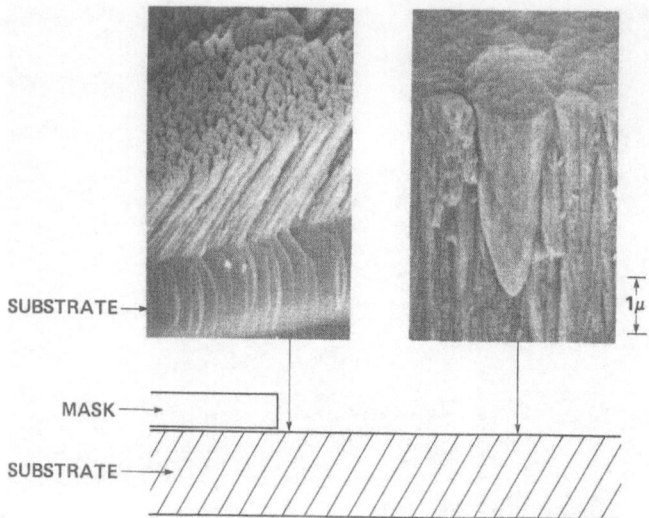

Fig. 2.18. Scanning electron micrographs of fracture surfaces of a plasma-deposited a-Si : H film (room temperature substrate) showing the effect of position relative to a metal mask on microstructure

substrate, an alternative explanation may be that an ion flux which normally scours the surface of weakly bonded species [2.46] is prevented by shadowing from scouring the interstitial regions.

2.4.2 Conclusion

It is clear that there is pervasive evidence for structural inhomogeneity on a number of different scales in much a-Si : H deposited by both plasma-deposition and sputtering. There is, however, material, often that with the best electronic properties, which exhibits no structural inhomogeneity under any of the probes used to date. The unanswered question is: is the "best" material free of inhomogeneity or is it simply a problem of inadequate means of detection? Although there will be evidence presented later in this chapter which suggests the latter, perhaps the best response is to turn the question around. Given the discussion earlier on nucleation and growth, the question arises, is there any way that a rigid covalently-bound solid can form a thin film by vapor deposition that does not have inhomogeneity?

I would like to suggest this topic of cluster contact in nucleating films as a prime target for theoretical modeling; much information would be gained about surface mobility constraints; etc. and the validity of the mechanistic approach to film growth determined.

2.5 Hydrogen Content

The presence of hydrogen in hydrogenated amorphous silicon is easily demonstrated – a few milligrams of material heated to ~ 700 °C will evolve hydrogen producing a pressure rise measured in torr in a 50 ml volume. The fact that at least some of this hydrogen is covalently bonded to silicon in the as-deposited material is equally easily demonstrated – an infrared transmission spectrum of a few micron thick layer shows the characteristic Si–H vibrational absorption lines in the ~ 2000 and ~ 600 cm^{-1} regions. The ease with which these measurements can be made belies the considerable effort of a number of researchers using a range of techniques to make accurate measurements of both average hydrogen content and hydrogen depth profiles. This section will discuss the techniques and the results of their application.

2.5.1 Techniques

a) Nuclear Reaction Analysis

This technique belongs to the general class of ion beam reaction techniques in which a high energy ion beam is incident on the sample, and some product of the reaction of the ion beam with the atomic constituents of the sample is emitted and analyzed. In the specific case of hydrogen, two reactions have been used:

$$H + {}^{19}F \rightarrow {}^{4}He + {}^{16}O + \gamma \tag{2.1}$$

$$H + {}^{15}N \rightarrow {}^{4}He + {}^{12}C + \gamma . \tag{2.2}$$

Both reactions yield γ-rays which are detected with standard sodium iodide scintillators. Of the two reactions, the second has been more widely applied since there is a narrow isolated resonance obviating the need for corrections from off-resonance contributions [2.52]. The energy of this resonance is 6.385 MeV and the γ-ray energy is 4.43 MeV. If the energy of the incident beam is exactly 6.385 MeV, only hydrogen on the surface of the sample will be detected, since the rapid loss of ion energy on entering the sample reduces the reaction cross section by ~ 3 orders of magnitude. Since the cross section above resonance is similarly reduced, a beam of energy higher than 6.385 MeV will only react with a thin layer of material. This permits depth profiling with a depth resolution in the case of a-Si:H ~ 40–100 Å. The accuracy of this technique is limited only by the calibration which, using hydrogen implants into crystalline silicon, is $\sim 8\%$. The detection limit under typical operating conditions is $\sim 10^{20}$ cm^{-3}, i.e., ~ 0.2 at.%. It is worth noting that characteristics of this technique include a beam diameter of ~ 4 mm, a rate of depth profiling ~ 0.4 h/µm and a deposition of about 0.2 W of energy into the probed area, the latter being potentially a cause of hydrogen diffusion during profiling [2.53 a].

b) Secondary Ion Mass Spectroscopy (SIMS)

SIMS is another member of the class of ion beam reaction techniques [2.53]. In this case the incident energy is much lower than for nuclear reaction analysis ~ 5–20 keV and the detected particles are in fact the sputtered protons themselves. SIMS analysis is more sensitive than nuclear reaction analysis by a factor ~ 1000 but has a number of drawbacks, some specific to hydrogen and some more general. The most general limitation comes from the nature of the process itself, namely, secondary ion emission under primary ion bombardment. Only a small fraction of the sputtered atoms are ionized and the magnitude of this fraction, the ionization probability, is known to be sensitive to the matrix from which the ion is ejected. This matrix sensitivity is not at present a calculable quantity and can only be factored out by careful cross comparison with other techniques such as nuclear resonance analysis. Another limitation related to the matrix sensitivity is that polyatomic clusters are also ejected; species such as SiH, SiH_2, etc., and to the extent that the distribution of hydrogen over the various species changes as a function of sample composition, this can introduce errors when, as is customary, only the ^1H ion is counted. The main drawback of SIMS specific to hydrogen stems from the fact that hydrogen both in a molecular form and in compounds such as H_2O and hydrocarbons is a pervasive contaminant in the vacuum systems and ion sources used for SIMS. As a result, unless special precautions are taken, the hydrogen background may be comparable to the true sample signal even when the concentration is in the atomic percent range. It is now standard practice to perform SIMS on hydrogen in UHV chambers and because of the importance of hydrogen in a number of material systems, considerable effort has been expended on optimizing performance for hydrogen [2.54, 55]. An argon or cesium beam is incident on the sample, the emitted ions are extracted electrostatically and then detected in a quadrupole mass spectrometer using either a Faraday cup or an electron multiplier.

c) Hydrogen Evolution

This technique, pioneered by *Fritzsche* [2.56] and his collaborators at the University of Chicago, is the simplest technique available for hydrogen content determination in a-Si : H. The technique relies on the fact that the material decomposes more or less completely into its constituent elements upon crystallization which typically occurs at about 650–700 °C. Heating a sample in a sealed volume to or beyond this temperature causes the hydrogen to evolve causing a pressure rise in a well-designed system of ~ 1 torr with a few hundred micrograms of material, a pressure that can be measured to an accuracy of 0.1 % with a capacitance manometer. If there is the possibility of other gases being evolved, e.g., argon incorporated in the sputtering process, the evolved gas must be analyzed quantitatively using mass spectroscopy. There is also the problem of the "more or less" complete decomposition; experiments using proton magnetic resonance [2.57] have shown that the crystalline material left after evolution can contain up to 1 at.% hydrogen. This represents the source of a systematic underestimation of hydrogen content by this technique, but expressed as a percentage of the typical 10–12 at.%, it is very similar to the errors inherent in SIMS and nuclear reaction analysis and is achieved at a fraction of the cost. In addition to absolute hydrogen content, the rate of hydrogen evolution plotted as a function of temperature has yielded valuable information about hydrogen bonding and diffusion. Typical times for analysis including a rate measurement are $\sim \frac{1}{2}$ h.

d) Proton Magnetic Resonance

Since the results of the application of this technique to hydrogenated amorphous silicon extend well beyond hydrogen content measurement, it will be discussed in detail later. For the moment it is sufficient to say that the absorption of radio-frequency energy by the Zeeman split nuclear spin energy levels of the protons produces an absorption line, the area under which can be integrated to yield, via appropriate calibration, an absolute hydrogen content. This technique's accuracy is limited primarily by sample mass and integration time constraints. Sample masses ~ 100 mg are desirable and yield accuracies $\sim \pm 0.5$ at.% using integrated pulsed signals over times of $\sim \frac{1}{2}$ h. The major limitation of the technique apart from mass is the requirement that the sample be either deposited on an electrically insulating proton-free substrate or removed from the substrate prior to measurement.

e) Infrared Absorption

As mentioned in the introduction to this section and as described elsewhere in this book, the SiH vibrational absorption bands are clearly identifiable in ir absorption spectra. Their existence has prompted and continues to prompt researchers to use the integrated oscillator strength under one or other mode as a measure of hydrogen content. The first to do this were *Connell* and *Pawlik* [2.58] (for a-Ge : H) and *Brodsky* et al. [2.20] for a-Si : H. Both groups used the known oscillator strengths for Si(Ge)H stretching vibrations in molecules and applied local field corrections to account for the solid matrix. While these and other groups obtained good agreement for limited sets of samples, *Fang* et al. [2.59] have clearly shown that ir measurements on the stretch modes do not constitute a reliable measure of hydrogen content over the range of hydrogen content obtained in reactively sputtered a-Si : H with errors $\sim 50\%$ at hydrogen contents of 5 at.%. Furthermore, *Oguz* et al. [2.60] have shown that damage of a sputtered film by ion implantation without any hydrogen evolution causes substantial changes in certain vibrational mode absorptions indicating directly that the local field correction is a function of the defect structure in the material. These results, combined with other reports of unusual distributions of oscillator strength between the different vibrational modes [2.61], indicate that ir spectroscopy, while valuable in determining bonding configurations, is not the technique of choice for routine hydrogen content measurements.

2.5.2 Results

The results will be presented in two sections, one for undoped material and one for doped material since in the latter case the results are complicated by hydrogen evolution occurring during film growth. It is also important to mention at this point that none of the techniques described could resolve compositional inhomogeneity commensurate with the type of structural inhomogeneity described in the previous section. As such, all the results are spatial averages with minimum lateral dimensions $\sim 50\ \mu m \times 50\ \mu m$ being obtained using SIMS.

a) Undoped a-Si : H

Sputtered Material. Since hydrogen is an additive in this instance, the absolute hydrogen content can be controlled up to some saturation value indepen-

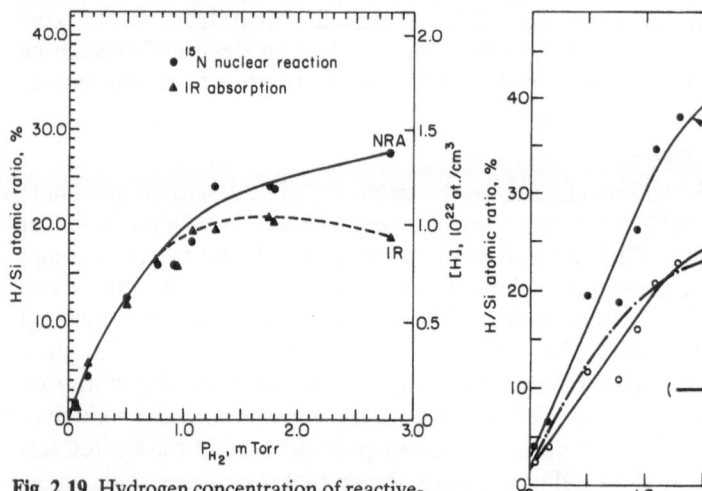

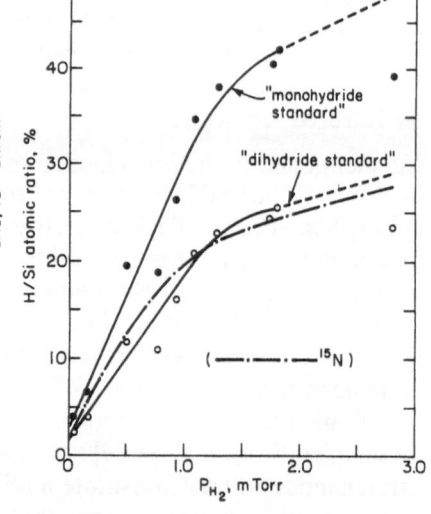

Fig. 2.19. Hydrogen concentration of reactively sputtered a-SiH$_x$ films as a function of hydrogen partial pressure, P_{H_2}. The left-hand y-axis is for ^{15}N nuclear reaction and the right-hand y-axis for ir absorption spectrometry. The two axes agree for the hypothetical Si atomic density of 5×10^{22} atoms cm^{-3} [2.62]

Fig. 2.20. H/Si atomic ratio determined by SIMS using a monohydride and a dihydride standard as a function of hydrogen partial pressure [2.62]

dent of other deposition conditions such as substrate temperature, dc electrical bias and argon partial pressure. The relationship between hydrogen content and hydrogen partial pressure for a typical sputtered film determined both by NRA and ir spectroscopy [2.62] is shown in Fig. 2.19. This illustrates both this relationship and the pitfalls of using ir spectroscopy without cross-calibration. Figure 2.20 shows the results of SIMS analysis on the same sample using either an a-Si : H standard with predominantly monohydride (SiH) vibrational spectra or one with dihydride (SiH$_2$) vibrational spectra, both cross-calibrated with NRA. The ~ 17 at.% discrepancy at $P_H \sim 1.7$ mtorr indicates that the matrix sensitivity issue is real and must be resolved by comparisons such as this prior to relying on SIMS as a routine technique. The result of Fig. 2.20 indicates that hydrogen content measurements in sputtered material cannot be considered in isolation of other deposition parameters. As such, no attempt will be made to review all the reported hydrogen content measurements. It is sufficient to mention that hydrogen contents in the range 8–15 at.% are commonly reported for sputtered material with "good" electronic properties with no discernable correlation between electronic "quality" and absolute content.

The majority of published reports on hydrogen content in sputtered a-Si : H do not describe depth profiling. Where it is described it is found that the hydrogen content is relatively uniform, except when approaching the film substrate interface. Figure 2.21 shows results of SIMS analysis on dc magnetron sputtered material [2.40]; nuclear reaction analysis on similar samples supports the observation of an interface enhancement and a comparable sample prepared using bias sputtering shows that this interface peak can be reduced considerably, suggesting that contamination from system contaminants such as residual water vapor may be responsible.

Plasma-deposited Material. Since hydrogen is subtracted from the starting material both in the initial electron impact dissociation of silane and in the

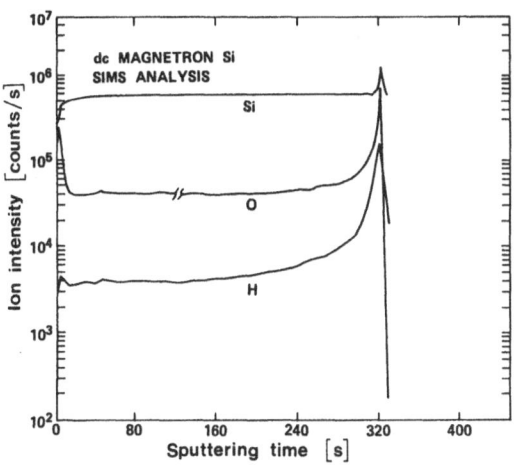

Fig. 2.21. SIMS depth profile of hydrogen and oxygen in dc magnetron sputtered a-Si : H [2.40]

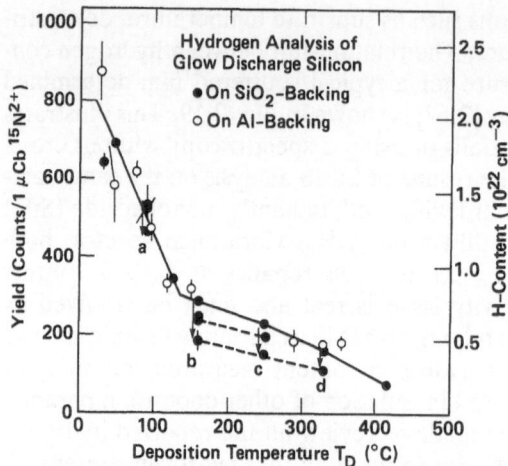

Fig. 2.22. Hydrogen concentration of undoped glow-discharge a-Si : H films versus deposition temperature T_D. Arrows indicate the influence of 15 min anneals at (*a*) 160 and 250 °C, (*b*) 250, 330 and 415 °C, (*c*) 330 and 415 °C, and (*d*) 415 °C [2.63]

subsequent surface reaction of free radicals, the hydrogen content is a strong function of deposition parameters. The most notable trend is that with substrate temperature. Figure 2.22 shows results from *Milleville* et al. [2.63]; the clear conclusion being that the hydrogen content, with a considerable amount of scatter, decreases monotonically with substrate temperature during deposition. The highest reported content is 58 at.% [2.64] achieved by cooling the substrate to − 120 °C; this material is an ~ 3.1 eV bandgap SiH$_{1.4}$ polymer. Increasing the deposition rate by any combination of parameters tends to increase hydrogen content. Of interest is the range of hydrogen content associated with "good" electronic properties. The indication is that the range 5–12 at.% is favored and it is worth noting that the "optimum" content reported by a number of groups has shifted downward with the passage of time, suggesting that as the deposition process is refined, lower contents may be expected.

Hydrogen depth profiling in plasma-deposited material indicates generally good vertical uniformity (± 1 at.%) but with deviations near surfaces and interfaces. These deviations can be of either sign and reach magnitudes ≥ 50% of the bulk value. Work by the groups at Dundee and Heidelberg [2.65] has demonstrated depletion of hydrogen at both top surface and substrate interface extending into the film ~ 0.1 μm for undoped material. The explanation proposed for this depletion is hydrogen effusion during growth and agreement on expected widths of the depletion layers is obtained with calculations using meaured diffusivities. Results by *Carlson* et al. [2.66] on multiple layer films indicate both depletion and enhancement at the interfaces even when deposition occurs at low temperature. In contrast, photo-emission studies [2.67] indicate hydrogen saturated surfaces; clearly there is a disagreement with the nuclear reaction analysis that needs to be resolved.

Finally, two caveats relative to the analysis techniques emerge from this literature. *Jones* et al. [2.68] found that hydrogen evolution measurements can be in error by a factor of 2 depending on the substrate material used. Cross-calibration with nuclear reaction analysis indicates that ~ 50% of the hydrogen in films deposited on 7059 glass can diffuse into the substrate during thermal evolution while films on aluminum, due to rapid reaction of the silicon with aluminum at ~ 500 °C, release all their hydrogen. Secondly, nuclear reaction analysis has to be employed with care due to the heat generated during the analysis. A sputtered sample containing 25 at.% hydrogen was found by *Krishnaswamy* et al. [2.69] to be physically degraded ("massive cracks and film shrinkage") after nuclear reaction analysis.

b) Doped a-Si : H

Although a number of hydrogen content measurements on doped material had been made in isolation prior to the 1980 paper of *Müller* et al. [2.70], this paper provided the first systematic study of hydrogen profiles in device structures such as *pn* diodes and produced the result that the hydrogen content scaled linearly with Fermi-level position over a range of about 6 at.%. This result is shown in Fig. 2.23 for a graded pn junction. For different junction structures the results are shifted on an absolute scale but are still linear. The extremes of behavior are clearly that high boron content material shows very low, ~ 1 at.% contents, while 1% phosphorus doped material is similar to undoped material at ~ 7 at.%. Subsequent refinement of these studies by *Demond* et al. [2.65] revealed some very interesting results. Firstly they showed that the surface depletion of hydrogen mentioned in the previous section varied both in magnitude and spatial profile as a function of doping. This is illustrated in Fig. 2.24. Of particular interest here is the contrast between the 0.1% B_2H_6 and undoped samples, the former showing an

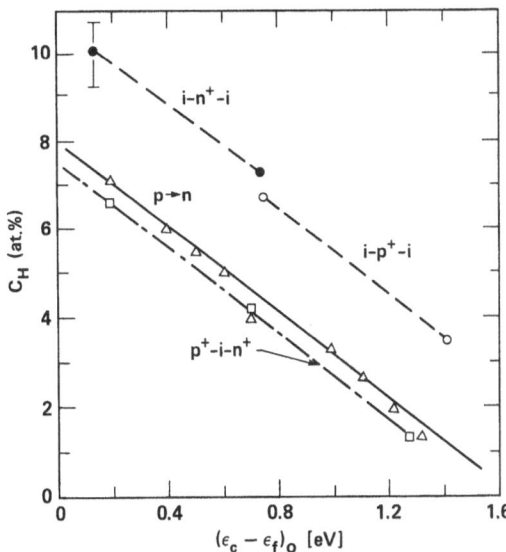

Fig. 2.23. Hydrogen content versus Fermi level position obtained by nuclear reaction analysis profiling of various junctions formed between doped and undoped plasma-deposited a-Si : H [2.70]

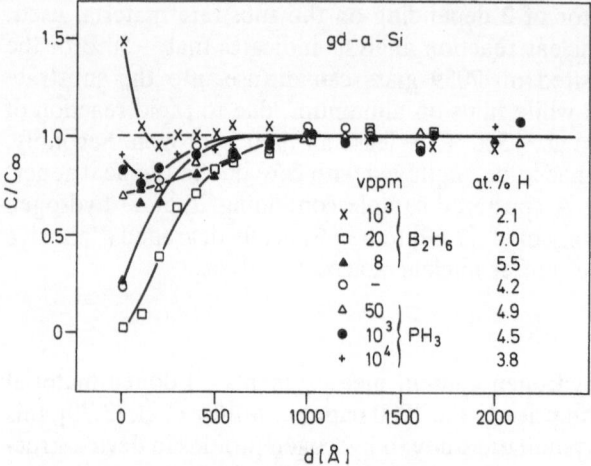

Fig. 2.24. Normalized hydrogen concentration determined by nuclear reaction analysis versus depth with the doping ratio as parameter [2.65]

enhancement of hydrogen at the surface ~ 50% of the bulk, the latter showing a depletion ~ 70%. Secondly, using the results of Fig. 2.24 to establish a bulk sampling depth of ~ 1000 Å, these authors remeasured the hydrogen content versus doping curve of Fig. 2.23; the new result is shown in Fig. 2.25. The major differences between these curves lie in the two peaks at ~ 0.1% PH_3/SiH_4 and ~ 0.01% B_2H_6/SiH_4. In this case C_H is clearly not a linear function of E_F. These results also indicate that the diffusivity of hydrogen is a function of doping. Yet another question regarding the hydrogen content of boron doped material is raised by the results of *Beyer* and *Wagner* [2.71] on hydrogen evolution. Their results are that (i) hydrogen evolves from heavily boron-doped material at much lower temperatures ($\leq 300\,°C$) than for undoped material and that (ii) as a consequence, the hydrogen content (measured by hydrogen evolution) of boron-doped material deposited at 300 °C

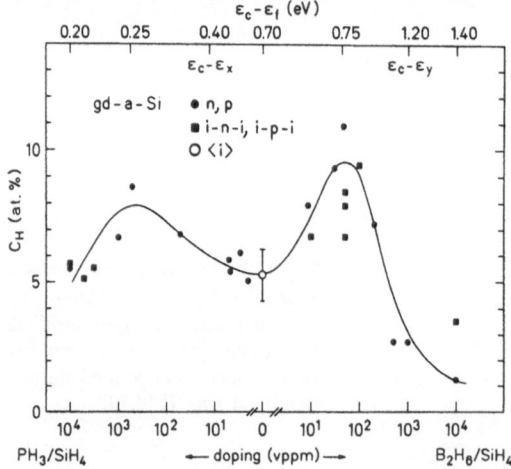

Fig. 2.25. Hydrogen concentration versus doping ratio of gd-a-Si samples. *n, p* means bulk material of the respective type, whereas *i-n-i, i-p-i* refers to a three-layer structure in which a central zone of *n* or *p*-type material is sandwiched between two intrinsic layers. Fermi-level positions corresponding to the doping ratios are given with reference to the conduction band edge [2.65]

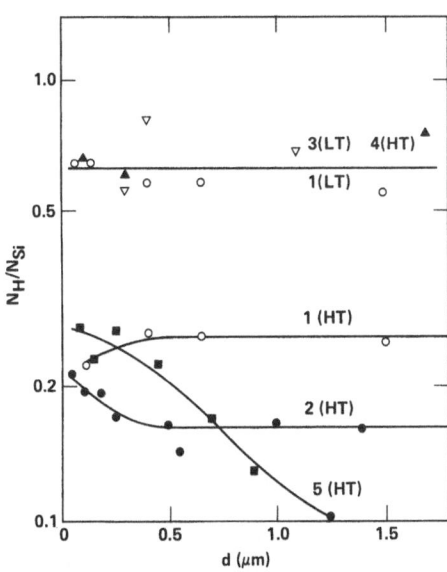

Fig. 2.26. Hydrogen content in the main evolution peaks as a function of film thickness for different a-Si:H films. Curve (*1*) undoped, $T_s = 25$ °C; (*2*) undoped, $T_s = 300$ °C; (*3*) 1% B_2H_6, $T_s = 25$ °C; (*4*) 1% B_2H_6, $T_s = 100$ °C; (*5*) 1% B_2H_6, $T_s = 300$ °C. HT and LT refer to high and low temperature evolution peaks, respectively [2.71]

actually decreases with increasing thickness of the film; this is illustrated by the curve labelled 5 (HT) in Fig. 2.26. Taken in conjunction with the results of *Demond* et al. [2.65], these results indicate that the absolute hydrogen content of doped plasma-deposited a-Si:H, particularly where the dopant is boron, is not a uniquely defined quantity. It depends on the thermal history both during and subsequent to deposition, and the fact that hydrogen effusion is taking place during deposition suggests that the hydrogen diffusivity in the materials on both sides of the doped layer may well influence the final hydrogen content.

Summarizing the results on hydrogen content measurement, it is clear that the scatter in data, even that taken from a single laboratory as shown in Fig. 2.23, is quite large and, within the 2–15 at.% range, there is to a first approximation no correlation with the electronic quality of the material. The question can then be asked why measure hydrogen content? As far as bulk measurements are concerned, the most appropriate answer is probably borrowed from mountaineering – "because it's there". As far as profiling is concerned, however, there are clearly issues relating to surfaces, interfaces and doping where hydrogen content measurements have and will continue to have considerable importance in building a coherent picture of material characteristics and behavior.

2.6 Impurities Other than Hydrogen

In order not to exceed the space restrictions, the discussion of impurities will be confined to those that are either deliberately added to produce substitutional doping or are present either as a result of contaminated source material, e.g., sputtering target, silane gas, etc. or due to their presence in the deposition ambient, e.g., argon in sputtering, chlorine from SiH_3Cl in SiH_4.

2.6.1 Dopant Impurities

The most commonly used dopants are boron, phosphorus and arsenic. Detection and spatial profiling of these dopants can be achieved by a number of techniques, the most sensitive and frequently employed being SIMS.

The issue addressed by most of the studies on plasma-deposited material has been the relationship between the dopant/silicon ratio in the gas phase compared to that in the film. The dopant incorporated shows, like the hydrogen content, considerable scatter from laboratory to laboratory, typical solid to gas phase ratios being ~ 0.5–2. Largely due to the fact that the absolute sensitivity of the majority of techniques is being stretched even at high dopant concentrations, there is little information about the linearity of the fractional incorporation with absolute dopant concentration. *Zesch* et al. [2.72] using SIMS and GDOS (glow discharge optical spectroscopy) have shown that the fraction of boron incorporated is linear to within a factor of two over the range of B/Si in the gas of 2×10^{-5} to 2×10^{-3}. This is illustrated in Fig. 2.27. *Chevallier* and *Beyer* [2.73] have reported a similar degree of linearity for boron in the 2×10^{-3} to 2×10^{-2} range using nuclear reaction analysis. They, along with *Street* et al. [2.74], have also studied compensated films where both boron and phosphorus are present. The results of Street et al. shown in Table 2.2, indicate a remarkable enhancement of both the boron and phosphorus incorporation ratio when the other dopant is present in excess, the apparent trend being toward "chemical" compensation, i.e., 1 : 1 dopant ratios. In contrast, Chevallier and Beyer found no such enhancement

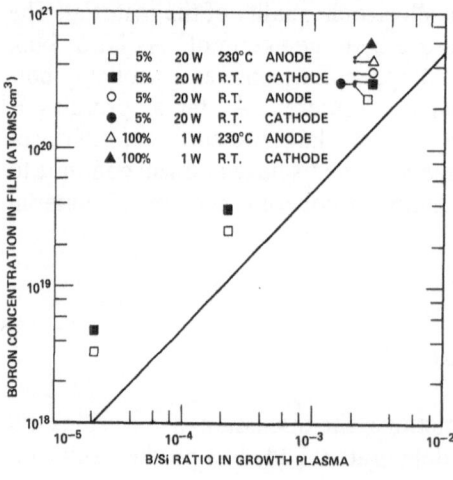

Fig. 2.27. Boron concentration in plasma-deposited a-Si : H films versus ratio of B to Si atoms in the growth plasma. A point representing a film whose B/Si ratio was the same as that in the plasma would fall on the solid line [2.72]

Table 2.2. Result of SIMS measurements of ^{11}B and ^{31}P concentrations. The enhancement factors E_P and E_B are the ratios of the measured to the nominal concentrations [2.74]

| Nominal concentrations | | Enhancement | |
P	B	E_P	E_B
10^{-3}		5.2	
10^{-3}	10^{-4}	4.3	23
10^{-3}	10^{-3}	3.5	4.2
10^{-3}	4×10^{-3}	9.4	4.1
	10^{-3}		3
3×10^{-4}	10^{-3}	9.3	7.1
3×10^{-3}	10^{-3}	4.8	13.3
10^{-4}	10^{-4}	8	4.5
3×10^{-4}	3×10^{-4}	5.8	6.2

over a similar range of dopant concentrations. Several possible explanations exist for this discrepancy. The matrix sensitivity of SIMS could result in higher yields of one species if, for instance, it was indeed complexed with another as, for example, boron-phosphorus groupings. Another more general explanation that can also help explain the scatter in fractional incorporation for boron involves gas-phase depletion. If all the gas fed into the plasma is converted into film, then clearly the average incorporation fraction will be unity at all concentrations. If on the other hand only a small fraction of the gas is consumed, there is much more latitude for selective chemistry to occur. Situations between these extremes will occur within a given deposition system as a function of gas flow rate and the placing of the substrate holder relative to the gas inlet and exhaust ports. Examination of the above results suggests that this explanation is not inconsistent with the data. Street et al. used a flow rate of ≥ 100 sccm, whereas Chevallier and Beyer used flow rates

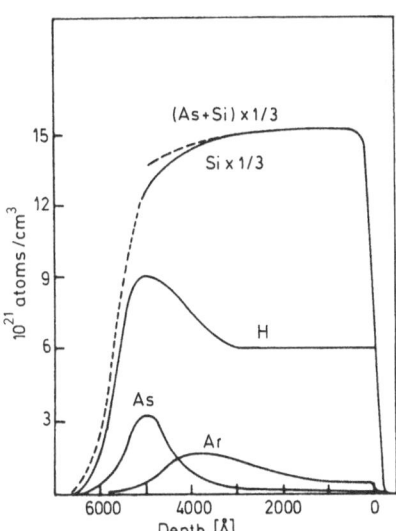

Fig. 2.28. Concentration profiles of arsenic, hydrogen and argon in sputtered a-Si:H with the bottom layer doped with arsenic [2.75]

~ 6 sccm and a deposition rate ~ 50% higher than Street et al. These factors would lead to a depletion of gas, assuming approximately equal plasma volumes, that would be ~ 30 times greater in the case of the work of Chevallier and Beyer.

Boron has attracted the most attention both because of its importance as the only easily available *p*-type dopant and because of the relatively high absolute sensitivity in SIMS and nuclear reaction analysis. Phosphorus can be detected by SIMS, but due to mass interferences from silicon isotopes, the sensitivity is poor and only the highest mass resolution instruments can be used. Arsenic has received relatively little attention due to its inferior doping characteristics but it has been studied in sputtered a-Si : H by *Deneuville* et al. [2.75]; their result is shown in Fig. 2.28. Since the arsenic is introduced as AsH_3 and the silicon is sputtered, the As/Si ratio in the film can be arbitrarily controlled and as such is not a particularly significant parameter in itself. It is, however, interesting to note the large influence that the presence of arsenic has on the argon and hydrogen contents, lowering one and raising the other by substantial fractions. This result, the authors suggest, is a manifestation of the microstructural properties of the material similar to that observed in plasma deposited material.

2.6.2 Contaminant Impurities

Since the contamination sources are different for plasma-deposited and sputtered material, they will be considered separately.

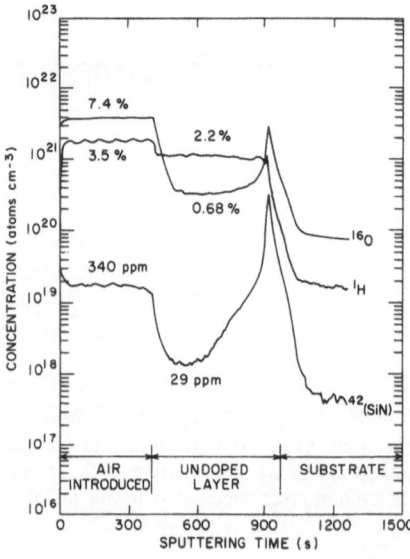

Fig. 2.29. SIMS concentration profiles of oxygen, hydrogen and SiN (for nitrogen) in plasma deposited a-Si : H with and without a deliberate air leak [2.76]

a) Plasma Deposited Material

The most extensive study of contaminant impurities and their effect upon electronic properties has been undertaken by *Delahoy* and *Griffith* [2.76] and coworkers at Brookhaven National Laboratory. Figure 2.29 illustrates SIMS results on material deposited in a "clean" system followed by a deliberate air leak, showing that the unintentional oxygen incorporation is ~ 1 at.%. Using deliberately added nitrogen and oxygen, this group has shown that simultaneous incorporation of nitrogen and oxygen is far more deleterious to electronic properties than either added singly. The source of these oxygen impurities in the absence of air leaks has been addressed by *Knights* et al. [2.77] using $H_2^{18}O$ to isotopically label the adsorbed water vapor on the walls of the reaction chamber and associated tubing. It is found that $\geq 50\%$ of the oxygen originates from this adsorbed water vapor and that system baking prior to deposition is an effective means of reducing the concentration. The Brookhaven group has also demonstrated from both SIMS and optical emission spectroscopy that chlorine from SiH_3Cl impurities in the silane tanks is a pervasive contaminant at the ~ 500 ppm level. They demonstrated that chlorine is electrically active in that solar cell properties are degraded, and have shown that by using special filters, the SiH_3Cl can be removed from the silane prior to depositon. Carbon is also a contaminant at levels similar to nitrogen (Fig. 2.1), i.e., ~ 10^{18}–10^{19} cm^{-3} with large interface peaks ($\geq 10^{20}$ cm^{-3}). It is presumed that hydrocarbons adsorbed on chamber walls are the major source although no studies have been published.

In a class by themselves are the inert gases used as diluents of silane by a number of groups. *Tanaka* et al. [2.78] and *Tsai* and *Fritzsche* [2.79] have shown that argon in concentrations up to ~ 8 at.% is incorporated into material deposited onto the cathode of asymmetric rf diode deposition systems when silane is diluted 10% in argon. This material is known (see section on structural inhomogeneity) to have isotropic small angle scattering indicated small voids which are not interconnected and it is believed that the argon is trapped in these defects. Other inert gases will presumably show the same behavior. It is important to note that material deposited on the anode of such systems, i.e., substrate potential close to plasma potential, does not show detectable argon content (≤ 0.01 at.%).

Finally, it should be mentioned that the material deposited at low substrate temperature, particularly that with columnar morphology, can undergo rapid post-deposition surface (internal or external) oxidation leading to average oxygen concentrations in the alloy range.

b) Sputtered Material

The fact that the sputtering process requires the presence of an inert gas of relatively high mass (Ne, Ar, Kr) suggests that these are likely to be impurities in sputtered material. This is confirmed directly by the work of *Ross* and *Messier* [2.80] illustrated in Fig. 2.30; concentrations of ~ 5 at.%

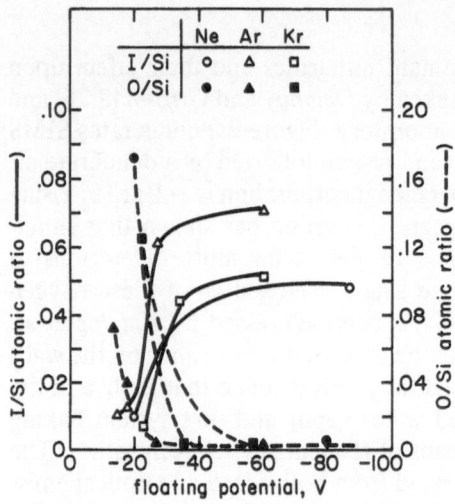

Fig. 2.30. Oxygen content and entrapped Ne, Ar and Kr contents in sputtered a-Si:H as a function of floating potential [2.80]

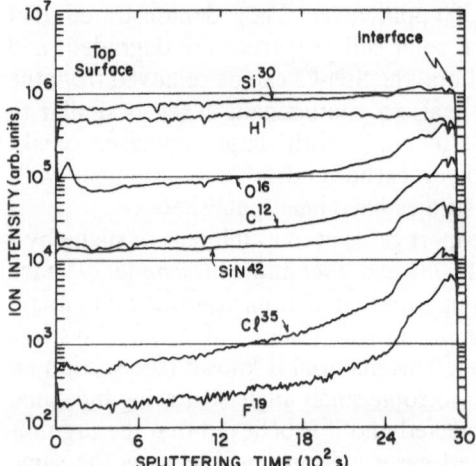

Fig. 2.31. SIMS profile of impurities in diode sputtered a-Si:H from the Harvard group. Note that the ion intensities are not normalized and do not permit relative concentrations of impurities to be gauged [2.82]

are found when, as in the case of plasma-deposited material, the substrate is held at a significant negative potential with respect to the plasma. The incorporation is linked by these authors to the suppression of columnar microstructure by ion bombardment of the growing film. This reduction in morphology is in turn reflected in a reduction in post-deposition oxidation to levels below the detection limit of electron microprobe (~ 0.01 at.%). These results are confirmed by several groups, notably those of *Tanaka* [2.78], *Deneuville* [2.75] and *Maruyama* [2.81]. The latter have expanded the study of neon incorporation using thermal evolution and Auger spectroscopy to show that there is a unique site for neon in hydrogenated material which they propose is neon trapped in a hydrogenated vacancy or void.

The other impurities found in plasma-deposited material are also present in sputtered material. Figure 2.31 shows results of SIMS profiling from the

Harvard group [2.82] indicating the characteristic interfacial enhancement and subsequent decay in the concentrations of oxygen and carbon. Noteworthy is the presence of chlorine and fluorine indicating their presence either in the sputtering gases or in adsorbates. It is important to recognize that there are considerably more opportunities for impurity incorporation in sputtering than plasma-deposition due both to the additional need to make a high purity target (and keep it pure) and to the fact that the reduced pressure of the sputtering process makes the relative partial pressure of contaminant gases emitted from the chamber walls higher.

2.7 Compositional Inhomogeneity

Ever since the hydrogen content of a-Si:H was first measured, it has been remarked that the amount of hydrogen present is far in excess of that needed to saturate the dangling-bond density generated when constructing a continuous random network model. The latter figure is $\sim 10^{17}$ cm^{-3}, the hydrogen content $\sim 10^{21}$ cm^{-3}. The implications of this discrepancy are either that there is a homogeneous hydrogen distribution (SiH$_2$ and SiH$_3$ clusters expected) with a far greater density of dangling bonds than a random network when the hydrogen is removed, or that the hydrogen is inhomogeneously distributed in some microscopic sense. It is the purpose of this section to review the experimental evidence that addresses this issue.

2.7.1 Infrared Spectroscopy and Hydrogen Evolution

Prior to the observation of structural inhomogeneity, *Knights* and coworkers [2.83, 84] had noted that a feature at 840 cm^{-1} in the Si–H vibrational spectrum could only be explained satisfactorily by the postulate that chains of SiH$_2$ groups (SiH$_2$)$_n$ or polysilane (more accurately polysilylene) existed in the material. The arguments behind this postulate are developed elsewhere in this book and are now generally accepted. Upon the observation of columnar morphology in SEM, it was found that the trends leading to pronounced columnar morphology were paralleled by increased intensity in the 840 cm^{-1} band. The proposition was then made that the polysilane-like chains existed in interstitial regions between columns of less-hydrogenated material. This proposition was supported by the observation that the pronounced internal surface oxidation of material with well-developed columnar morphology, observed via the growth of Si–O vibrational modes, occurred at the expense of the (SiH$_2$)$_n$ vibrational features. Subsequently, more detailed measurements of hydrogen evolution and transmission electron microscopy of thin samples before and after hydrogen evolution have provided strong support for the idea that in material with a visible columnar morphology, there is hydrogen clustering in the form of (SiH$_2$)$_n$ at or near the column surfaces.

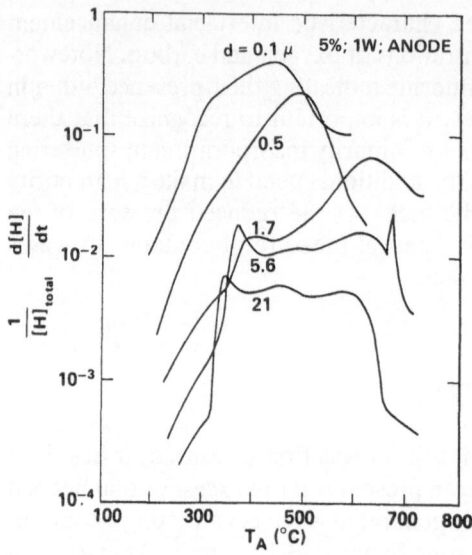

Fig. 2.32. Derivative evolution spectra normalized to total hydrogen evolved for samples of plasma-deposited a-Si:H of various thicknesses. All samples have columnar amorphology. Curves are offset by factors of e^{-1} [2.85]

Figure 2.32 shows the results of hydrogen evolution measurements by *Biegelsen* et al. [2.85] on material with columnar morphology (not particularly pronounced) as a function of film thickness. The conclusion that there is a peak in the evolution rate at $\sim 350\,°C$ that is thickness independent implies rapid diffusion of hydrogen through some form of connected internal microvoids. The observation that the $(SiH_2)_n$ vibrational modes are reduced sharply in intensity during the 350 °C evolution step suggests that the evolving hydrogen is indeed clustered. Finally, a combination of TEM and Raman measurements on annealed and etched material with columnar morphology by *Nemanich* et al. [2.35] indicates directly that the intercolumnar spaces become more electron transparent, i.e., the clustered hydrogen is indeed located in these spaces prior to evolution. Neutron scattering experiments on both plasma-deposited and sputtered material produce conflicting evidence on this issue. *Leadbetter* et al. [2.86] note that an increase in small angle scattering occurs on evolution through the 350 °C step and suggest that the intercolumnar spaces are in fact empty, i.e., true voids and that, while the surfaces of the columns may be more heavily hydrogenated, the columns are essentially homogeneous. *Postol* et al. [2.12], using deuterium substitution, also conclude that the hydrogen is predominantly located on or near void surfaces rather than as a separate phase. Clearly more work in this area would be useful although the sample mass requirements are onerous.

Two final results that bear directly on the question of homogeneity also come from hydrogen evolution measurements. First is the observation by a number of groups [2.85, 87–89] that the electron spin density measured after substantial (atomic percentages) amounts of hydrogen has been evolved during the $\sim 350\,°C$ anneal step is only increased to the $10^{17}–10^{19}\ cm^{-3}$ range. This factor of ≥ 100 between hydrogen atoms evolved and dangling bonds created is direct evidence for surface reconstruction of dangling bonds to preserve spin pairing. Second is the evolution temperature itself; the SiH bond is so strong that the lowest temperature at which silane will decompose is ~ 600 °C and even then the energy of bond scission is partially recovered by the formation of an H_2 molecule. In order for covalently bonded hydrogen to evolve at 350 °C or below, there has to be a substantial recovery of energy by Si–Si bond formation which can only occur if hydrogens are clustered on adjacent silicon atoms.

2.7.2 Proton Magnetic Resonance

Since nuclear magnetic resonance is a technique infrequently used on amorphous solids and since some of the more recent advances in technique have been used on a-Si : H, it is useful to briefly describe the theoretical and experimental background.

The Hamiltonian for a nuclear spin in an external magnetic field is the sum of a number of terms

$$H = H_Z + H_{DT} + H_{DM} + H_{CS} + H(t) \, ,$$

where H_Z is the Zeeman Hamiltonian due to the external magnetic field, H_{DT} is the heteropolar dipolar Hamiltonian due to dipole-dipole interaction between unlike spins, H_{DM} is the homopolar Hamiltonian which arises from dipole-dipole interactions between like spins, H_{CS} is the chemical shift Hamiltonian arising from magnetic fields at the nucleus due to orbital motion of the electrons and $H(t)$ are fluctuating magnetic fields due to, e.g., motion of nuclear or electron spins or applied by the experimenter. The Zeeman term determines the position of the resonance, perturbed slightly by the chemical shift term. In the case of a-Si : H, the heteropolar dipole term is negligible compared to the homopolar dipole term but both contribute to a random static field that determines the width of the resonance line. Rapid motion of either nuclear or electron spins causes the average dipolar field to be reduced and hence the linewidth decreased; this motional narrowing can typically be separated from the other terms by studying the temperature dependence of the linewidth. Finally, by applying pulsed magnetic fields in suitable sequences and/or spinning the sample at high speeds, $H(t)$ can be made equal and opposite to the dipolar Hamiltonians permitting the chemical shift term to be studied.

As in many spectroscopies, the resonance can be studied either in the time or frequency domain. Both have been used on a-Si : H, but due to a relatively long spin-lattice relaxation time (T_1), there are signal/noise advantages to working in the time domain and then Fourier transforming to the frequency domain. Thus, the typical experiment consists of applying a sequence of magnetic field pulses spaced $\geq 5 \ T_1$ apart via an external coil which

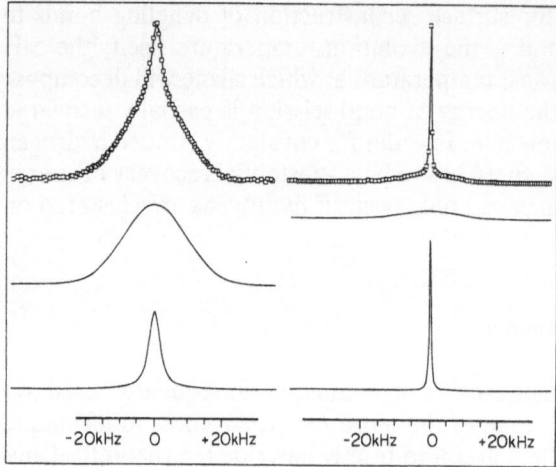

Fig. 2.33. Experimental proton magnetic resonance spectra in a-Si:H with Gaussian and Lorentzian fits showing the quality of the fits. On the left is the spectrum of an as-deposited (room temperature) sample; on the right is that of the same sample annealed at 400 °C [2.99]

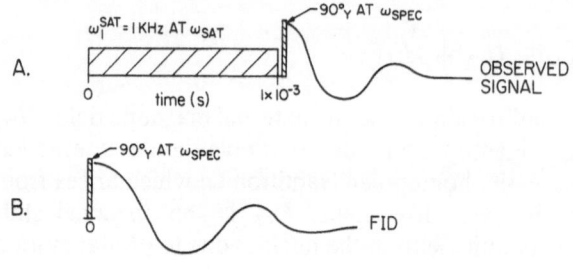

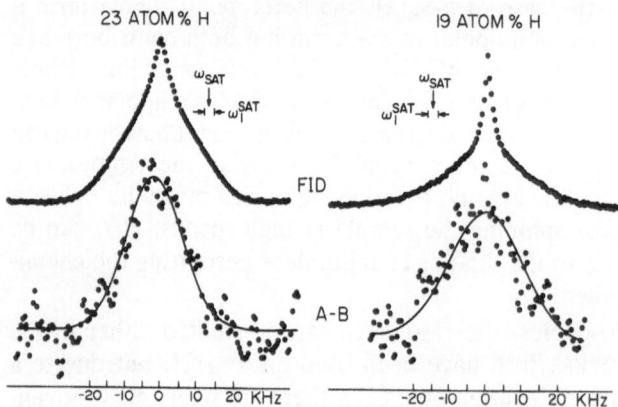

Fig. 2.34. Schematic diagram (*top*) of the pulse sequences used in the "hole-burning" experiments. Below, the results for two samples. The dark line in the difference spectra ($A - B$) are Gaussians best fit to the broad component of the FID spectra [2.95]

places the magnetization perpendicular to the main field and then measuring the decay of the magnetization. This free induction decay is governed by the spin-spin interactions and in a homogeneous system would yield a single characteristic decay time, T_2. On Fourier transformation this translates into the linewidth. Digital transient recorders and computerized signal averaging are typically used to acquire the data.

A number of groups have now studied proton NMR [2.90–92] on both plasma-deposited and sputtered material and have obtained essentially the same results. The resonance line shape typical of these studies is shown on the top left-hand side of Fig. 2.33. It can be decomposed into two lines as shown, one a Gaussian with, in this case, a ~ 25 kHz width and one a Lorentzian with a width ~ 3 kHz. A multiple-pulse sequence experiment designed to remove the dipolar contribution to the linewidth causes this line to narrow to ~ 500 Hz, a value consistent with expected chemical shift interactions. Thus, the implication of this result is that there are two distinct environments for protons, one giving rise to the narrow Lorentzian line having a slower T_2, i.e., experiencing weaker homonuclear dipole interactions, while the other has a faster T_2, i.e., stronger dipolar interaction. The "distinctness" of the two environments has been an issue primarily because the measured spin-lattice relaxation time T_1 has been found to be identical to within $\pm 10\%$ for the two environments. There are, however, two experiments which indicate that the environments are indeed separate. The first, a hole-burning experiment [2.95], is illustrated in Fig. 2.34. A weak rf pulse is applied to the sample at a frequency shifted from the line center sufficiently far to be absorbed only by protons contributing to the broad line. In the absence of any spin (spectral) diffusion, this pulse would literally "burn" a hole in the line. If there is spin diffusion, then the absorbed energy will be distributed over the communicating spins. This distribution can be detected by measuring the free induction decay (FID) before and after the saturating pulse, the difference between the two FIDs being those spins whose populations on the Zeeman split levels have been equalized (saturated) by the off-resonance pulse. The result clearly indicates that there is no spin diffusion between the two environments during the measurement time ~ 1 ms. *Carlos* and *Taylor* [2.96], using a dipolar echo measurement, have confirmed this result in the 25–250 μs regime.

Calculations to determine what microscopic distributions of hydrogen can give rise to the observed lines have been made by most of the experimentalists involved in the measurements. For the broad Gaussian line the FWHM is equal to $3.36 \sqrt{M_2}$ where M_2 is the second moment of the line given by

$$M_2 = 3/5 \; \gamma^4 h^2 I(I + 1) \sum_{ij} r_{ij}^{-6} \; ,$$

where I is the nuclear spin ($\frac{1}{2}$ in this case), γ is the nuclear gyromagnetic ratio and r_{ij} is the spacing between spins i and j. Thus for simple systems, e.g., local clusters such as SiH_2 and SiH_3 and cubic lattices, the linewidth is calculable.

In the limit where the spin density is less than 0.001 of the available site density, the system is termed dilute, the line is Lorentzian and a linewidth can be calculated for a random distribution. Even though, as will be shown later, the content of hydrogen in the narrow line is typically larger than the dilute limit on an atomic percent basis, the fact that the amorphous network has a much larger number of potential sites than the corresponding crystal is used to argue that the narrow line does indeed arise from a dilute random distribution. Table 2.3 shows the results of linewidth calculations by *Carlos* and *Taylor* [2.96] for a number of potential hydrogen clustering configurations. Three broadly accepted conclusions have been drawn from a comparison of these calculations with the results. The first is that the narrow line can only arise from a random distribution of protons bound in monohydride (SiH) configurations. The second is that given suitable spatial distributions, a number of configurations [SiH_2, $(SiH_2)_n$, heavily clustered SiH such as found on a surface] can give rise to the broad line and that comparison with other hydrogen-bonding specific measurements such as vibrational spectroscopy has to be made to identify in any specific instance which might be the major contributor. The third, arrived at by different reasoning and from different results by three groups, is that there are regions of many samples that have essentially zero hydrogen. *Reimer* et al. [2.94] concluded that this was universally necessary to provide buffer layers between protons in the two environments to prevent spin diffusion occurring during the hole-burning experiment. *Jeffrey* et al. [2.97] concluded this from the result that the linewidths of both the narrow and broad lines were independent of total hydrogen content over a factor of five in hydrogen content in sputtered a-Si:H. *Carlos* and *Taylor* [2.96] compared the measured linewidth of the narrow line with that calculated from the measured hydrogen content assuming a random distribution and concluded that clustering was occurring in certain samples even in this environment and as a result that there were empty or near empty regions.

Table 2.3. Calculated ^{1}H NMR linewidths for different proton distributions [2.96]

Bonding arrangement	σ [kHz]*
SiH_2	13.4[a]
SiH_3	19[b]
S(111):H	8.5[b]
C–Si hydrogenated monovacancy	114[b]
C–Si hydrogenated divacancy	94[b]
Uniform array	0.28/at. %[c]
Random array	0.46/at. %

* Full width at half maximum   [b] Assumes a Gaussian shape
[a] This is the width of a Pake doublet due to SiH_2 [c] Assumes a cubic lattice

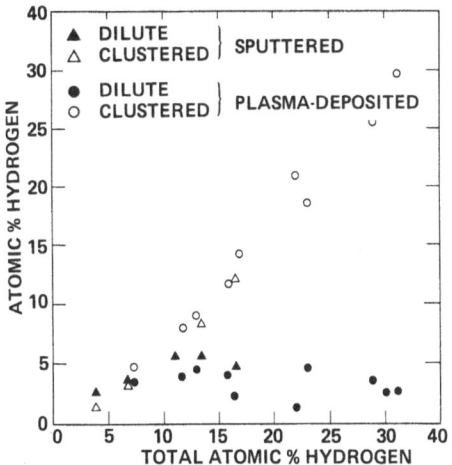

Further information concerning the role of the hydrogen in the broad line has come from studies of the distribution of hydrogen content between the two lines and of the linewidths as a function of total hydrogen content. Figure 2.35 shows the results of *Reimer* and *Knights* [2.98] on a broad range of plasma-deposited samples together with results from *Jeffrey* et al. [2.97] on sputtered samples. As the hydrogen content (C_H) is increased from zero to ~ 10 at.% in the sputtered material, the proportion going into the broad line increases from ~ $\frac{1}{2}$ that in the narrow line to approximately the same. Increasing C_H above this level causes the amount in the broad line to grow but the amount in the narrow line remains constant. Over this whole range the linewidths do not change. In the as-deposited plasma-deposited material at the lowest concentration (~ 8 at.%), the amount of hydrogen in the two lines is approximately the same and increasing the overall hydrogen content causes all excess hydrogen to go into the broad line. Interestingly, as will be discussed below, annealing the plasma-deposited material causes preferential evolution from the broad line, i.e., annealed plasma-deposited material has a similar distribution of hydrogen to low hydrogen content sputtered material. Figure 2.35 also illustrates the overall similarity in behavior, as far as proton NMR is concerned, between sputtered and plasma-deposited material.

The potential of a combination of NMR measurements and hydrogen evolution has been recognized by several groups. *Reimer* et al., studying material deposited at room temperature with a high hydrogen content [2.99] found the following: (i) the hydrogen evolving first (at low temperature) came exclusively from the heavily clustered environment, some increase in the content of the dilute environment actually being observed as hydrogen evolved; (ii) the linewidth of the dilute environment narrowed sharply, starting to decrease even before significant evolution from the other environment had commenced, indicating directly that there was clustering even in this "dilute" environment in the as-deposited material; (iii) after the major loss of

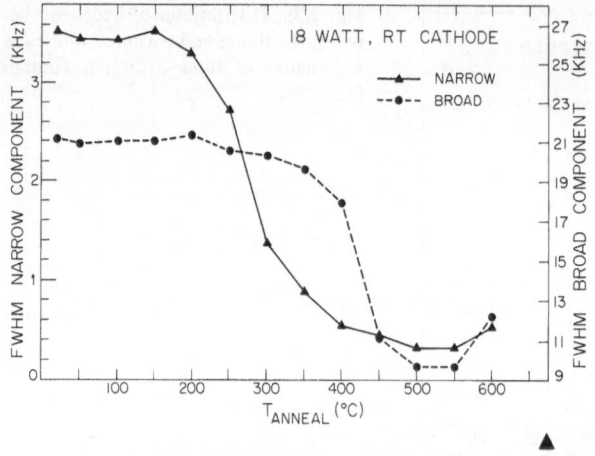

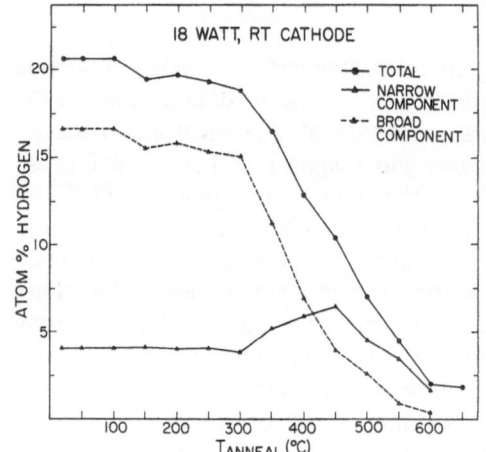

▲

Fig. 2.36. The full width a half maximum (FWHM) for both high and low density phases of hydrogen in the sample shown in Fig. 2.1. Note the two different scales for the two components [2.99]

◀ **Fig. 2.37.** At.% hydrogen for high and low density hydrogen phases in a sample deposited at room temperature from 5% SiH₄/Ar onto a cathode substrate. T_{anneal} refers to the temperature the sample was annealed to [2.99]

hydrogen at ∼ 350°–450 °C from the clustered environment, hydrogen evolved at an approximately equal rate from both types of-site, this being reflected in both linewidths. Figures 2.36, 37 illustrate these results. *Ueda* et al. [2.100] reported similar results as far as the hydrogen loss is concerned (i.e., from the broad line) in sputtered a-Si : F : H. *Carlos* and *Taylor* [2.101], starting with a high temperature plasma-deposited sample, observed only a slight preference for evolution from the broad line and saw no lineshape changes except those accounted for by hydrogen loss.

The overall picture that develops from these linewidth studies and the hydrogen evolution results is that under all conditions studied in both plasma-deposited and sputtered material, there is strong evidence for an inhomogeneous distribution of hydrogen that is strongly correlated, at least in the case of plasma-deposited material, with structural inhomogeneity. As mentioned earlier, there are still some unresolved issues from the neutron scattering results as to whether the two inhomogeneities are commensurate or even of the same scale, but the balance of results suggests that this is the most probable relationship.

2.7.3 Proton Spin-Lattice Relaxation and Other NMR Measurements

The richness of techniques encompassed by NMR spectroscopy and the speed with which they have been applied to a-Si are such that reviewing all the results cannot be attempted in the space available. The reader is referred to [2.90–101] for a more complete coverage. There are, however, several important results concerning proton spin-lattice relaxation and boron and fluorine NMR which deserve mention.

Spin-lattice relaxation is the process by which the excited spin population returns to the equilibrium thermal distribution following rf irradiation. Energy is transferred to the lattice via fluctuating magnetic fields at or near the frequency corresponding to the Zeeman splitting. In most proton containing solids, these arise from rapid motion of nuclei, e.g., rotating CH_3 groups in polymers or from electrons in paramagnetic centers. The issue in a-Si : H is that material with very low densities of paramagnetic centers and with no apparent motion of the protons (from the temperature independent linewidth) is found to have a rather short spin-lattice relaxation time ($T_1 \sim$ a few seconds). The results of *Carlos* et al. [2.102] on the temperature dependence of T_1 for both plasma-deposited and sputtered a-Si : H are shown in Fig. 2.38. The low temperature minimum has also been observed in sputtered a-Si : H by *Jeffrey* et al. [2.97]. The power law dependence on temperature shown here for sputtered material has been attributed directly to relaxation via paramagnetic dangling bonds. The low temperature minimum, on the other hand, cannot be so simply explained. Other measurements show that from sample to sample, the minimum stays at approximately the same temperature while the magnitude of T_1 varies in the range ~ 0.2–10 s. Below the minimum, T_1 is frequency dependent whilst above it is frequency inde-

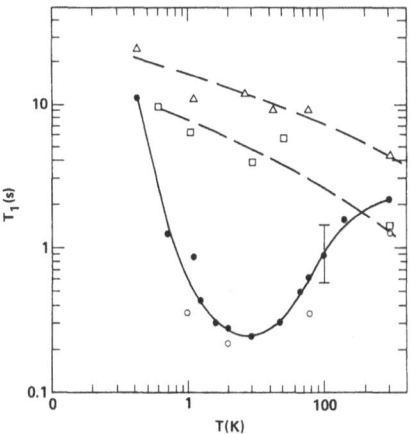

Fig. 2.38. Spin lattice relaxation time (T_1) for protons for the glow-discharge film (●) and the sputtered films [(A △), (B □), (C ○)]; from [2.102]

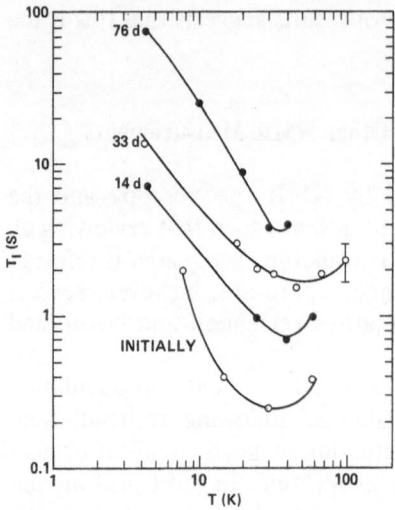

Fig. 2.39. Spin lattice relaxation time (T_1) in plasma-deposited a-Si : H as a function of temperature; the different curves represent measurements after the indicated number of days at liquid helium temperature [2.96]

pendent. On annealing above 250 °C, *Reimer* et al. [2.99] observed a decrease in the room temperature magnitude of T_1 and eventually, at ~530 °C, *Carlos* and *Taylor* [2.101] observed a transition from the "minimum" behavior to the power-law type curve.

The first attempt to explain this behavior by *Carlos* and *Taylor* [2.92], invoked a two-step relaxation process whereby spins diffused to two-level disorder modes associated with hydrogen atoms at which the relaxation process itself took place. *Movaghar* and *Schweitzer* [2.103], in a more detailed analysis, pointed out that in order for this model to produce the correct magnitude of T_1, the number of hydrogen atoms associated with disorder modes would have to be ~ 10^{19}–10^{20} and since these atoms would be in rapid motion, a motionally narrowed component should be readily observable in the resonance line itself. Subsequently, *Conradi* and *Norberg* [2.104] proposed a model based on relaxation via trapped molecular hydrogen that has passed an important experimental test. The mechanism in this model involves spin diffusion to "sinks" to the lattice in a similar fashion to *Carlos* and *Taylor* [2.92], the difference being that the relaxing centers are orthohydrogen molecules which are coupled to the lattice via transitions between rotational energy levels. This model requires only a small number (~ 1% of the total hydrogen) of hydrogen molecules and is essentially the same as that used to explain the relaxation of molecular hydrogen in rare gas solids [2.105]. The prediction of this model is that if the orthohydrogen (spin 1) is converted to parahydrogen (spin 0) by holding the sample at liquid helium temperatures, T_1 should increase. *Carlos* and *Taylor* [2.96] have demonstrated that this is indeed the case. Figure 2.39 shows the results of holding a sample for a total of 76 days at liquid helium temperatures confirming that T_1 increases substantially; the effect is reversed on annealing to room temperature. This trapped molecular hydrogen is expected to produce a motionally narrowed line which is not observed by Carlos and Taylor, a possible explanation being that site symmetry reduces the required concentration to below that detectable. *Reimer* et al. [2.95] have, however, observed an additional motionally narrowed line in "aged" high-hydrogen content a-Si:H, in boron-doped a-Si:H and in materials such as a-C:H, a-Si:N:H [2.106] which might be attributable to hydrogen molecules. Having shown that spin diffusion and relaxation through orthohydrogen are the essential elements in the T_1 process, it is clear that deuterium substitution, which will modify both processes for protons and provide additional information via the deuterium resonance, is an attractive area of research. Work has commenced in this area and readers are referred to the papers of *Reimer* et al. [2.107] and *Leopold* et al. [2.108] for the results.

NMR has also been used to address the question of both homogeneity and bonding configuration in heavily boron-doped material and the issue of fluorine homogeneity in both plasma-deposited and sputtered a-Si:F:H and a-Si:F. In the case of boron doping, *Reimer* et al. [2.95] noted that the proton resonance line in a 1% B_2H_6/SiH_4 sample had a small additional broad line (FWHM ~50 kHz) which was similar in width to that observed for protons in a-B:H and attributed this to heteropolar ($^{10}B/^{11}B$) broadening. These authors deduced that there was spatial clustering of boron in the same regions in which the heavily clustered hydrogen exists. *Greenbaum* et al. [2.109] studied the boron (^{11}B) resonance in plasma-deposited material with ~10% boron incorporated. They found that, like the protons, there were two sites for boron giving rise to different lines although in this case the linewidth is primarily determined by quadrupolar interactions. From the

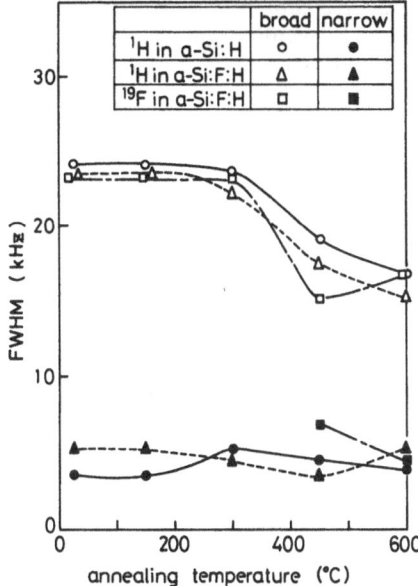

Fig. 2.40. Changes with annealing contents of H and F contributing to the broad and narrow NMR lines for a-Si:H and a-Si:F:H [2.93]

magnitude of the quadrupolar interaction extracted from the frequency dependence of the linewidths, the authors concluded that at least 90% of the boron is in fact threefold coordinated and suggested that the two sites are BSi_3 and Si_2BH. Comparison of the measured T_2 with a random distribution of boron in a glass ($3 B_2S_3 7 As_2S_3$) suggests that the boron is significantly clustered, although comparison with a-B and boron carbide leads the authors to rule out large numbers of B–B bonds.

Ueda et al. [2.93, 100] have studied fluorine NMR in sputtered a-Si : F and a-Si : F : H. They observed two dipolar broadened lines in both the as-deposited a-Si : F and on annealing in a-Si : F : H. Their results on annealing are shown in Fig. 2.40. The linewidths, observed in a-Si : F ~ 23 kHz and ~ 4.4 kHz, are motional narrowed at room temperature but are still clearly separated at liquid helium temperatures indicating, as in the case of protons, two spatially separated reservoirs. In contrast, the two-line behavior observed only after annealing in a-Si : F : H is only pronounced at 77 K with the narrow line being a motionally-narrowed portion of the total, disappearing at 4.2 K and dominating at 300 K. The authors attribute the motion to $(SiF_2)_n$ or SiF_4 units that are more mobile in a-Si : F : H than in a-Si : F.

2.8 Surfaces and Interfaces

The properties of surfaces and interfaces will certainly be important for any device applications of a-Si : H. As has been the case for crystalline thin film semiconductors like GaAs, after initial characterization and optimization, the interfacial properties often determine device capabilities. Because of the high pressure deposition conditions of a-Si : H, examination of the free surface using traditional surface analysis techniques is difficult. Furthermore, the amorphous structure makes interpretation difficult or the techniques inapplicable, as in the case of LEED.

Despite the difficulties with obtaining clean surfaces, the properties of interfaces can still be studied. While interface structures obtained through standard fabrication procedures are important, "clean" interfaces are also obtainable. In the following, the properties of the a-Si : H surface and oxidation properties are considered as are the interactions at the metal/a-Si : H interface.

2.8.1 Surface and Oxidation Properties

Some of the properties of the free surface and substrate interface have been addressed in previous sections. For instance, the nucleation process may cause structural changes at the substrate interface. In addition to the hydrogen profiling and photo-emission results mentioned previously, Raman measurements have been carried out on 100 Å thick films and it was found that the spectra resembled that of thick (~ 1 μm) films [2.110]. Since the spectra obtained from thick films averages over the top 4000 Å, it has been suggested that the 100 Å films are similar to the average thick films. Thus, the Raman and photoemission [2.67] results seem to contradict the nuclear-reaction profiling results [2.65] and more work must be done to sort out the apparent disagreement.

Another measurement of surface or substrate interface properties is the subgap absorption measured by photo-thermal deflection spectroscopy (PDS). It has recently been shown that the sub-gap absorption of low defect films does not scale with thickness [2.111]. This result has been interpreted as indicating that significant defect density exists on either the free surface or the substrate interface.

The oxidation properties of a-Si:H have been studied by ellipsometry and quartz oscillator frequency change. In the frequency change experiment, a-Si:H was deposited on a quartz oscillator microbalance which was then exposed to air in situ [2.112]. The results are shown in Fig. 2.41. The conditions of curve b exclude effects due to the pressure change and aerodynamic loading. Two components of mass increase were observed, both of which had approximately equal magnitude. One component could be removed by vacuum annealing at 140 °C which was attributed to absorbed gas while the remaining component was attributed to silicon-oxide formation. These measurements also showed an almost instantaneous initial component of approximately 10 to 20% of the long exposure limit, and the long exposure oxide was found to be ~ 5 Å thick. Furthermore, the oxidation properties did not change with film thickness indicating the effects occurred at the surface.

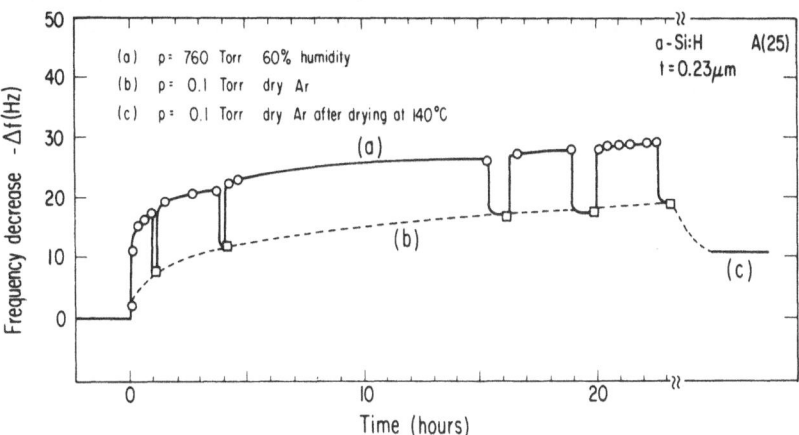

Fig. 2.41. Frequency decrease of an a-Si:H film upon exposure to air of 60% humidity at 760 torr [curve (a)]. Curves (b) and (c) were measured in the 0.1 torr dry Ar environment which surrounded the sample before air was admitted. Curve (c) was obtained after drying the sample at 140 °C for one hour in dry Ar flow at 0.1 torr [2.112]

The ellipsometry results found a similar oxide thickness for long exposure to air [2.113]. However, these results also showed there was considerable anisotropy in the oxidation process. This was attributed to the structural anisotropy of the film. This effect has also been observed by *Street* and *Knights* [2.114]. It should be emphasized that the oxidation rate of a-Si : H is significantly less than that of crystalline Si surfaces [2.113]. This is apparently even true for a-Si : H which is less than ideal.

2.8.2 Metal/a-Si : H Interfaces

Because of the initial considerations of Schottky barrier configurations for solar cell structures and because of the need for metal ohmic contacts to any device, the properties of the metal/a-Si : H interface have received the most attention to date. It is clear that a complete understanding of the interface requires both electrical and structural characterization. A most sensitive electrical probe is examination of current-voltage characteristics produced in a Schottky diode configuration. The structural analysis is, however, more difficult. Standard structural probes like TEM, SEM and x-ray diffraction are of limited but important use. A technique which has proved especially important is Raman scattering utilizing the interference enhanced Raman scattering (IERS) configuration [2.115]. These structural probes are further supported by the compositional sensitive probes of Rutherford ion-backscattering (RBS) and scanning Auger microprobe (SAM). Certainly for the complicated problem of interfacial interactions at the metal/a-Si : H interface, the correlation of structural, compositional and electrical measurements is required.

At the time of this writing the interactions at the interface of a-Si : H and four different metals have been studied in some detail. These are Pd, Pt, Au and Al. The Pd, Pt and Au metals have been selected for the high metal work functions, while Au and Al which have high conductivities are usually considered for ohmic contacts. Another benefit of the materials is that they resist rapid oxidation. Thus, if reasonable care is taken in transferring the a-Si : H film before metalization, oxide formation can be minimized.

Before embarking on a discussion of the results, it is interesting to attempt to anticipate what interactions may occur. For Pd and Pt deposited on crystalline Si, it has been shown that different silicide compounds form after annealing [2.116]. The first compound formed is M_2Si (M = Pd, Pt) which is near the composition of the stable metallic glass phases of these materials. It has been suggested that an amorphous membrane may form at the interface [2.117], but this has not been observed for Pd on c-Si [2.118]. The removal of the steric limitations due to the underlying c-Si lattice may, however, allow formation of an amorphous "silicide" region. While silicide compounds are known for Pd and Pt, there are no stable silicides with Au and Al. Thus it might be anticipated that thermal processing will cause gradual intermixing and atomic interdiffusion.

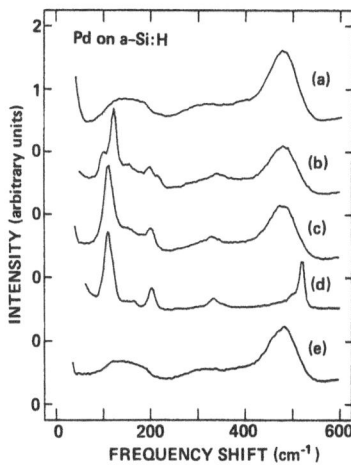

Fig. 2.42. The Raman spectrum of a-Si : H (*a*); compared to that of Pd on a-Si : H after various preparation sequences (*b–e*). All spectra were obtained using the IERS configuration. The Pd on a-Si : H spectra were obtained (*b*) from Pd deposited on freshly prepared a-Si : H, (*c*) from the sample annealed to 300 °C, (*d*) at 500 °C, and for (*e*) Pd deposited on an aged (oxidized) a-Si : H film. A spectral slitwidth of 10 cm^{-1} was used [2.121]

Consider now the results for Pd deposited on a-Si : H [2.119–121]. Typical Raman spectra obtained using the IERS technique are shown in Fig. 2.42. It should be noted that the IERS technique enhances the Raman intensity near the metal/a-Si : H interface. The relatively sharp spectral features that appear after Pd deposition indicate the formation of a crystalline silicide which has been identified as a form of Pd$_2$Si. It has been found that approximately 20 Å of Pd is initially consumed to form the silicide. Annealing to temperatures below 180 °C causes the silicide to grow while annealing to above that temperature causes a change in the sharp spectral features (Fig. 2.42) due to the silicide. This change indicates the formation of a second Pd$_2$Si phase. Further annealing to 550 °C causes no change in the features due to the silicide, but as shown in Fig. 2.42, a sharp feature appears at ~ 520 cm^{-1}, indicating the formation of crystalline Si.

Before the Pd was deposited on the a-Si : H, care was taken to limit exposure to air. For the ~ 15 min exposure to air, an incomplete oxide of < 5 Å would result and this does not impede silicide formation. Similar samples were exposed to air for two weeks before the Pd deposition [2.121]. The spectrum obtained from this sample which is shown in Fig. 2.42 indicates that no silicide formation occurs when the sample is oxidized.

Because the initial formation of the silicide occurs at similar temperatures for Pt or Pd deposited on c-Si, it is anticipated that Pt on a-Si : H will also be similar. This, however, is not the case. The Raman spectra obtained after deposition showed no evidence of sharp spectral features due to crystalline silicide formation [2.122]. However, a weak broad low frequency component was observed and annealing at 150 °C caused growth of this feature. This component has been ascribed to the formation of an amorphous Pt-Si phase. Annealing above 200 °C causes the appearance of two sharp spectral features which are assigned to the simultaneous formation of PtSi and Pt$_2$Si [2.123]. This is unlike crystalline Si where annealing to 300 °C causes first the forma-

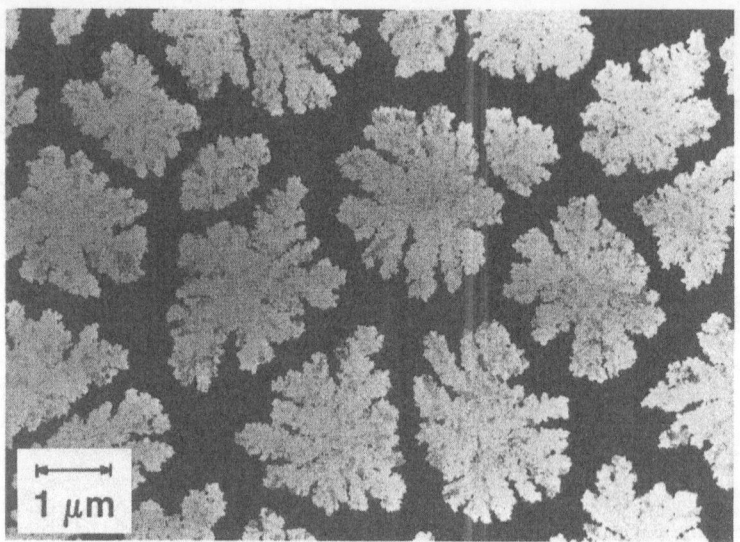

Fig. 2.43. Bright-field TEM micrograph (9800 ×) of 8.0 nm Au on 35 nm of a-Si:H which shows the dendritic growth of islands after in situ annealing to ~ 180 °C [2.124]

tion of Pt_2Si, and only after all the Pt is in that phase does the PtSi phase begin to form [2.116].

While the Pd or Pt interfacial interactions on a-Si:H differ somewhat from those on c-Si, dramatic differences are observed for Au [2.123, 124] or Al [2.125] deposited on a-Si:H which have been annealed to 200 or 250 °C, respectively. In both cases the Raman spectra indicate the formation of crystalline Si. The details of the process can be further elucidated by TEM. As shown in Fig. 2.43, "snow flake" like islands with dimensions ~ 1 μm have formed near the Au/a-Si:H interface. A line scan using SAM has indicated that the islands are Si rich and are the crystalline Si regions observed in the Raman spectra. It should be noted that similar effects have been seen for metals deposited on sputtered a-Si without H [2.126].

While this chapter is concerned with structural aspects, the electrical properties of the interface are important because they are so directly affected and because they can be used to identify structural changes. The current voltage (J-V) characteristics of Schottky diodes have been measured and fit with the expression

$$J = J_0 \exp(qV/\eta kT) ,$$

where J_0 is the reverse bias saturation current and η is the ideality parameter [2.127]. For Schottky barriers on crystalline Si, ideality parameters of < 1.1 are obtained where 1.0 is the theoretical value for the thermionic emission model. For all the metals described here, rectification was observed over at

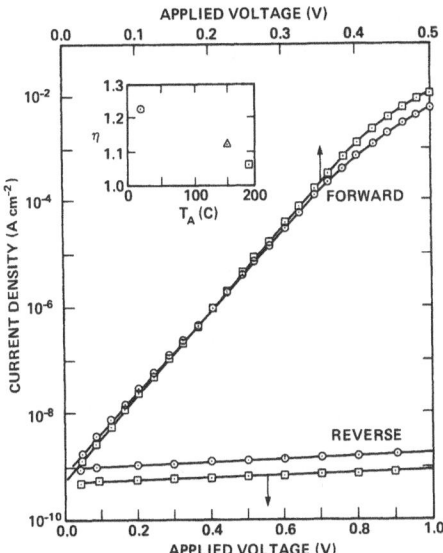

Fig. 2.44. Characterization of Pd Schottky-barrier contacts on a-Si:H: current density versus voltage characteristic before (○) and after (□) annealing at 180 °C for 15 min in a vacuum. The insert shows the ideality factor η obtained at select anneal temperatures T_A [2.127]

least 5 orders of magnitude of current density [2.123, 127]. Typical *J-V* curves for a Pd-a-Si:H diode are shown in Fig. 2.44. For all four metals, the ideality parameter of the as-deposited diodes exhibited a value between 1.2 and 1.3. However, upon annealing to ~200 °C, this value, was lowered to ~1.05 indicating almost ideal behavior. The only samples which showed strong deviations from rectifying characteristics were the Al/a-Si:H diodes after annealing to 250 °C, and here nearly ohmic properties were obtained. Thus, this group of experiments demonstrates clearly the need for correlated electrical and structural measurements of interface properties.

2.9 Conclusions/Future Directions

The major conclusion that one is forced to draw from a review of virtually any aspect of a-Si:H is that there is an incredible richness of effects that are either being investigated or worthy of further investigation. Structural and chemical characterization is no exception, from impurity-defect complexes through disclinations revealed by etching to orthohydrogen "heatsinks" for "hot" spatially separated proton "baths", the diversity of issues, tools for addressing them and implications for other measurements are amazing.

Unfortunately, a plate full to the overflowing of tasty morsels can obscure the need to identify and address the real "meat and potatoes" issues that remain. This section will therefore focus on two of these issues that are (i) unresolved, (ii) potentially resolvable and (iii) have major implications for both fundamental understanding and technological application.

The first is the interrelationship between impurities and defects. Clear experimental evidence has been presented here for a connection: the hydrogen evolution measurements of *Beyer* and *Wagner* [2.70], the study of the effects of doping and compensation on electronic properties by *Street* et al. [2.74] and the work of *Schiff* et al. [2.49] on doping-precipitated morphology. These connections are, however, neither satisfactorily explored nor adequately explained. *Street* [2.128] has recently proposed a model for doping that invokes satisfaction of the (8-*N*) rule for valence electrons on the dopant as the mechanism whereby, once the Fermi level is close to either the donor or acceptor levels, additional dopant atoms will always be threefold coordinated. If the dopant atoms are primarily threefold coordinated with silicon, then a topological constraint will exist for the surrounding silicon lattice that will in most cases produce a dangling bond either hydrogenated or unhydrogenated. This association between single dopant atoms and single defects can explain the electronic consequences, but it does not offer an obvious explanation for the microstructural effects. Clearly further studies of the precise relationship between defects, both electronic and structural, hydrogen content and dopants is a major area for investigation. In a more general sense it is also obvious, as mentioned before, that the issue of causal relationships between electronic and structural defects needs further investigation, but this is a less tractable problem than that of the dopants due to the lower degree of control over defect concentration.

The second issue is the broad understanding and control of the vertical homogeneity, both compositional and structural in a-Si : H films. Conflicting evidence has been presented in this article over the existence of steep gradients in hydrogen concentration at or near interfaces. This aspect can probably be resolved by careful experiments to eliminate technique-dependent artifacts; I personally suspect that defect-creation or beam-heating during nuclear reaction analysis may be causing enhanced hydrogen out-diffusion but if this explanation is incorrect, then several other techniques have to be looked at very carefully. There is no conflict over the presence of large interface "spikes" of contaminant impurities; here the issue is control – cleaner source materials, cleaner chambers, load-locks and lots of (unfortunately) expensive analysis. I fear that in this area, the high cost (in equipment) of going to the next step will be dwarfed by the cost of SIMS, etc., needed to ascertain what the first expenditure has achieved and together, these costs may be prohibitive for many institutions now in this field. There is then the combination of the developing scale in microstructure described by *Ross* and *Messier* [2.42] and several recent reports [2.126] that a number of electronic properties including solar cell fill factor improve with increasing film thickness. Are smaller defects coalescing and segregating into larger defects as the film gets thicker? Finally, there are the questions relating to the interfaces between amorphous Si : H and both metals and insulators. Can they be adequately characterized for undoped material both when the a-Si : H is deposited first and the other material subsequently and vice versa?

Further, and particularly relevant to the technological applications, does doped a-Si : H behave radically differently from undoped material in its interface behavior? Certainly the strong structural effects of doping suggest that this is a good possibility.

Having described two very important issues (and certainly missed out many more), it is appropriate to pick out a theme from these that was touched on in the introduction and that offers a natural opportunity to conclude the review as a whole. In both the defect-structure and interface-impurity structure issues, it is obvious that correlations exist between the results of the different measurements, yet explanations that are put forward are still frequently one-sided, i.e., the explanation for the electronic effects neglects to even address the structural effects and vice versa. In the introduction, I mentioned a number of good reasons why the electronic and structural sides had to work together. Let me point to the metal-a-Si : H interface work, where some of that is happening, as a good example of what I hope will be the future and exhort all who are considering entering this field to follow this example – it won't be easy but the end will be worthwhile.

Acknowledgments. I would like to gratefully acknowledge Bob Nemanich for contributing the section on surfaces and interfaces, Vi Moffat for her patience and skill in typing the many drafts of the manuscript and Jim Boyce for critically reading the section on magnetic resonance. I would also like to acknowledge the many colleagues at Xerox PARC who have contributed both to the substance of some of the work discussed and to the environment that made both the work and the review possible.

References

2.1   R. C. Chittick, J. H. Alexander, H. F. Sterling: J. Electrochem. Soc. **116**, 77 (1969);
      R. C. Chittick: J. Non-Cryst. Solids **3**, 255 (1970)
2.2   W. E. Spear, P. G. LeComber: Solid State Commun. **17**, 1193 (1975)
2.3   A. J. Lewis, G. A. N. Connell, W. Paul, J. R. Pawlik, R. J. Temkin: AIP Conf. Proc. **20**, 27 (1974)
2.4   A. Triska, D. Dennison, H. Fritzsche: Bull. Am. Phys. Soc. **20**, 392 (1975)
2.5   J. C. Knights: AIP Conf. Proc. **31**, 296, AIP New York (1976)
2.6   A. Barna, P. B. Barna, G. Radnoczi, L. Toth, P. Thomas: Phys. Stat. Sol. (a) **41**, 81 (1977)
2.7   J. M. Ogier: Annales de Chimie et de Physique **20**, 5 (1880)
2.8   H. J. Eméleus, K. Stewart: Trans. Faraday Soc. **32**, 1577 (1936)
2.9   W. L. Jolly: Adv. Chemistry **80**, 156 (Am. Chem. Soc., New York 1969)
2.10 R. Mosseri, C. Sella, J. Dixmier: Phys. Stat. Sol. (a) **52**, 475 (1979)
2.11 J. F. Graczyk: Phys. Stat. Sol. (a) **55**, 231 (1979)
2.12 T. A. Postol, C. M. Falco, R. T. Kampwirth, I. K. Schuller, W. B. Yelon: Phys. Rev. Lett. **45**, 648 (1980)
2.13 D. E. Polk: J. Non-Cryst. Solids **5**, 365 (1971)
2.14 J. C. Knights, T. M. Hayes, J. C. Mikkelsen Jr.: Phys. Rev. Lett. **39**, 712 (1977)
2.15 A. Pines, M. G. Gibby, J. Waugh: J. Chem. Phys. **59**, 569 (1973)

2.16 J. A. Reimer, P. Dubois-Murphy, B. C. Gerstein, J. C. Knights: J. Chem. Phys. **74**, 1501 (1981)
2.17 B. Lamotte, R. Rousseau, A. Chenevas-Paule: J. Physique **42**, p. C4–839 (1981)
2.18 F. R. Jeffrey, P. Dubois Murphy, B. C. Gerstein: Phys. Rev. B **23**, 2099 (1981)
2.19 L. Guttman, W. Y. Ching, J. Rath: Phys. Rev. Lett. **44**, 1513 (1980)
2.20 M. H. Brodsky, M. Cardona, J. J. Cuomo: Phys. Rev. B **16**, 3556 (1977)
2.21 J. E. Smith Jr., M. H. Brodsky, B. L. Crowder, M. I. Nathan: In *Proc. 2nd Intern. Conf. on Light Scattering in Solids,* ed. by M. Balkanski (Flammarion, Paris 1971) p. 330
2.22 R. J. Nemanich, S. A. Solin: Phys. Rev. B **20**, 392 (1979)
2.23 R. J. Nemanich, S. A. Solin, R. M. Martin: Phys. Rev. B **23**, 6348 (1981)
2.24 S. Veprek, Z. Iqbal, H. R. Oswald, F. A. Sarotl, J. J. Wagner: J. Physique **42**, C4–251 (1981)
2.25 W. E. Spear, P. Willeke, P. G. LeComber, A. G. Fitzgerald: J. Physique **42**, C4–257 (1981)
2.26 R. Tsu, M. Izu, S. R. Ovshinsky, F. H. Pollak: Solid State Commun. **36**, 817 (1980)
2.27 S. Usui, M. Kikuchi: J. Non-Cryst. Solids **34**, 1 (1979)
2.28 Z. Iqbal, S. Veprek, A. P. Webb, P. Capezzuto: Sol. State Commun. **37**, 993 (1981)
2.29 A. Barna, P. B. Barna, G. Radnoczi, H. Sugawara, P. Thomas: Thin Solid Films **48**, 163 (1978)
2.30 G. Abowitz, L. B. Leder: J. Vac. Sci. Tech. **15**, 1746 (1978)
2.31 G. S. Cargill III: Phys. Rev. Lett. **28**, 1372 (1972)
2.32 R. Messier, S. V. Krishnaswamy, L. R. Gilbert, P. Schwab: J. Appl. Phys. **51**, 1611 (1980)
2.33 A. G. Dirks, H. J. Leamy: IEEE Trans. Mag.-**14**, 835 (1978)
2.34 H. J. Leamy, A. G. Dirks: J. Phys. D **10**, L95 (1977)
2.35 R. J. Nemanich, D. K. Biegelsen, M. Rosenblum: J. Phys. Soc. Japan **49** A, 1189 (1980)
2.36 P. Swab, S. V. Krishnaswamy, R. Messier: J. Vac. Sci. Techn. **17**, 362 (1980)
2.37 S. R. Herd: Phys. Stat. Sol. A **44**, 363 (1977)
2.38 J. C. Knights, R. A. Lujan: Appl. Phys. Lett. **35**, 244 (1979)
2.39 R. C. Ross, R. Messier: J. Appl. Phys. **52**, 5329 (1981)
2.40 T. M. Donovan: Final Technical Report, Contract DS-9-8142-4 (Naval Weapons Center, China Lake, CA 1981)
2.41 A. Chenevas-Paule: Private communication
2.42 R. Messier, R. C. Ross: J. Appl. Phys., to be published
2.43 P. D'Antonio, J. H. Konnert: Phys. Rev. Lett. **43**, 1161 (1979)
2.44 A. J. Leadbetter, A. A. M. Rashid, R. M. Richardson, A. F. Wright, J. C. Knights: Solid State Commun. **33**, 973 (1980)
2.45 J. C. Knights: J. Non-Cryst. Solids **35/36**, 159 (1980)
2.46 J. C. Knights, R. A. Lujan, M. P. Rosenblum, R. A. Street, D. K. Biegelsen: Appl. Phys. Lett. **38**, 331 (1981)
2.47 A. Chenevas-Paule: Private communication
2.48 J. F. Sadoc, R. Mosseri: J. Physique **42**, C4–189 (1981)
2.49 E. A. Schiff, P. D. Persans, H. Fritzsche, V. Akopyan: Appl. Phys. Lett. **38**, 92 (1981)
2.50 R. F. Bunshah, R. S. Juntz: J. Vac. Sci. Tech. **9**, 1404 (1972)
2.51 J. A. Thornton: Ann. Rev. Mat. Sci. **7**, 239 (1977)
2.52 W. A. Lanford, H. P. Trautvetter, J. F. Ziegler, J. Keller: Appl. Phys. Lett. **28**, 566 (1976)
2.53 J. F. Ziegler et al.: Nucl. Inst. Meth. **149**, 19 (1978)
       A. Benninghoven, C. A. Evans Jr., R. A. Powell, R. Shimizu, H. A. Storms (eds.): *Secondary Ion Mass Spectrometry* SIMS II, Springer Ser. Chem. Phys., Vol. 9 (Springer, Berlin, Heidelberg, New York 1979)
       A. Benninghoven, J. Giber, J. László, M. Riedel, H. W. Werner (eds.): *Secondary Ion Mass Spectrometry* SIMS III, Springer Ser. Chem. Phys., Vol. 19 (Springer, Berlin, Heidelberg, New York 1982)

2.54 G. C. Clark, C. W. White, D. D. Allred, B. R. Appleton, C. We Magee, D. G. Carlson: Appl. Phys. Lett. **31**, 582 (1977)
2.55 C. W. Magee, E. M. Botnick: J. Vac. Sci. Tech. **19**, 47 (1981)
2.56 H. Fritzsche, M. Tanielian, C. C. Tsai, P. J. Gaczi: J. Appl. Phys. **50**, 3368 (1979)
2.57 J. Reimer, J. Knights: Unpublished work
2.58 G. A. N. Connell, J. R. Pawlik: Phys. Rev. B **13**, 787 (1976)
2.59 C. J. Fang, K. J. Gruntz, L. Ley, M. Cardona, F. J. Demond, G. Müller, S. Kalbitzer: J. Non-Cryst. Solids **35/36**, 255 (1980)
2.60 S. Oguz, D. A. Anderson, W. Paul, H. J. Stein: Phys. Rev. B **22**, 880 (1980)
2.61 M. Taniguchi, M. Hirose, Y. Osaka: J. Non-Cryst. Solids **35/36**, 189 (1980)
2.62 R. C. Ross, I. S. T. Tsong, R. Messier, W. A. Lanford, C. Burman: J. Vac. Sci. Tech. **20**, 406 (1982)
2.63 M. Milleville, W. Fuhs, F. J. Demond, H. Mannsperger, G. Müller, S. Kalbitzer: Appl. Phys. Lett. **34**, 173 (1979)
2.64 J. C. Knights, G. Lucovsky, R. Nemanich: Phil. Mag. B **37**, 467 (1977)
2.65 F. J. Demond, G. Müller, H. Damjantschitsch, H. Mannsperger, S. Kalbitzer, P. G. LeComber, W. E. Spear: J. Physique **42**, C4–779 (1981)
2.66 D. E. Carlson, C. W. Magee, A. R. Triano: J. Electrochem. Soc. **126**, 688 (1979)
2.67 B. von Roedern, L. Ley, M. Cardona, F. W. Smith: Phil. Mag. B **40**, 433 (1979)
2.68 D. I. Jones, R. A. Gibson, P. G. LeComber, W. E. Spear: Solar Energy Mat. **2**, 93 (1979)
2.69 S. V. Krishnaswamy, R. Messier, C. S. Wu, S. B. McLane, T. T. Tsong: J. Vac. Sci. Tech. **18**, 309 (1981)
2.70 G. Müller, F. Demond, S. Kalbitzer, H. Damjantschitsch, H. Mannsperger, W. E. Spear, P. G. LeComber, R. A. Gibson: Phil. Mag. B **41**, 571 (1980)
2.71 W. Beyer, H. Wagner: J. Physique **42**, C4–783 (1981)
2.72 J. C. Zesch, R. A. Lujan, V. R. Deline: J. Non-Cryst. Solids **35/36**, 273 (1980)
2.73 J. Chevallier, W. Beyer: Phys. Stat. Sol. (a) in press
2.74 R. A. Street, D. K. Biegelsen, J. C. Knights: Phys. Rev. B **24**, 969 (1981)
2.75 A. Deneuville, J. C. Bruyère, M. Toulemonde, J. J. Grob, P. Siffert: AIP Conf. Proc. **73**, 120 (1981)
2.76 A. Delahoy, R. W. Griffith: J. Appl. Phys. **52**, 6337 (1981)
2.77 J. Knights, J. C. Mikkelsen Jr., B. Wacker: Unpublished results
2.78 K. Tanaka, S. Yamasaki, K. Nakagawa, A. Matsuda, H. Okushi, M. Matsumura, S. Iizima: J. Non-Cryst. Solids **35/36**, 475 (1980)
2.79 C. C. Tsai, H. Fritzsche: Solar Energy Mat. **1**, 29 (1979)
2.80 R. C. Ross, R. Messier: J. Appl. Phys., to be published;
      R. C. Ross: PhD Thesis, The Pennsylvania State University (1981)
2.81 Y. Katayama, T. Shimada, K. Usami, E. Maruyama: J. Physique **42**, C4–787 (1981)
2.82 W. Paul, D. A. Anderson: Solar Energy Mat. **5**, 229 (1981)
2.83 J. C. Knights: Japan J. Appl. Phys. **18**, Supp. 18-1, 101 (1979)
2.84 G. Lucovsky, R. J. Nemanich, J. C. Knights: Phys. Rev. B **19**, 2064 (1979)
2.85 D. K. Biegelsen, R. A. Street, C. C. Tsai, J. C. Knights: Phys. Rev. B **20**, 4839 (1979)
2.86 A. J. Leadbetter, A. A. M. Rashid, N. Colenutt, A. F. Wright, J. C. Knights: Solid State Commun. **38**, 957 (1981)
2.87 C. C. Tsai, H. Fritzsche, M. H. Tanielian, P. J. Gaczi, P. D. Persans, M. A. Vesaghi: Proc. 7th Intern. Conf. on Amorphous and Liquid Semiconductors, ed. by W. E. Spear (CICL Edinburgh 1977) p. 339
2.88 P. John, I. M. Odeh, M. J. K. Thomas, M. J. Tricker, F. Riddoch, J. I. B. Wilson: Phil. Mag. B **42**, 671 (1980)
2.89 K. Zellama, P. Germain, S. Squelard, J. Monge, E. Ligeon: J. Non-Cryst. Solids **35/36**, 225 (1980)
2.90 J. A. Reimer: J. Physique **42**, C4–715 (1981)
2.91 M. E. Lowry, R. G. Barnes, D. R. Torgeson, F. R. Jeffrey: Solid State Commun. **38**, 113 (1981)
2.92 W. E. Carlos, P. C. Taylor: Phys. Rev. Lett. **45**, 358 (1980)

2.93   S. Ueda, M. Kumeda, T. Shimizu: Japan J. Appl. Phys. **20**, 6 (1981)
2.94   J. A. Reimer, R. W. Vaughan, J. C. Knights: Phys. Rev. Lett. **44**, 193 (1980)
2.95   J. A. Reimer, R. W. Vaughan, J. C. Knights: Phys. Rev. B **24**, 3360 (1981)
2.96   W. E. Carlos, P. C. Taylor: Phys. Rev. B **26**, 3605–3616 (1982);
       W. E. Carlos, P. C. Taylor: Phys. Rev. B **25**, 1435 (American Institute of Physics 1982)
2.97   F. R. Jeffrey, M. E. Lowry, M. L. S. Garcia, R. G. Barnes, D. R. Torgeson: AIP Conf.
       Proc. **73**, 83 (1981)
2.98   J. A. Reimer, J. C. Knights: AIP Conf. Proc. **73**, 78 (1981)
2.99   J. A. Reimer, R. W. Vaughan, J. C. Knights: Solid State Commun. **37**, 161 (1981)
2.100 S. Ueda, M. Kumeda, T. Shimizu: J. Physique **42**, C4–729 (1981)
2.101 W. E. Carlos, P. C. Taylor: J. Physique **42**, C4–725 (1981)
2.102 W. E. Carlos, P. C. Taylor, S. Oguz, W. Paul: AIP Conf. Proc. **73**, 67 (1981)
2.103 B. Movaghar, L. Schweitzer: AIP Conf. Proc. **73**, 73 (1981)
2.104 M. Conradi, R. Norberg: Phys. Rev. B **24**, 2285 (1981)
2.105 M. S. Conradi, K. Luszczynski, R. E. Norberg: Phys. Rev. B **20**, 2594 (1979)
2.106 J. A. Reimer, R. W. Vaughan, J. C. Knights, R. A. Lujan: J. Vac. Sci. Tech. **19**, 53
       (1981)
2.107 J. A. Reimer, R. W. Vaughan, J. C. Knights: Phys. Rev. B **23**, 2567 (1981)
2.108 D. J. Leopold, J. B. Boyce, P. A. Fedders, R. E. Norberg: Phys. Rev. B **26**, 6053 (1982)
2.109 S. G. Greenbaum, W. E. Carlos, P. C. Taylor: Phys. Rev. B **43**, 663 (1982)
2.110 C. C. Tsai, R. J. Nemanich: J. Non-Cryst. Solids **35 and 36**, 1203 (1980)
2.111 W. B. Jackson, D. K. Biegelsen, R. J. Nemanich, J. C. Knights: Appl. Phys. Lett. **42**, 105
       (1983)
2.112 H. Fritzsche, C. C. Tsai: Solar Energy Mat. **1**, 471 (1979)
2.113 D. E. Aspnes, B. G. Bagley, A. A. Studna, A. C. Adams, F. B. Alexander Jr.: In
       Tetrahedrally Bonded Amorphous Semiconductors, ed. by R. A. Street, D. K. Biegelsen,
       J. C. Knights (AIP, New York 1981) p. 307
2.114 R. A. Street, J. C. Knights: Phil. Mag. B **43**, 1091 (1981)
2.115 G. A. N. Connell, R. J. Nemanich, C. C. Tsai: Appl. Phys. Lett. **36**, 31 (1981)
2.116 G. Ottaviani: J. Vac. Sci. Tech. **16**, 1112 (1979)
2.117 R. W. Bené, R. M. Walser: In *Thin Film Phenomena-Interfaces and Interactions,* ed. by J.
       E. E. Baglin, J. M. Poate (Electrochemical Soc., Princeton, NJ 1978) p. 21
2.118 P. S. Ho, P. E. Schmid, H. Föll: Phys. Rev. Lett. **46**, 782 (1981)
2.119 C. C. Tsai, R. J. Nemanich, T. W. Sigmon: J. Phys. Soc. Japan **49** Suppl. A, 1265 (1980)
2.120 R. J. Nemanich, C. C. Tsai, T. W. Sigmon: Phys. Rev. B **23**, 6828 (1981)
2.121 R. J. Nemanich, C. C. Tsai, M. J. Thompson, T. W. Sigmon: J. Vac. Sci. Tech. **19**, 685
       (1981)
2.122 R. J. Nemanich, C. C. Tsai: J. Physique **42**, C6–822 (1981)
2.123 C. C. Tsai, M. J. Thompson, R. J. Nemanich: J. Physique **42**, C4–1077 (1981)
2.124 C. C. Tsai, R. J. Nemanich, M. J. Thompson: J. Vac. Sci. Tech. **21**, 632 (1982)
2.125 C. C. Tsai, R. J. Nemanich, M. J. Thompson, B. L. Stafford: Physica B, in press
2.126 S. R. Herd, P. Chaudhari, M. H. Brodsky: J. Non-Cryst. Solids **7**, 309 (1972)
2.127 M. J. Thompson, N. M. Johnson, R. J. Nemanich, C. C. Tsai: Appl. Phys. Lett. **39**, 274
       (1982)
2.128 R. A. Street: Phys. Rev. Lett. **51** (to be published)

3. Fundamental and Applied Work on Glow Discharge Material

Walter E. Spear and Peter G. LeComber

With 40 Figures

This chapter is concerned specifically with the properties of amorphous silicon (a-Si) prepared in a radio frequency (rf) glow discharge by decomposing silane which may contain small amounts of doping gases. The subject has now reached an important, perhaps even critical stage: the fundamental developments during the 1970s in the field of glow discharge Si have established the considerable applied potential of this material and we are now beginning to see a rapid growth of applied developments, in some cases already on an industrial scale. It is therefore essential that a review written at the present time should deal with both these aspects and should attempt to relate the fundamental insight into the material properties with the applied possibilities which have led to the present worldwide activity in the a-Si field.

This aim is reflected in the general layout of the chapter, which is centred on two aspects of basic importance to the present developments. The first concerns the density of states in the mobility gap of glow discharge Si, studied by the field-effect and more recently by other techniques. Already in 1972 the early field-effect experiments had led to the realisation that a-Si possessed a remarkably low level of gap states; this is probably the most important single factor which makes a-Si into an electronically viable material for fundamental and applied work. A direct result on the applied side has been the a-Si field effect transistor discussed in Sect. 3.3, which could find application as a switching element in large area addressable displays and in other devices.

The second crucial development was the discovery in 1975 that the electronic properties of the glow discharge material could be controlled very effectively by substitutional doping from the gas phase; again, this is closely related to the low gap state density of the material. The possibility of doping has opened up the rapidly growing field of a-junction devices which will be discussed in Sect. 3.5 with particular reference to the materials aspect.

We shall begin the chapter with the subject common to all following sections, namely, the glow discharge technique itself. Deposition systems, on a laboratory and industrial scale, will be discussed as well as some of the recent work on deposition rate and the silane plasma.

3.1 The Glow Discharge Technique

With the limited information presently available on the processes taking place in the glow discharge plasma, it is not surprising that the techniques for the preparation of a-Si from a silane glow discharge were developed largely on an empirical basis. Accordingly, we shall begin with a discussion of reactors, techniques and deposition conditions which have been found to produce electronically viable a-Si specimens. In Sect. 3.1.4 some of the recent work aimed at a more detailed understanding of the plasma will be summarised.

3.1.1 Radio Frequency Deposition Systems

Glow discharge preparation units can be divided into two groups, depending on the method of coupling the radio frequency (rf) excitation into the plasma. The first glow discharge deposition of a-Si was carried out in an inductively coupled system by *Sterling* and his collaborators [3.1, 2] at the S.T.L. Laboratories. The same approach was adopted in the early work of the Dundee group on the electronic properties of the material [3.3, 4] and inductively coupled reactors are still used as a convenient method for depositing small area specimens. Figure 3.1 illustrates the method. The gas *G* (silane or an appropriate gas mixture) is passed through a fused silica reaction tube *Q*, between 5 and 10 cm in diameter. The pressure lies in the range from 0.1 to 1 torr at gas flow rates between 0.1 and 10 cm^3/min at STP. A rotary pump *RP*, sometimes fitted with a Roots blower, maintains the gas flow through the system. The substrate *S* is held on the heated pedestal *H*, which is immersed in the lower part of the glow discharge plasma *P*. The latter is excited by an external coupling coil connected to an rf generator, normally operated at a frequency of 13.56 MHz.

It is important to note that the film grows in close contact with the plasma and that its electronic properties are critically dependent on the interactions taking place at the interface. In spite of the simplicity of the basic arrangement of Fig. 3.1, the reproducibility and quality of the specimens can be maintained only by careful attention to all the parameters involved. Elec-

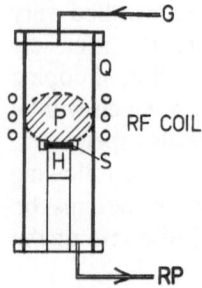

Fig. 3.1. Inductively coupled glow discharge deposition system. [(*G*) gas; (*Q*) fused silica reaction tube; (*S*) substrate; (*H*) heated pedestal; (*RP*) rotary pump; (*P*) plasma]

tronic properties as well as hydrogen content are also dependent on the dimensions of the system, presumably as a result of floating potentials on surrounding surfaces. In our experience the height of the rf coil above the substrate is a sensitive variable, as are the substrate temperature T_d and the rf power level. It has generally been found (see Sect. 3.3) that for the lowest densities of gap states, T_d values between 250 °C and 330 °C are required, together with a low level of rf power (1–10 W), just sufficient to maintain a weak glow discharge.

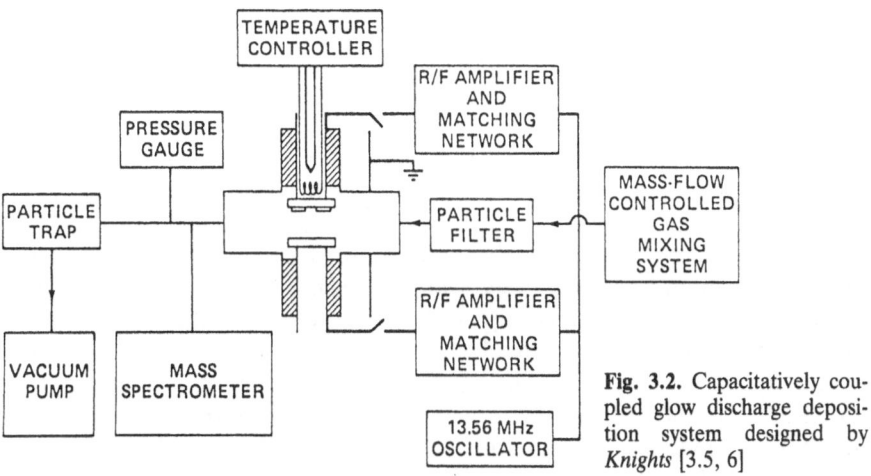

Fig. 3.2. Capacitatively coupled glow discharge deposition system designed by *Knights* [3.5, 6]

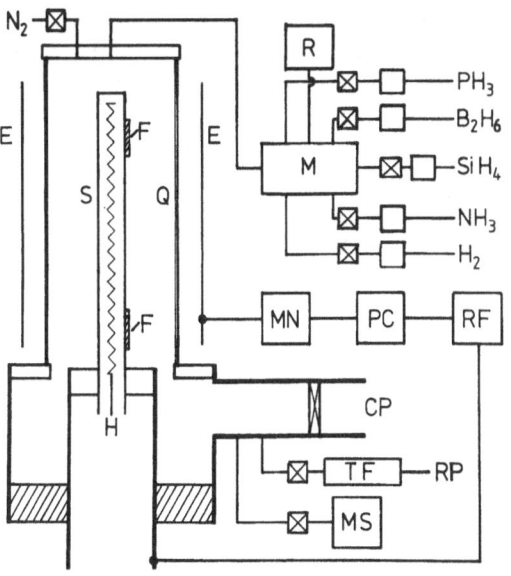

Fig. 3.3. Vertical glow discharge deposition system used at the Dundee laboratory. [(S) specimen holder; (H) electrically insulated heater; (Q) quartz enclosure; (E) external rf electrodes; (F) rotatable "flaps" for shading during run; (M) mixing chamber; (R) reservoir for pre-mixing of gases; (MN) matching network; (PC) power controller; (RF) rf generator; (CP) cryo-pump; (TF) tubular furnace; (RP) rotary pump; (MS) mass spectrometer]

The second type of deposition system employs capacitative coupling. Figure 3.2 shows a versatile unit of this kind designed by *Knights* [3.5, 6]. An obvious advantage over the inductive system lies in the relative ease with which the parallel plate geometry can be scaled up for the uniform deposition of larger area specimens (Sect. 3.1.2). It can be seen that the upper electrode holding the substrates can either be connected to the rf ground potential (i.e., that of the surrounding metal box) or to the "hot" rf electrode via the matching network and a coupling capacitor. In the latter case the substrate surface will acquire a steady negative bias which, as shown by *Knights* [3.6], can influence the material properties. Typical values of parameters such as pressure, flow rate, substrate temperature and rf power mentioned above apply equally well to the capacitative system.

Figure 3.3 is a schematic diagram of a system that has been used in the Dundee laboratories during the last few years. The main aims of the design have been to reduce the contamination problem and to achieve optimum versatility, particularly for the deposition of multilayer specimens. The stainless steel specimen holder S and its electrically insulated heater H are mounted vertically to reduce the deposition of small particles which are difficult to eliminate completely from a plasma system. S is held in a demountable quartz enclosure Q and the rf field is applied between the external electrodes E which are electrically connected together and adjusted parallel to the surface of S. The whole of the specimen holder assembly is insulated from the ground and can easily be withdrawn from the apparatus for substrate loading. Two thermocouples are incorporated to monitor the temperature of the holder and of the substrate surface, respectively. Rotatable "flaps" F have been fitted which can be operated during a deposition run to shade certain regions of the substrate. Such a facility has proved to be very useful in the development of junctions and other multi-layer devices. An obvious advantage of the vertical arrangement over that shown in Fig. 3.2 is that both sides of the holder S can be used; in our largest unit this gives a total useable area of 300 cm^2.

rf generators, operating at 5.0, 13.56 and 40.68 MHz, can be connected through a power controller PC and a matching network MN between the external plates E and the holder assembly S. The gas handling system is made entirely from stainless steel and consists of the five channels indicated in Fig. 3.3, all connected to the mixing chamber M. Each channel is fitted with a mass flowmeter and a piezoelectric valve for electronic flow and ratio control. In addition, several 1 l reservoirs such as R are connected to M, enabling the operator to flow pre-mixed gases through the system. Nitrogen flushing lines, essential for safe operation, are provided at several parts of the apparatus.

The whole system is normally kept under high vacuum produced by the 3000 l s^{-1} cryopump CP. We believe that this helps considerably in reducing cross-contamination from gases that have been used in previous runs. During a deposition run CP is shut off and the gas flow is maintained by the rotary pump *RP*. The tubular furnace *TF* mounted in front of the pump decomposes the hydrides which reduces contamination of the pump oil and adds to the safety of the system. The mass spectrometer *MS* is a useful part of the unit allowing examination of a gas sample before deposition. Finally, to comply with safety regulations, the whole apparatus, including the gas supplies, are kept in a continuously ventilated enclosure which can be completely isolated from the laboratory.

3.1.2 Large Scale Deposition

The reactors described in the previous section are useful in the research or development laboratory for preparing a limited batch of specimens, but they are hardly suitable for industrial production on a larger scale. The rapid growth during the last few years in the field of a-Si photovoltaic devices has stimulated interest in the industrialisation of the glow discharge technique. At the 1981 Grenoble Conference, *Kuwano* and *Ohnishi* [3.7] of the Sanyo laboratories described the design of what is probably the first industrial plant for the production of a-Si devices.

As shown schematically in Fig. 3.4, the system consists of five separate vacuum chambers connected by vacuum locks. Trays loaded with substrates are placed into the first chamber on the right and then pass in a continuous process through the three central chambers in which boron doped, undoped and phosphorus doped layers are deposited to form the *p-i-n* photovoltaic junction. The main advantage of separating these stages is that problems arising from dopant contamination are completely avoided. The latter can partly be overcome in the smaller systems described earlier, but would seriously affect reproducibility in a continuous industrial process. Another important point is that the central deposition chambers are not exposed to air, except for occasional maintenance.

In spite of the greater cost and complexity of the Sanyo system over possible single chamber designs in which the gases are cycled, its advantages

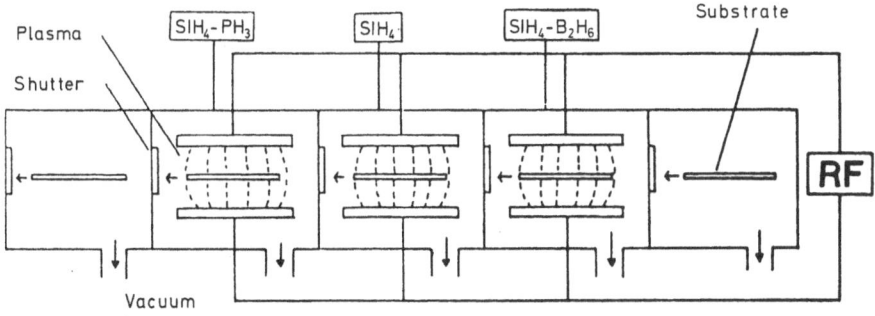

Fig. 3.4. Industrial glow discharge reactor designed by the Sanyo laboratories [3.7]

are considerable and it is likely that future industrial developments will follow along similar lines.

3.1.3 The Deposition Rate and the Use of Higher Silanes

Experience has shown that high quality a-Si films have to be prepared in the silane glow discharge at deposition rates of less than about 3 Å s^{-1} and with the minimum rf power. Under these conditions ion and electron energies in the plasma remain sufficiently low to reduce defect formation at the growing surface to an acceptable level. The present limitation imposed by the deposition rate is not too serious in laboratory work involving specimens a fraction of a μm in thickness. However, for thicker films such as those discussed in Sect. 3.5.4, deposition times tend to become unacceptably long.

Can this limitation be overcome? Recently *Scott* et al. [3.8, 9] have investigated the deposition and electronic properties of glow discharge a-Si specimens produced from *di*-silane (Si_2H_6) and *tri*-silane (Si_3H_8), both of which form less stable molecules than the mono-silane used in the previous work. The encouraging result was that with a low rf power (2 W in an inductively coupled system), the deposition rate from the higher silanes was more than 20 times larger than that obtainable from SiH_4. Moreover, films up to 20 μm thick could be deposited, presumably because of reduced film-substrate strain. Such thicknesses are difficult to achieve in monohydride depositions without cracking or flaking-off and suggest that structural and/or morphological differences may exist between the films produced by the two methods.

According to the information published so far, the electrical and photoconductive properties of films rapidly deposited from Si_2H_6 appear to be substantially the same as those obtained in specimens from SiH_4 but produced at a much smaller rate. For example, in Fig. 3.5 [3.9] the dark conductivity activation energy ε_σ is plotted against the deposition temperature T_d for

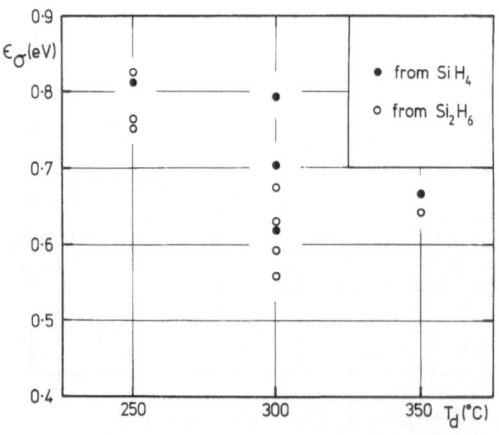

Fig. 3.5. Plot of the activation energy ε_σ of the dark conductivity against the deposition temperature T_d for a-Si specimens prepared from SiH_4 and from Si_2H_6 [3.9]

undoped specimens prepared from mono and disilane. In spite of the scatter in the results, it can be seen that average values of ε_σ tend to be somewhat smaller for the Si_2H_6 films, suggesting some differences in the energetic distribution of defect states in the central region of the mobility gap (Sect. 3.2, Fig. 3.12). The reported increase in photoconductivity of these specimens [3.8] over the SiH_4-films is most probably connected with the same effect. As shown by *Anderson* and *Spear* [3.10], the movement of ε_f towards ε_c (in their experiments produced by phosphorus doping) leads to a rapid increase in the recombination lifetime of photogenerated electrons as ε_σ approaches 0.6 eV. The observed sensitisation was ascribed to the filling up of recombination centres situated between 0.75 eV and 0.6 eV and it is likely that the same mechanism applies to the di-silane specimens.

In their discussion of the deposition mechanism, *Scott* et al. [3.9] suggest, that the initial rate-determining step in the decomposition of silanes involves formation of the SiH_2 diradical:

$$SiH_4 \rightarrow SiH_2 + H_2 \tag{3.1}$$

$$Si_2H_6 \rightarrow SiH_2 + SiH_4 . \tag{3.2}$$

The significant point is that the reaction in (3.2) is about 100 times faster (at 400 °C) than that given by (3.1), leading to the conclusion that the higher deposition rate obtainable in Si_2H_6 decomposition is closely related with the greater rate of SiH_2 production in the plasma.

In view of the comparatively high cost of commercially available di-silane, it is tempting to explore the possibility of producing higher silanes directly within the present experimental setup. Figure 3.6 shows the method adopted by *Ogawa* et al. [3.11] in their recent study of enhanced deposition rate. Higher silanes are formed by circulating mono-silane through an ozoniser

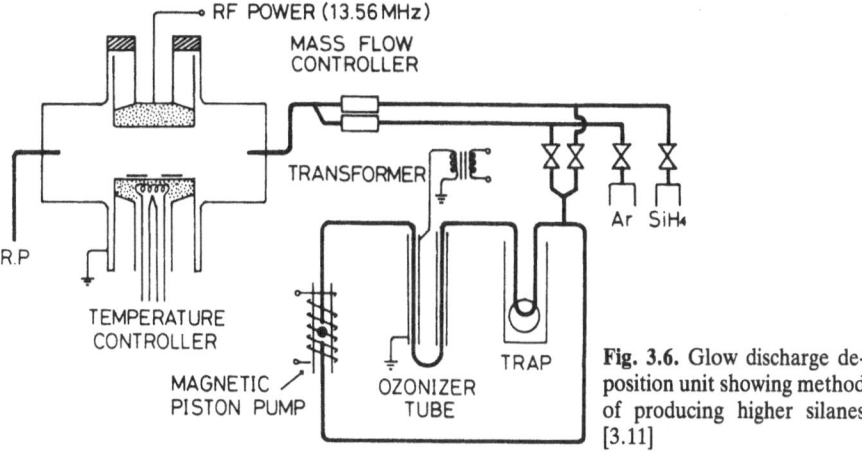

Fig. 3.6. Glow discharge deposition unit showing method of producing higher silanes [3.11]

tube connected to a 6 kV transformer and condensed into a cold trap. Remaining gaseous products, such as hydrogen or SiH_4, are then pumped away. The mixture of higher silanes is introduced into the glow discharge by allowing the trap to warm up. Ogawa et al. deposited specimens in this apparatus at a rate exceeding 60 Å s^{-1} and a substrate temperature of 350 °C. Measurements showed that the optical gap, the activation energy and the dark and photoconductivities are substantially the same as those obtained from mono-silane at appreciably lower deposition rates.

The recent developments described in this section are most encouraging and, provided the performance of a-Si devices can be maintained at the higher deposition rates, could prove to be of considerable importance in large scale applications of a-Si.

3.1.4 Recent Studies of the Silane Plasma

In spite of the widespread use of the glow discharge techniques described in the previous sections, the reactions taking place in the silane plasma, and particularly at the plasma-substrate interface, are still largely unknown. In the following we shall discuss briefly the nature of the plasma and summarise some of the present work aimed at a closer understanding of the processes involved in the deposition of a-Si.

The weak glow discharge employed here produces a dilute plasma state which comprises a wide variety of species: electrons, ions, excited neutrals (both free radicals and gas molecules) and photons. All these participate in the interactions to varying degrees. The electrons possess average energies between 0.5 and 5 eV, which means that their effective temperature is 10 to 100 times that of the gas ($\simeq T_d$). Electron densities are about 10^{10} cm^{-3}, approximately equal to that of the positive ions. The rf glow discharge is maintained by inelastic electron collisions, which in the case of the silane plasma lead to reactive neutral species such as SiH, SiH_2, SiH_3, Si_2H_6, H, H_2 and ionised species SiH^+, SiH_2^+, SiH_3^+, etc. More detailed information on the silane glow discharge is given in the review article by *Griffith* [3.12], who also considered likely electron impact processes which lead to the above species.

Experimental work on the silane plasma has been started by a number of groups during the last few years. The aim is twofold: first, to identify the species and reactions involved in the growth of the film and secondly, to relate well-characterised plasma conditions to the electronic properties of the a-Si specimen. The experimental method generally used is optical emission spectroscopy by which the radiation from the de-excitation of atoms and molecules in the plasma can be studied. The main advantage of this approach is that it does not perturb the plasma and can also be used as a permanent monitor to achieve reproducible plasma conditions. However, the spectroscopy is limited to emitting species and it is essential to complement it by other techniques, such as mass spectroscopy.

Figure 3.7 is an example of a typical optical emission spectrum from a pure silane glow discharge plasma obtained by *Matsuda* and *Tanaka* [3.13]. The rf power was 20 W and, by means of an optical fibre system, the emission from a region close to the substrate could be recorded. The spectrum shows primarily neutral species such as SiH, H_2 and the H_α and H_β lines of the Balmer series. The prominent SiH band contributes to the violet appearance of the silane glow discharge. Species such as SiH_2 and SiH_3 have no known emission spectrum.

The absence of lines from ionised species (Si^+, SiH^+, SiH_2^+, etc.) is explained by the fact that the threshold energy for dissociation of SiH_4 is about 8 eV, whereas electron energies in excess of 11.9 eV are required to ionise the molecule [3.14]. Recent mass-spectrometric results [3.13, 15] suggest that under typical a-Si deposition conditions, the total density of ionised species lies 4 to 5 orders of magnitude below that of the neutrals. SiH_3^+ and SiH_2^+ appear to be the dominant ionised species at low rf power.

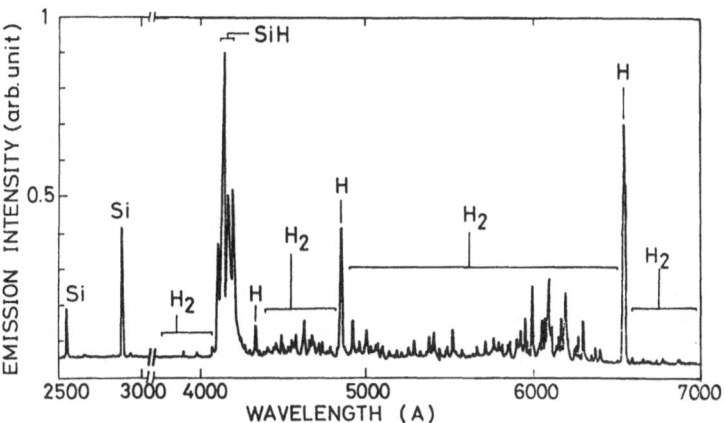

Fig. 3.7. Optical emission spectrum from a pure silane glow discharge plasma run at a power of 20 W [3.13]

In the work published so far, the optical emission intensities (Fig. 3.7) have been investigated as a function of deposition parameters such as flow rate, growth rate, pressure and rf power [3.13, 16, 17]. The experiments have also been extended to plasmas containing dopant gases [3.18], SiH_4-SiF_4 mixtures [3.13] and gaseous impurities [3.12]. Some of the results [3.13, 16] suggest a relationship between the intensities of the H and SiH lines and the ir absorption in the Si-H vibrational band of the film.

There is no doubt that even at this early stage the work summarised here has already made a useful contribution to our understanding of the silane plasma and has also stimulated interest in the problems involved. It is likely to be followed by further interesting developments in the field.

3.2 Field Effect and the Density of States in Amorphous Silicon

The work on a-Si during the last decade has demonstrated the critical dependence of the electronic properties of this material on the density and distribution of localised states. The *independent* measurement of the distribution function $g(\varepsilon)$, defined as the number of states per unit volume in unit energy interval, is therefore of considerable importance. In fact, the $g(\varepsilon)$ measurements on a-Si in the early 1970s, using the field effect technique [3.19–22], provided a wealth of information on the effect of preparation technique and deposition conditions on $g(\varepsilon)$; this contributed greatly to the subsequent development of the subject. In spite of the difficulties of the experiment and particularly the analysis of the data, the field effect experiment still appears to be the most promising approach to the determination of $g(\varepsilon)$. Recently, for example, it has been used to study the effect of hydrogen evolution and of illumination on the creation of defects in a-Si (Sect. 3.2.4). Furthermore, the field effect experiments on a-Si have led to the development of the thin-film transistor which is the subject of discussion in Sect. 3.3.

The aim of this section is to review the principle and analysis of the field effect experiment, discuss the calculated density of state distribution and finally, compare these with the results from independent measurements.

3.2.1 Principle and Experimental Details of the Field Effect Experiment

In the experiment, the localised state distribution is investigated by displacing it with respect to the Fermi level ε_F. This is done by means of an external electric field, applied in a direction normal to the surface of the film to be studied. The energy range moving past ε_F as the field is increased can be calculated from the measured changes in the specimen conductivity. The major experimental problem consists of applying as large a field as possible without any significant leakage current into the specimen. Figure 3.8 shows two arrangements that have been widely used [3.20, 21]. In (a) the a-Si film is deposited directly onto a thin ($\simeq 170$ μm) quartz substrate Q, which also acts as the dielectric for applying the external field. The other surface contains the narrow gate electrode G which is carefully aligned with the gap between the source and drain electrodes S and D deposited on the free surface of the a-Si film. The measurement consists of applying a relatively small constant voltage V_{SD} between the source-drain contacts and recording the specimen current I_D as voltages V_G of up to 10–15 kV are applied between the gate electrode and the free surface of the a-Si, which will remain essentially at earth potential. Measurements are made for both positive and negative V_G.

The main disadvantage of using the quartz substrate as the dielectric is the necessity for large gate voltages. This can be overcome by the thin-film insulator shown in Fig. 3.8b [3.20, 21]. A film ($\lesssim 1$ μm) of amorphous sili-

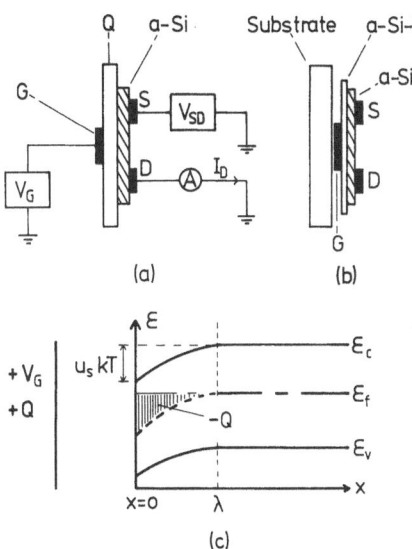

Fig. 3.8. (a, b) specimen geometries used in the field effect experiments. (S, D) and (G) represent the source, drain and gate electrodes. (Q) is the thin quartz dielectric. **(c)** electron energy ε plotted as a function of distance from the a-Si/dielectric interface at $x = 0$. $u_s kT$ denotes the magnitude of the band bending at the interface and λ its spatial extent into the semiconductor. Q and $- Q$ represent the total charge induced on the gate and in the a-Si, respectively, by the positive gate voltage V_G

con nitride (a-Si-N) is deposited by the glow discharge process onto a substrate which carries the evaporated gate electrode G. The a-Si is deposited directly onto this Si-N film and source and drain electrodes are then evaporated as before. In this way much higher fields can be applied with comparatively low voltages, significantly extending the range of measurements.

In addition to providing important information on the localised state density, the field effect experiment also gives the sign of the majority carriers in the amorphous sample. For instance, consider Fig. 3.8c in which the positive V_G induces a negative charge distribution in the a-Si and produces an electron accumulation layer close to the interface. If the sample is n-type, then the conductivity in the a-Si channel will increase. In this case, the specimen current I_D between source and drain will also increase, leading to the type of characteristics shown, for example, in Fig. 3.9. For negative values of V_G, the bands will bend up giving a small decrease in I_D for the n-type specimen before the onset of inversion. Clearly the characteristics expected for a p-type sample (Fig. 3.10 a) will be the reverse of those for an n-type specimen.

The magnitude of the changes in I_D are, of course, critically dependent on the magnitude of $g(\varepsilon)$ in the material, with a low $g(\varepsilon)$ giving the largest changes. Evidently, useful information about the general features of $g(\varepsilon)$ can be obtained directly from the experimental data.

The field effect experiment can, however, suffer from one major limitation which in extreme cases could invalidate the above general conclusions. Underlying the interpretation of the data is the basic assumption that the localised states which screen the field are representative of a volume distribution rather than interface states. Little is known about surface states on

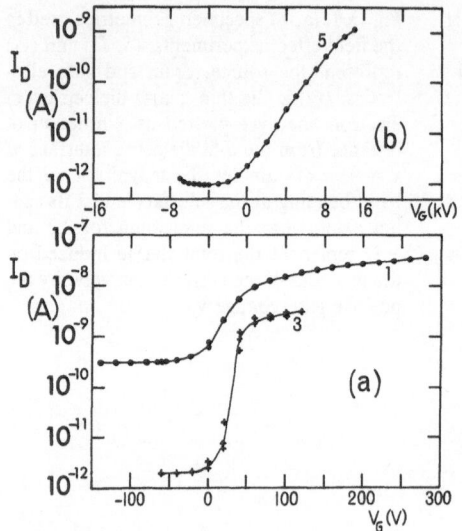

Fig. 3.9 a, b. The dependence of the specimen source-drain current I_D on the gate potential V_G for a number of glow discharge specimens deposited at about 250 °C. (a) Si–N dielectric and (b) quartz dielectric [3.20]

amorphous semiconductors and we have always taken the view that the only reasonable way to test the above assumption is to assess critically the deduced distribution in the light of measurements of other volume properties [3.23]. We shall return to this point in Sect. 3.2.5.

3.2.2 Field Effect Results on Amorphous Silicon

a) Glow Discharge Samples

Figure 3.9 shows a number of field effect curves for glow discharge Si specimens deposited at $T_d \simeq 250$ °C [3.20]. The results in Fig. 3.9 a were obtained using Si–N as the dielectric, those in Fig. 3.9 b with quartz. The most obvious feature is the large rise in I_D when V_G increases in the positive direction. This leads to the conclusion that the overall density of states is relatively small and also that the current in these specimens is carried predominantly by electrons. In contrast, samples produced at $T_d < 80$ °C invariably indicated *p*-type conduction as shown in Fig. 3.10 a for a specimen deposited at $T_d = 40$ °C on to a Si–N dielectric [3.20]. In addition, the overall change in I_D is minute compared to that in high T_d samples, indicating a significantly larger density of states.

Figures 3.10 b, c illustrate the effect of annealing the low T_d specimen [3.20]. After the first stage of annealing at $T_A = 140$ °C (Fig. 3.10 b), the specimen has undoubtedly been converted to predominant electron conduction and its state density at the Fermi level has been reduced. Further annealing at $T_A = 210$ °C (Fig. 3.10 c) continues this process very effectively, as is apparent from the large increase in the range of I_D which now extends over three orders of magnitude.

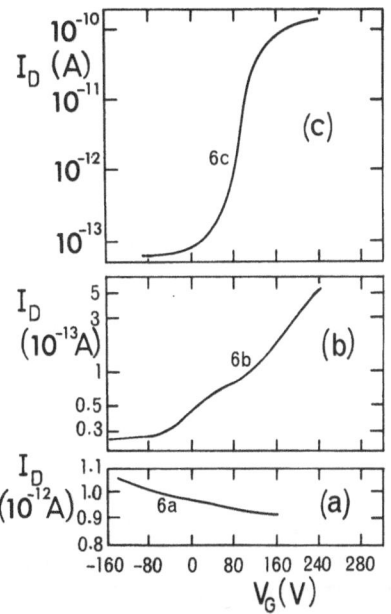

Fig. 3.10 a–c. Field effect results for a specimen deposited at a low deposition temperature ($T = 40\,°C$). (**a**) as deposited, (**b**) annealed to $140\,°C$, and (**c**) annealed to $210\,°C$. After [3.20]

b) Evaporated and Sputtered Samples

In general, field effect experiments on evaporated specimens either gave no observable change in I_D or a very small p-type response, even at the highest gate voltages used, unless the samples were subsequently annealed or were evaporated in a poor vacuum [3.20]. Similar results have also been obtained on samples prepared by sputtering. As discussed in Chap. 4 it is possible significantly to decrease $g(\varepsilon)$ of this material by sputtering in a gas mixture containing hydrogen. A large field effect response was, however, only observed after samples were deposited onto quartz substrates coated with a thin layer (150–200 Å) of SiO_x or Si-N [3.24]. The field effect characteristic of such a specimen (see Chap. 4 for more details) is then similar to that for the glow discharge material shown in Fig. 3.9 b.

c) General Conclusions from Experimental Field Effect Data

The experiments support the suggestion, based on electrical conductivity, drift mobility and optical data [3.4, 20], that evaporated and sputtered films have a much higher localised state density than glow discharge specimens, leading to significant differences in the optical and electrical properties. In addition, the decrease of $g(\varepsilon)$ of sputtered material prepared in hydrogen can also be directly demonstrated using the field effect technique.

In the next section we summarise the approaches that have been used to calculate the density of state distribution from the experimental field effect data.

3.2.3 Analysis of the Field Effect Results

The calculation of $g(\varepsilon)$ from the experimental field effect data is a complex problem which has attracted quite a lot of attention during the last few years. All the methods of analysis, discussed in the following, involve two common assumptions; first, the semiconductor is homogeneous and secondly, the effect of interface states can be neglected except that they may, if present, produce band bending at zero gate voltage. The validity of these assumptions will be discussed in Sect. 3.2.5.

We begin by considering the step-by-step method of analysis in which $g(\varepsilon)$ is progressively deduced from the observed current I_D as V_G is increased in equal steps ΔV_G. This approach, developed by the authors for the analysis of the early field effect results on a-Si [3.19], involved in its original form three simplifying assumptions. The first of these was concerned with the potential profile $V(x)$ in the semiconductor, which was assumed to have the form of a Schottky barrier or an exponential barrier; this neglected the effect of the localised state distribution on the shape of the barrier (Sect. 3.5.1). Poisson's equation can then be solved to relate the band bending $u_s kT$ at the surface (Fig. 3.8 c) with the spatial extent λ of the barrier. By integrating the current contributions throughout the sample thickness it is possible, at any step in the experiment, to determine u_s and λ from the measured I_D. Finally, the total excess charge eN_s induced in the semiconductor at the given V_G is required. This can readily be calculated from the known capacitance between gate and semiconductor.

The next problem consists of deducing $g(\varepsilon)$ from eN_s, u_s and λ as the distribution is moved past ε_F by the ΔV_G steps. A simple solution, adopted in the earlier work, rested on the assumption that the induced charge carriers occupied previously empty states within $u_s kT$ of ε_F over a distance λ. This implied (i) zero temperature statistics and (ii) abrupt band bending, inconsistent with the above form of $V(x)$. In a subsequent publication [3.22], these two approximations were removed; the simple procedure was initially used to obtain $g(\varepsilon)$ to a first approximation. Based on this solution, the form of the field effect response curve was calculated by numerical methods. Small changes in $g(\varepsilon)$ were then made until agreement between predicted and experimental curves was achieved. The work showed that the main quantitative change produced by this refinement was to increase $g(\varepsilon)$ by a factor of between two and three.

Even the refined step-by-step method still relies on an assumed form of $V(x)$. This limitation was removed by *Goodman* and *Fritzsche* [3.25] who started the analysis with a trial solution for $g(\varepsilon)$ and obtained a numerical solution of Poisson's equation with a self-consistent $V(x)$. $g(\varepsilon)$ [and $V(x)$] were then modified until the predicted and experimental field effect characteristics agreed. Although more exact than the step-by-step procedure, this approach involves considerable computing time and produces results that differ from the simpler method by less than a factor of two (see, for example,

[Ref. 3.25, Fig. 3]). Subsequently, it has been pointed out and demonstrated by a number of authors [3.26–28] that the Goodman-Fritzsche method can be simplified by a change of the independent variable in Poisson's equation, with a significant reduction in computing time.

A basically different approach was followed by *Grünewald* et al. [3.29] who published a method for calculating the induced space charge density $\varrho(v)$ *directly* from the experimental data. The density of states has then to be found by deconvoluting

$$\varrho(v) = e \int_{-\infty}^{+\infty} g(\varepsilon)[f(\varepsilon - eV) - f(\varepsilon)]d\varepsilon , \tag{3.3}$$

where $f(\varepsilon)$ is the Fermi distribution function. If correlation effects are important, then it is not possible to use conventional Fermi-Dirac statistics. Under these conditions it has been suggested [3.30] that the calculated $g(\varepsilon)$ may be significantly modified.

We mentioned earlier that all the methods of analysis assume that the only effect of interface states might be to produce some band bending at zero gate voltage. The calculated values of $g(\varepsilon)$, at least near ε_F are sensitive (within a factor of 2 to 3) to the choice of the flat-band position on the V_G-axis. This value cannot be determined unambiguously from field effect data at a single temperature. However, it can be deduced, at least in principle, from experiments at different temperatures [3.27, 28]. In fact, measurements on a-Si at various temperatures were used to provide a check on the band bending calculated from the refined step-by-step procedure with that deduced from the temperature dependence of the field effect mobility [3.22]. Considering the approximations involved, the agreement was found to be very satisfactory.

In summary, the determination of $g(\varepsilon)$ from field effect data is a complex problem that can only be solved either by making a number of simplifying assumptions, as in the step-by-step method, or by numerical techniques involving a trial solution for $g(\varepsilon)$ which is then adjusted until agreement between the calculated and experimental field effect conductance is obtained. Generally, these two approaches lead to similar results, although *small* features in $g(\varepsilon)$, sometimes found by the former, are not always uniquely determined by the latter. In this connection it may be significant that the more direct method of *Grünewald* et al. [3.29] reproduces the features in $g(\varepsilon)$ obtained by the step-by-step method [3.19, 22] when similar $I_D(V_G)$ data are used [3.29–31]. Finally, before discussing some of the calculated $g(\varepsilon)$ results, it is worth noting that the neglect of interface states and the difficulties associated with the choice of the flat-band voltage probably introduce larger uncertainties in $g(\varepsilon)$ than those produced by the different methods of analysis!

3.2.4 Discussion of Calculated $g(\varepsilon)$ Data

Figure 3.11 shows the calculated density of states [3.20, 21] for a number of glow discharge a-Si films, deposited at the indicated values of T_d. $g(\varepsilon)$ is plotted against the energy normalised to the electron mobility edge at ε_c; the position of the Fermi level for each sample is indicated by the arrow. The full-line portions were calculated from the experimental data and the dotted parts were based on an extrapolation procedure. The results range from ε_A, situated about 0.2 eV below ε_c, to almost ε_y which was previously associated with hole hopping transport [3.4]. The curves in Fig. 3.11 lead to a number of important conclusions [3.20]:

(i)   the overall level of the localised state density, at least between ε_y and ε_A, is a strong function of the deposition temperature;
(ii) a minimum in $g(\varepsilon)$ exists close to the centre of the gap;
(iii) $g(\varepsilon)$ increases as $\varepsilon_c-\varepsilon$ approaches ε_y which, in agreement with conductivity and photoconductivity results [3.32], is situated at $\varepsilon_c-\varepsilon_y \simeq 1.2$ eV;
(iv) between ε_A and ε_c there is a rapid increase in $g(\varepsilon)$ which has been identified with a range of electron tail states, about 0.2 eV wide.

Sample 1 in Fig. 3.11, deposited at $T_d = 300$ °C, is representative of the high-T_d specimens. It is clearly *n*-type and the density of states near the centre of the gap is remarkably low, $\lesssim 10^{17}$ cm^{-3} eV^{-1}. As the deposition temperature is lowered, the general level of $g(\varepsilon)$ increases and ε_F moves closer towards the centre of the gap. Eventually for specimens deposited at $T_d < 350$ K (sample 6a, for example), the Fermi level has moved so close to ε_y that the dominant conduction mechanism is now by holes hopping at this energy.

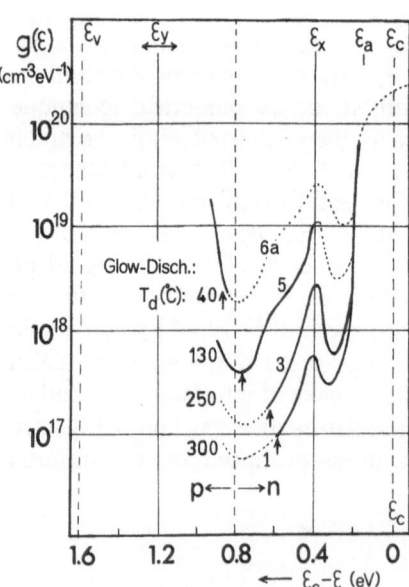

Fig. 3.11. Deduced density of states for glow discharge specimens deposited at the given temperatures T_d. The position of the Fermi level is indicated by the arrow. Full lines were calculated from experimental data and the dotted parts were based on an extrapolation procedure [3.20]

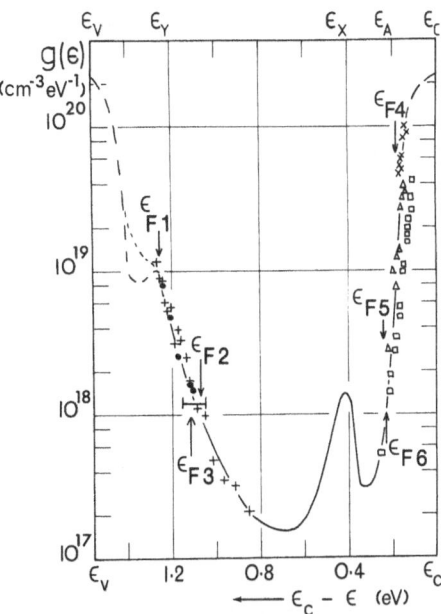

Fig. 3.12. Density of states of high T_d glow discharge samples. The centre region refers to undoped specimens and the points to comparatively lightly doped specimens: ε_{F6}, ε_{F5} and ε_{F4}: Fermi level positions of P-doped samples; ε_{F3}, ε_{F2} and ε_{F1}: Fermi level positions of B-doped samples

Substitutional doping in a-Si (Sect. 3.4) has made it possible to shift ε_F throughout most of the mobility gap so that $g(\varepsilon)$ can be determined for high-T_d specimens over a larger energy range than covered in Fig. 3.11. Figure 3.12 includes the results for a number of undoped and comparatively lightly doped samples. The $g(\varepsilon)$ data in the centre of the gap were determined as before on undoped material. The Fermi level positions denoted by ε_{F6}, ε_{F5} and ε_{F4} refer to specimens doped with an increasing level of P, and the results give more information on $g(\varepsilon)$ in the region of the electron tail states. On the other hand, light B-doping moves ε_F towards the valence band and extends the $g(\varepsilon)$ range to energies of over 1.2 eV below ε_c. It must, however, be emphasised that the results obtained in Fig. 3.12 with the doped specimens give no information on the effect of doping on $g(\varepsilon)$ in the central region of the gap. It is possible that moderate or heavy doping can produce an increase in $g(\varepsilon)$ in this energy range. To investigate this point, measurements on a series of compensated specimens are being carried out.

The field effect technique has also been used to study the effect of hydrogen evolution on $g(\varepsilon_F)$ in a-Si [3.33]. Figure 3.13 shows a plot of $g(\varepsilon_F)$ as a function of the annealing temperature T_A. Once T_A exceeds about 400 °C, the temperature at which appreciable hydrogen evolution begins [3.34], $g(\varepsilon_F)$ starts to increase rapidly. Eventually, little or no field effect response is observed, setting a lower limit to $g(\varepsilon_F)$ of about 10^{20} cm^{-3} eV^{-1}. The photoconductivity σ_{PC}, measured at a standard photon flux of about 10^{15} s^{-1} cm^{-2} and normalised to the value before annealing, is also shown as a function of T_A in Fig. 3.13. In these samples the photoconductivity depended linearly on

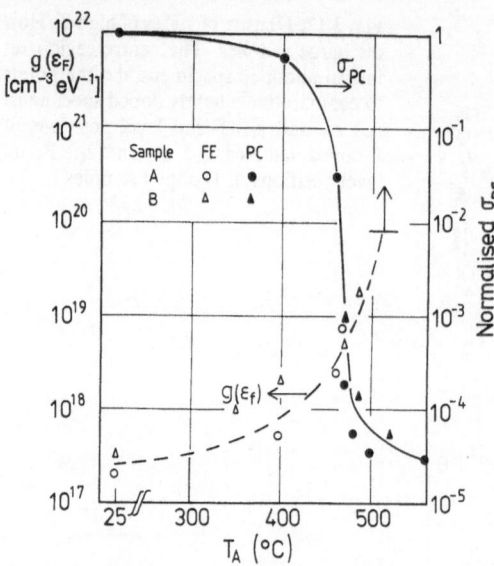

Fig. 3.13. Density of states $g(\varepsilon_F)$ at the Fermi level and the photoconductivity σ_{PC} of two specimens as a function of the annealing temperature T_A. The magnitude of σ_{PC} has been normalised to the value before annealing [3.33]

light intensity and in previous work this has been associated with recombination via states at ε_F [3.32]. The data in Fig. 3.13 suggests that σ_{PC} depends almost inversely on $g(\varepsilon_F)$ and therefore supports this interpretation.

Finally we shall look briefly at the experiments reported recently by *Grünewald* et al. [3.29], *Tanielian* et al. [3.35] and *Powell* et al. [3.36], who used the field effect technique to investigate the formation of defects under illumination. The data [3.29] shown in Fig. 3.14 is typical of the results obtained. Curve 1 represents $g(\varepsilon)$ of an undoped a-Si specimen annealed at

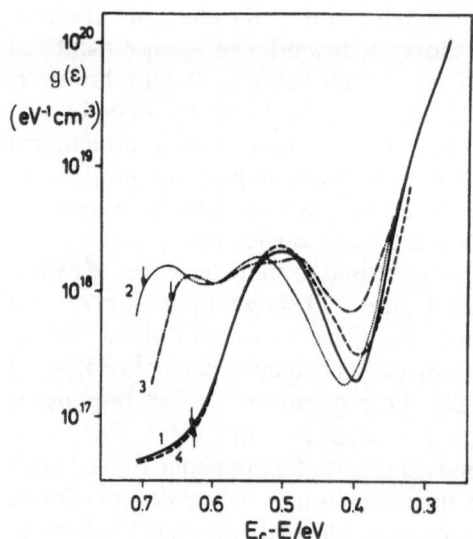

Fig. 3.14. Density of states of undoped a-Si: (*1*) annealed at 170 °C, (*2*) after strong illumination, (*3*) after 30 min anneal at 170 °C, and (*4*) after 4 h anneal at 170 °C. Arrow indicates position of ε_F [3.29]

170 °C. After strong illumination (curve 2), $g(\varepsilon)$ has increased by about an order of magnitude in the range investigated. Annealing in the dark at 170 °C for half an hour (curve 3) and for four hours (curve 4) recovers the initial low $g(\varepsilon)$. The corresponding changes in conductance and photoconductivity observed after illumination have been interpreted in terms of bulk changes in $g(\varepsilon)$ and ε_F [3.37–39], or in terms of surface space charge regions [3.40, 41]. *Powell* et al. [3.36] concluded from their results that a major part of the effect is likely to be associated with a bulk shift of ε_F.

3.2.5 Comparison of the Deduced $g(\varepsilon)$ with Other Measurements

The calculation of $g(\varepsilon)$ from the field effect experiment rests on the basic assumptions that the effect is representative of a homogeneous bulk property of the a-Si and is not dominated by interface states at the gate insulator. In our view the best test of this assumption is to compare the $g(\varepsilon)$ data with independent measurements of the bulk properties [3.19–22]. The evidence from extensive electrical and optical measurements has been reviewed in some detail previously [3.23] and we shall therefore refer here only briefly to this work.

Measurements of electrical conductivity [3.4], optical absorption [3.32], photoconductivity [3.42] and drift mobility [3.3, 4] all provide independent support for the general features in $g(\varepsilon)$ shown in Figs. 3.11, 12. Furthermore, the effect of doping, either from the gas phase [3.43, 44] or by ion implantation [3.45, 46], also provides evidence for the broad minimum in $g(\varepsilon)$ near the centre of the gap and the rise in $g(\varepsilon)$ near ε_A and ε_y. It can therefore be concluded that, in spite of the difficulties associated with the field effect and its analysis, it does provide us in a-Si with a density of states distribution that is consistent with independent measurements of bulk properties.

The technique of deep level transient spectroscopy (DLTS) [3.47, 48] has recently been applied to a-Si [3.49] and the results were used to deduce the density of localised states. As with the field effect, certain assumptions had to be made in order to calculate both the magnitude of $g(\varepsilon)$ and the energy scale. However, the above authors stress as a major advantage of DLTS over the field effect technique the fact that it leads to a true bulk density of states, unaffected by interface or surface states. In their original paper, *Cohen* et al. [3.47] deduced $g(\varepsilon)$ values for two samples in which the conductivity was considerably higher than normal, although these films were not intentionally doped. Values of $g(\varepsilon)$ as low as $2 \times 10^{14}\ \mathrm{cm}^{-3}\,\mathrm{eV}^{-1}$ were reported, but these values were increased in a subsequent publication [3.49] and are represented in Fig. 3.15 by the line marked DLTS. Comparison with a typical field effect $g(\varepsilon)$ denoted by F.E. in Fig. 3.15 shows that as well as giving significantly lower $g(\varepsilon)$ values, the DLTS technique also suggests a widely different distribution with a deep minimum in $g(\varepsilon)$, 0.4 to 0.5 eV below ε_c.

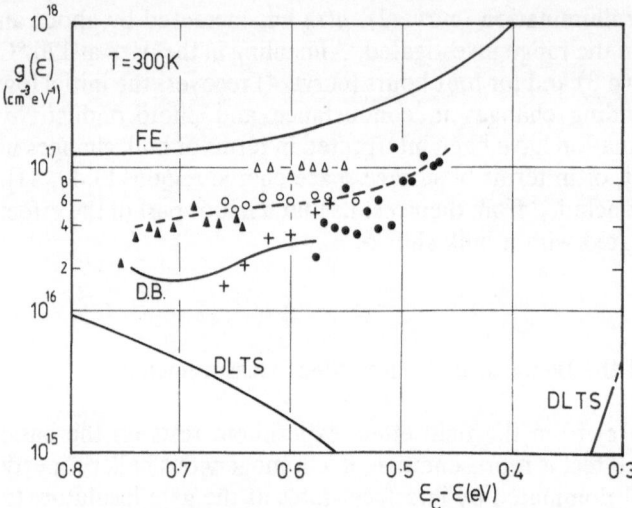

Fig. 3.15. Density of states in glow discharge a-Si deduced from different experiments: (FE) field effect [3.20, 22]; solid line (DB) space charge limited currents [3.50]; dashed line and experimental points: space charge limited currents [3.51]; (DLTS) deep level transient spectroscopy [3.49 c]

The analysis of space charge limited current flow in a-Si provides an independent alternative approach to the $g(\varepsilon)$ problem. It is particularly relevant in the present controversy because there is no doubt that the space-charge limited current (SCLC) is determined by the *volume distribution of localised states*. Extensive work on SCLCs in a-Si has recently been carried out by *Den Boer* [3.50] and by *Mackenzie* et al. [3.51] in our laboratory. A typical current-voltage characteristic for the highly injecting n^+-i-n^+ structures used is shown in Fig. 3.26 (Sect. 3.5.2). With relatively minor approximations these measurements lead to the bulk $g(\varepsilon)$ values shown by the points and the dashed line in Fig. 3.15 [3.51]. The full line marked D.B. represents the average of *Den Boer* [3.50] results. Clearly, there is satisfactory agreement between the results obtained in the two laboratories on different specimens. Furthermore, the values of $g(\varepsilon)$ in the range investigated lie between 2 $\times$ 10^{16} and 10^{17} cm^{-3} eV^{-1} and follow the shape of the field effect data rather than that from the DLTS studies. On this basis it would appear that the DLTS technique considerably underestimates the density of states in a-Si. In view of the uncertainty by a factor of two to three in the precise magnitude of $g(\varepsilon)$ from field effect and the somewhat smaller uncertainty in the data from SCLC, it also appears from Fig. 3.15 that interface states can have relatively little effect on the field effect results. Evidently, further work is required to resolve the present disagreement on this fundamentally important property of glow discharge a-Si.

3.3 The Amorphous Silicon Field Effect Transistor and Its Possible Applications

The field effect experiments described in the preceding section demonstrated that a-Si deposited by the glow discharge process has a remarkably low density of gap states, which is a basic requirement for most electronic applications. Another advantage of the glow discharge technique is that it can be used to deposit sequentially thin layers of different materials, simply by changing the gas composition. This procedure has found increasing use in the development of the a-Si field effect transistor (to be referred to as FET in the following), which forms the subject of this section. The possibility of this application was originally suggested by the field effect studies in the 1970s [3.19–21], which showed that large current changes could be obtained for moderate fields produced by the gate electrode.

In 1976 the Dundee group and colleagues at RSRE, Malvern, jointly proposed the use of a-Si field effect devices in the addressing of liquid crystal matrix displays as an alternative to the thin film CdSe transistors which had been developed by *Brody* and coworkers [3.52]. They reasoned that an elemental covalently bonded material such as a-Si should have distinct advantages over the more complex II–VI compounds as far as ease of preparation, reproducibility and stability were concerned. The design and characteristics of an a-Si insulated gate FET suitable for driving liquid crystal displays were published by *LeComber* et al. [3.53] and experiments at RSRE showed that such a device could switch liquid crystal elements satisfactorily.

In the following we shall consider the design, characteristics and fabrication of a-Si FETs and discuss their possible applications in addressable liquid crystal displays, logic circuits and image sensors [3.54].

3.3.1 Design and Characteristics of Elementary Device

The basic design of the a-Si FET element is shown schematically in the insert to Fig. 3.16. It represents a direct development of the arrangement originally used in the field effect experiments described in Sect. 3.2.1. A metal electrode, typically 30 µm wide, is first evaporated onto a glass substrate to form the gate electrode. A thin insulating layer, 0.3 µm or less, of amorphous silicon nitride (Si–N) is then deposited by the glow discharge technique to form the gate dielectric. This is followed by an a-Si layer, also a few tenths of a µm thick. The final step consists in depositing the required pattern of source and drain contacts on to the a-Si surface. It is an essential point for any future application of these devices that large area arrays can be produced reproduceably and cheaply. In this connection it is most important that conventional photolithographic techniques have been successfully applied to the fabrication of arrays of a-Si/Si–N FET [3.55a].

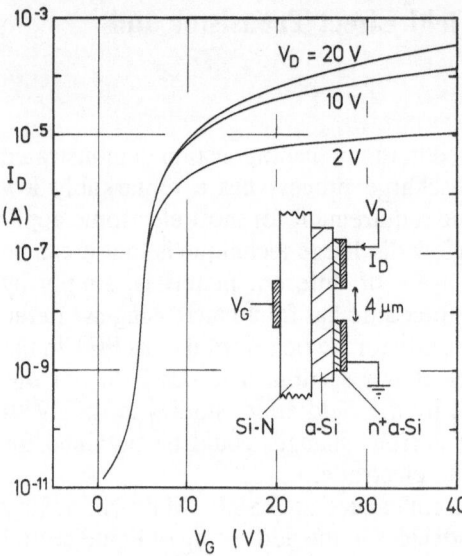

Fig. 3.16. Transfer characteristics of a-Si FET element. The drain current I_D is plotted against the gate voltage V_G for three drain potentials V_D. The insert shows the basic design of the FET [3.55 b]

The *dc* performance of an elementary device with a 4 μm channel length and a 500 μm channel width, produced by photolithographic techniques, is illustrated by the transfer characteristics shown in Fig. 3.16 [3.55 b]. The source-drain current I_D is plotted logarithmically against the gate voltage V_G for drain potentials of 2, 10, and 20 V. It can be seen that with + 15 V on the gate, drain currents in excess of 10 μA can be achieved for drain voltages as low as 10 V. In the off-condition with $V_G = 0$, the current through the device drops below 10^{-11} A. The remarkable rise in I_D is caused by an electron accumulation layer formed at the Si/Si–N interface which produces an efficient current path between the source and drain electrodes.

A number of different groups have reported results for similar a-Si FETs. To compare the data, which refer to different device geometries and gate dielectrics, we follow *Powell* et al. [3.56] and plot in Fig. 3.17 the sheet conductance G_s logarithmically against the field in the semiconductor at the insulator/a-Si interface [3.54]. The surface field has been calculated from $\varepsilon_i V_G / \varepsilon_{Si} d$, where ε_i is the relative permittivity of the insulator, ε_{Si} the relative permittivity of the a-Si and d the thickness of the gate insulator. The curve denoted by D is recent Dundee data obtained with a 0.3 μm thick glow discharge Si–N insulator [3.55 b]; curve P is from *Powell* et al. [3.56] who also used a glow discharge Si–N film of about 0.3 μm tickness; the curves marked T are from data published by *Matsumura* et al. [3.57 a]; curves C and F represent recent results from the *Canon* [3.57 b] and *Fujitsu* [3.57 c] laboratories, also in Japan; finally curve X are new results from the Xerox laboratories [3.58] for a double-gate device. In double-gate operation this device gives more current than is obtained by simply summing the currents when either gate is operated independently. However, even under these

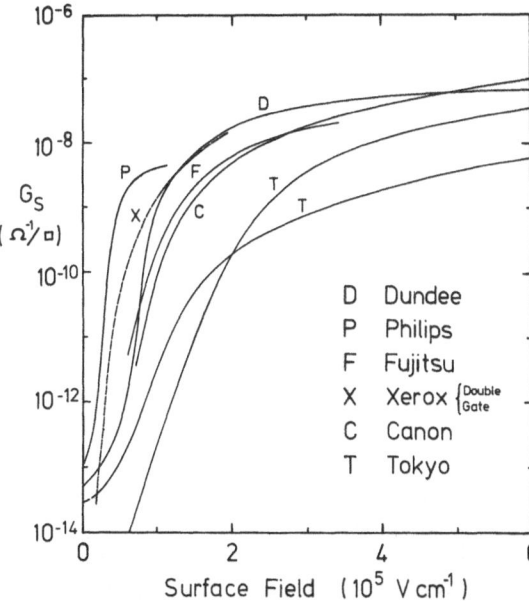

Fig. 3.17. A comparison of the performance of a-Si FET devices made by various research groups [3.55 b]

D   Dundee
P   Philips
F   Fujitsu
X   Xerox {Double Gate}
C   Canon
T   Tokyo

conditions it gives no more sheet conductance than our present single-gate devices. In plotting the data shown in Fig. 3.17 we have taken the dielectric constant of Si–N as 6.4, of SiO_2 as 3.9 and that of a-Si as 11.8. For device applications the transfer characteristics shown in Figs. 3.16, 17 should have the following properties: (i) low off-conductance, (ii) high on-conductance, and (iii) a rapid rise from the off to the on-state at gate voltages below 15 V so that devices are compatible with modern integrated circuit voltage levels. All the data in Fig. 3.17 satisfy the first of these criteria, but the achievement of high on-currents clearly presents one of the main problems.

It is perhaps worth emphasising that the formation with positive gate voltage of the highly conducting electron accumulation layer, which determines the properties of the on-state, is governed by two fundamental material properties: (i) a low density of localized gap states in the bulk of the a-Si material itself, and (ii) a low density of interface states at the Si–N boundary. The differences in the samples may arise, at least partly, for the former reason, since localized state densities tend to depend critically on the plasma condition during the glow discharge process. It is equally likely, however, that different laboratories produce devices with different interface state densities, particularly if different insulator materials are involved.

The radiation hardness of a-Si FETs could be significant in some applications. Results obtained with γ-radiation from Cobalt 60 indicate that doses up to 5 mrad, the maximum used, have relatively little effect on the dc characteristics of the devices [3.59].

3.3.2 Application to Addressable Liquid Crystal Displays

Liquid crystal displays are potentially very attractive for large-area applications as they are low power devices and have good visibility in high ambient light conditions. However, the difficulty of multiplexing large arrays of such devices has so far limited their use. If a small nonlinear device such as an FET is used to control each display element, much larger arrays could, in principle, be addressed. Liquid crystal devices typically require an ac drive signal of 3–6 V without dc component as the latter tends to degrade the material. The device responds to the rms value of the ac signal and has a threshold of 1–2 V rms. Voltages below this level will not turn on the display.

In the following we shall begin by looking at the fabrication of the a-Si FET array and then review its dynamic behaviour in a liquid crystal display. A more extensive discussion is given in [3.54, 55].

Figure 3.18 shows a schematic diagram of a number of elements in the display panel. A transistor is incorporated in a corner of each element of the array. The FETs are interconnected by means of X and Y buses, G_1, G_2...and S_1, S_2..., linking gate and source contacts, respectively. The drain contact of each transistor is connected to the ITO squares, D. From the section through the panel in Fig. 3.18b it can be seen that the liquid crystal material is sandwiched between the substrate carrying the FETs and an ITO coated glass top plate which is normally returned to ground. The liquid crystal element is therefore in series with the drain circuit and behaves electrically as a capacitor C_{LC} with some leakage resistance R_{LC}.

Figure 3.19a shows a section through an individual a-Si device and Fig. 3.19b illustrates the design of the FET in part of the matrix array [3.55]. As stressed above, all processing is carried out by standard photolithographic techniques and with conventional etches.

The transient behaviour of an FET in the matrix is mainly determined by the capacitive loading of the liquid crystal element (C_{LC} in Fig. 3.18b).

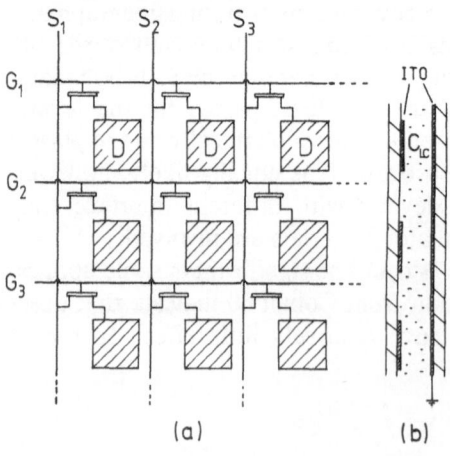

(a)   (b)

Fig. 3.18a, b. Schematic lay-out of an addressable liquid crystal panel. (a) shows gate and source connections G_1, G_2... and S_1, S_2..., and the drain connected to the ITO squares D. (b) section through panel. C_{LC} denotes the capacitance of a liquid crystal element [3.55]

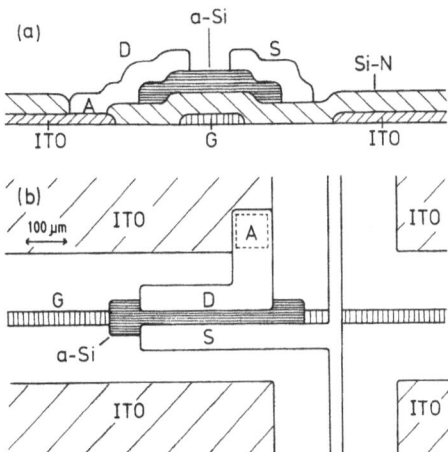

Fig. 3.19 a, b. Design of a-Si field effect transistor element. (a) section through device, (b) FET in part of the matrix array [(ITO) indium-tin oxide squares; (G) gate electrode; (D) drain electrode; (S) source electrode; (A) contact hole etched through silicon nitride film] [3.55]

This aspect has been investigated with C_{LC} = 10 pf, corresponding to ITO elements of 1 mm sidelength. When the FET was switched on, the required drive potential across C_{LC} of 3 V was reached in less than 10 μs [3.55 b]. On the other hand, the charge retention on the elementary capacitor in the off-state of the FET is limited by leakage through the liquid crystal material to about 40 ms. It is therefore possible, even with the present devices, to address about 1000 lines in this time [3.55 b].

We are presently assessing the performance of panels containing 20 × 25 elements, each 1 mm² in area. The good uniformity of the transfer charac-teristics of elements within a panel and their reproducibility in different depositions is encouraging. So far, the information on long term stability of the a-Si devices is still somewhat limited. In a test an element has now been run at 80 Hz with V_s = ± 10 V and V_G = 15 V for 28 months, i.e., for about 6×10^9 switching operations. There has been no sign of deterioration or change in transfer characteristics of this device which is unpassivated and unencapsulated.

3.3.3 Application of Amorphous Silicon FETs in Logic Circuits

Recently, *Matsumura, Hayama* and coworkers showed that an integrated inverter circuit [3.60] and an image sensor [3.61] can be fabricated from a-Si FETs. The devices used in their work were made from a-Si deposited in a dc glow discharge and with low pressure CVD SiO_2 as the gate insulator [3.60]. Although these FETs were relatively leaky and had a slow turn-on, they demonstrated the feasibility of using the a-Si technology in these circuits. At the same time work in our laboratory has been concerned with similar appli-cations [3.62], initially aimed at providing integrated a-Si drive circuits for

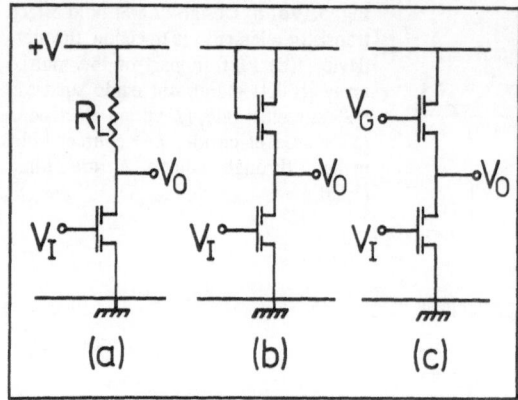

Fig. 3.20. Three examples of inverter circuits. Configuration (*a*) is used in the present work [3.54, 62]

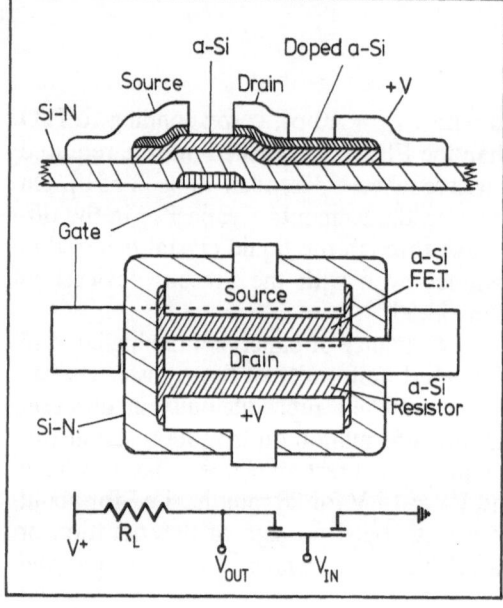

Fig. 3.21

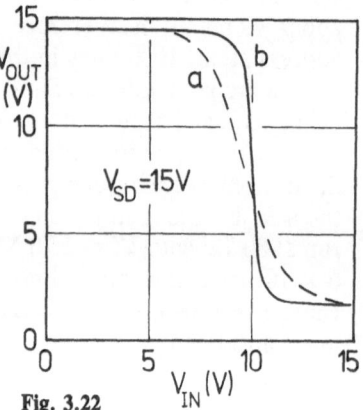

Fig. 3.22

Fig. 3.21. Section and plan view of the integrated a-Si inverter [3.62]

Fig. 3.22. Characteristics of (*a*) single a-Si inverter and (*b*) three inverters connected in series [3.54, 62]

the liquid crystal display panels discussed above. In the following we shall briefly review the work on the inverters and related devices [3.54, 62].

Three basic circuits for producing the inverter logic are shown in Fig. 3.20. *Matsumura* and *Hayama* [3.60] used circuit (c) which requires two FETs and an additional supply V_G. We chose circuit (a) and made the load resistor from an integrated a-Si gap cell. This is simpler and has the further advantage that the a-Si resistor requires less space than an FET load. Figure 3.21 shows a section through the integrated device and also its layout in plan view. The input voltage is applied to the gate of the FET, the source is connected to ground and the output taken from the drain connection. To obtain the appropriate value of load resistance, a doped a-Si layer about 250 Å thick and about 3×10^8 $\Omega/\square$ was deposited. This was etched away from the source-drain gap of the FET after the top Al metallisation leaving doped a-Si under the source and drain, which had the added advantage of improving the electrical contact at these electrodes.

The dashed curve in Fig. 3.22 shows the transfer characteristics of such a device having a load of approximately 30 MΩ and measured with a supply potential of 15 V. The inverter logic is clearly seen: the output changes from about 14.5 to 2 V as the input swings from about 5 to 15 V. The full line in Fig. 3.22 shows the characteristics of three such devices connected in series. As well as sharpening the characteristics, this demonstrates the important point that the output from one device can be successfully used as the input for successive stages.

By simple extension of the above basic design, logic circuits such as NAND and NOR gates as well as *bi*-stable multivibrators have been produced [3.54, 62]. As mentioned previously, one of the primary aims of our present work is to investigate the possibility of drive circuits for the liquid crystal panels. Of particular importance in this respect is the fabrication of a shift register to address sequentially the gate buses G_1, G_2, . . . in Fig. 3.18. Recently, a four stage shift register using a-Si FETs has been made [3.54, 62] in which each stage can shift a 20 ms pulse by 10 ms. All the circuits mentioned in this section are slow by crystalline standards. At present they are limited to the kHz range, although it must be emphasized that the exploratory a-Si devices described were in no way optimised for speed of response. Possible improvements are discussed in Sect. 3.3.5.

3.3.4 Addressable Image Sensing Elements

In this section we review the work of *Matsumura* et al. [3.61] and our results on the a-Si image sensors [3.62]. The circuit used by both groups is shown in the top part of Fig. 3.23 and is that originally suggested by *Weimer* et al. [3.63]. Light incident on the a-Si photoconductor charges up the capacitor whilst the FET is off. The FET is then pulsed on and the capacitor discharges rapidly, providing an output current pulse the magnitude of which is a function of the light intensity. Figure 3.23 shows the magnitude of the output current pulse as a function of light intensity for gate pulse frequencies of 2, 20 and 40 Hz [3.62]. The output current at 20 Hz, for example, varies by over an order of magnitude as the light intensity is varied from 10 to 1000 lux. These relatively large changes in output current suggest that operation with a grey scale should be feasible.

Further work on this potentially important application is presently being carried out in a number of laboratories throughout the world.

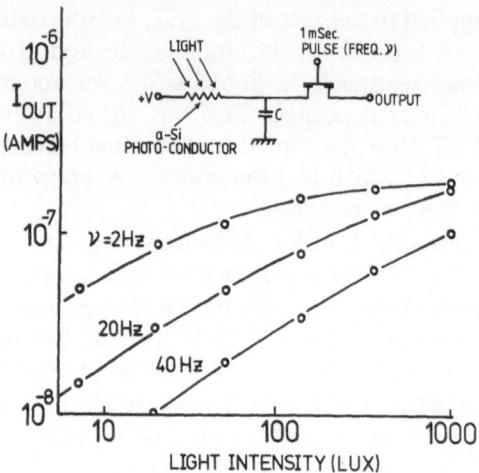

Fig. 3.23. Circuit of the a-Si image sensor and measured output current as a function of light intensity for various gate pulse frequencies [3.54, 62]

3.3.5 Limitations of the Present Amorphous Silicon FETs

The work described above has clearly demonstrated the feasibility of using a-Si FETs in a number of practical circuits. In particular, the devices presently available already offer a viable solution to the problem of multiplexing liquid crystal displays [3.54, 55]. However, the factor that appears most likely to limit their use in other applications is their relatively slow switching time. A figure of merit characterising the high-frequency performance of thin-film transistors is the gain-bandwidth product, which leads to the maximum operating frequency

$$f_\mathrm{m} \simeq \frac{\mu_\mathrm{FE} V_\mathrm{G}}{2\,\pi L^2}\;.$$

with $\mu_\mathrm{FE} \simeq 0.4\ \mathrm{cm}^2\,\mathrm{V}^{-1}\,\mathrm{s}^{-1}$, $V_\mathrm{G} = 15$ V and $L = 4\ \mu\mathrm{m}$, a value of $f_\mathrm{m} \simeq 6$ MHz is obtained [3.55 b]. In practical circuits this will often be further limited by the time required for the ON current to charge circuit capacitances. The considerable increase in ON currents in recent optimised devices [3.55 b] represents a significant advance in this direction and it is possible that a-Si FETs will be used in a growing number of applications, in a frequency range approaching 1 MHz.

3.4 Doping in the Amorphous Phase

This section is concerned with substitutional doping of a-Si in the glow discharge plasma and also by ion implantation. The possibility of efficient dop-

ing in the a-phase was demonstrated by the authors in 1975 [3.43, 44] and led
to the first a-Si p-n junction [3.64]. This development has removed one of the
main limitations of amorphous semiconductors and opened up exciting possibilities for both fundamental and applied work.

3.4.1 Control of Electrical Properties of Amorphous Silicon by Gas Phase Doping

The mechanism of substitutional doping in an amorphous semiconductor
differs basically from that in the crystalline material because changes in electrical properties are brought about primarily by changes in the gap state
occupation. If, for example, N_D substitutional donors are incorporated, practically all the excess electrons will condense into empty gap states near the
Fermi level, displacing the latter towards ε_c by an amount $\Delta\varepsilon_F$. For sensitive
doping $\Delta\varepsilon_F$, which is of the order of $N_D/g(\varepsilon_F)$, should be as large as possible;
a low level of gap states is therefore an essential condition. The field effect
measurements discussed in Sect. 3.2 suggest that preparation by the glow
discharge technique should be a most promising approach. It was, in fact, on
the basis of these results that we started the first doping experiments in 1975.
Reference to Sect. 3.3 explains the reason for the insensitivity to doping of
evaporated amorphous semiconductors. Although it may be possible to
incorporate donors or acceptors into the material, the overall $g(\varepsilon)$ is too high
to allow much change in the Fermi level position and consequently in the
electrical properties.

Gas phase doping in glow discharge Si or Ge is carried out by introducing
small but accurately controlled amounts of phosphine (PH_3) or diborane
(B_2H_6) into the reactor by means of the gas handling system shown in Fig. 3.3
(Sect. 3.1). The gaseous doping ratio D is defined as the ratio of the number
of phosphine (or diborane) to silane molecules in the mixture. Ratios as low
as 10^{-6} can be obtained by pre-mixing in the reservoir R (Fig. 3.3), whereas
for D values exceeding 3×10^{-4}, direct electronically controlled mixing
during the deposition run is sufficiently accurate.

Figure 3.24 summarises the doping dependence of the room temperature
conductivity σ_{RT} in glow discharge Si. σ_{RT} is plotted against the doping ratio-
on the right side of the diagram for phosphorus doping and on the left for
boron doping; the centre refers to the undoped material. The results illustrate the remarkable control of the conductivity that can be achieved in
carefully prepared specimens. Curve (a) represents the typical dependence of
specimens prepared in the deposition unit shown in Fig. 3.3. With $D \simeq 10^{-6}$,
σ_{RT} on the n-side has been increased by an order of magnitude over that of
the undoped specimen, and the highest doping ratio of 10^{-2} leads to $\sigma_{RT} \simeq$
2×10^{-2} $(\Omega\ cm)^{-1}$.

Light boron doping first moves ε_F into the centre of the mobility gap,
where σ_{RT} reaches its "intrinsic" value of about 10^{-12} $(\Omega\ cm)^{-1}$. The speci-

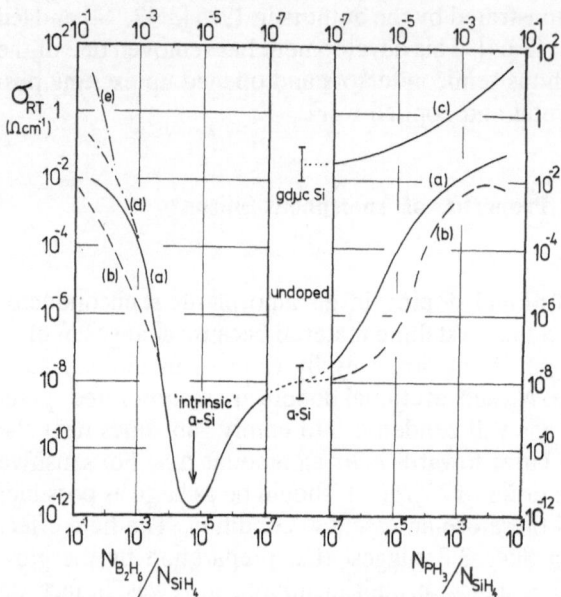

Fig. 3.24. Doping dependence of the room temperature conductivity σ_{RT} of glow discharge Si. On the right σ_{RT} is plotted against the phosphine doping ratio, on the left against the diborane doping ratio. The centre refers to the undoped material. (Curve a) refers to specimens prepared in the system shown in Fig. 3.3, (Curve b) to the doping results published originally in 1976 [3.44]. (Curve c–e) were obtained from microcrystalline Si specimens prepared in the glow discharge (Sect. 3.6.2)

men then turns p-type and the rapid rise of σ_{RT} with further boron doping indicates that hole conduction has taken over, which is also confirmed by the sign of the thermoelectric power and by the field effect.

From measurements of the temperature dependence of σ it is concluded that, by doping, the Fermi level position can be controlled over about 1.2 eV, essentially between the onset of the electron and hole tail states. This has also been confirmed by photoemission measurements [3.65]. The information summarised in Fig. 3.24 leaves little doubt that systematic control of the electronic properties of glow discharge Si by P or B doping is a very real possibility.

Curve (b) included in Fig. 3.24 shows the originally published doping results [3.44] obtained with specimens from a smaller capacitatively coupled system. Comparison with (a) suggests that doping in the larger deposition unit of Fig. 3.3 is more effective, particularly on the n-side. What are the possible reasons? The observed conductivity changes (taking phosphorus doping as an example) depend on three factors.

i) The density of phosphorus sites N_P incorporated into the growing structure. N_P is likely to be determined by the density of gaseous phosphorus-hydrogen fragments in the plasma at the interface of the growing film and should therefore depend on plasma parameters and also on the layout of the deposition system. From the previous work which led to curve (b), it was concluded that for the particular preparation unit $(N_P/N_{Si})_S \sim \frac{1}{2}(N_{PH_3}/N_{SiH_4})_G$.

ii) The fraction of the N_P phosphorus sites which will act as donors. Little is known about the factors which determine the formation of a donor configuration during the growth of the random network. In the above work [3.44], N_P was measured for a series of n-type specimens by ion probe analysis; from the movement of ε_F with doping and the known density of state distribution, the corresponding density of donors $N_D(\simeq N_D^+)$ was deduced. The results suggested that $N_D/N_P \simeq 0.3$ so that about two-thirds of the phosphorus sites incorporated into these specimens did not form a donor-like configuration.

iii) Finally, for a given N_D, σ_{RT} will be determined by the density of state distribution; that is, by the number of gap states that will have to be filled as ε_F moves towards ε_c.

To return to the comparison of curves (a, b), we believe that (i) and possibly (ii) are the most likely reasons for the increased doping efficiency. Further ion probe measurements would be necessary to distinguish between the two possibilities. On the basis of field effect measurements, (iii) is unlikely to account for the observed difference.

Figure 3.24 also includes results for microcrystalline Si specimens prepared directly in the glow discharge (curves c–e). These will be discussed in Sect. 3.6.2.

3.4.2 Doping by Ion Implantation

In the field of crystalline semiconductors, ion implantation has become an important doping technique. It appeared, therefore, of fundamental and applied interest to explore the feasibility of this technique in the doping of an amorphous semiconductor. In a paper presented at the 1977 Edinburgh conference [3.45], we were able to show (in collaboration with colleagues at the Max-Planck Institute in Heidelberg) that substitutional doping of a-Si by boron and phosphorus implantation is indeed possible. Figure 3.25 summarises the results of more recent work [3.46]. It shows the dependence of σ_{RT} of glow discharge a-Si on the implanted densities of phosphorus, boron and alkali ions (curves P_I, B_I and A_I, respectively). For comparison, curves P_G and B_G, representing gas-phase doping, have been included. Evidently, ion implantation can be used to control the electrical properties of a-Si over the same range as gas-phase doping, although Fig. 3.25 indicates that the latter is considerably more efficient. Similar results have been obtained with other group III and V ions [3.46].

The curve denoted by A_I represents the results of interstitial doping by the alkali ions Na, K, Rb and Cs [3.66]. It is interesting to note that in spite of the different ionic masses, the conductivity changes produced by these ions are practically identical. Interstitial doping by Li ions and by in-diffusion of an evaporated Li layer have been investigated by *Beyer* and *Fischer* [3.67]. The major disadvantage of Li as an interstitial donor lies in the limited thermal stability of the electrical properties of the doped specimen. This is overcome by the use of the heavier alkali ions which lead to stable properties up to 400 °C [3.66].

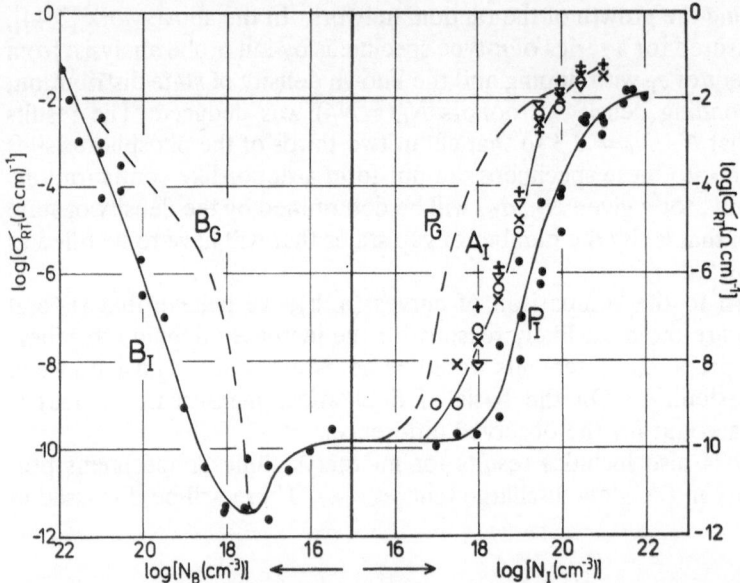

Fig. 3.25. Doping of a-Si by ion implantation. The room temperature conductivity σ_{RT} is plotted against the impurity concentration. (P$_I$) phosphorus ion implantation; (A$_I$) alkali ion implantations ($\bigcirc$Na, $+$ K, $\triangledown$ Rb, $\times$ Cs); (B$_I$) boron ion implantations. All implantations at a specimen temperature of 280 °C. P$_G$, B$_G$ represent the results for gas phase doping [3.46]

Compensation experiments using ion implantation have led to encouraging results and it has been shown that viable a-Si *p-n* junctions can be made by this technique [3.46]. Recent work on the use of implantation doping in the production of a-Si solar cells was reported in a paper at the 1981 Grenoble Conference [3.68]. The problem of implantation damage in a-Si is discussed in [3.46] and has also been investigated by *Müller* and *LeComber* [3.69] using irradiation with chemically inactive species.

3.5 The Amorphous Silicon Junction and Its Applications

The effective doping of a-Si discussed in the last section has opened up interesting possibilities for the application of this material in amorphous junction devices. A particular attraction lies in cheap, large-area devices and it is this aspect which has stimulated worldwide developments in the a-Si field. In this section we shall begin with an introduction to the basic properties of the amorphous barrier and junction and then go on to discuss some of the junction devices that have been investigated during the last few years.

3.5.1 Formation and Electronic Properties of the Metal/a-Si Barrier and the Amorphous Junction

The amorphous barrier differs basically from its crystalline counterparts because the net space charge in the barrier region is determined not only by ionised impurities but also by the density of state distribution in the mobility gap. Useful insight into the formation and properties of the metal/a-Si barrier and their dependence on doping levels has been gained from model calculations [3.70] based on the experimentally determined density of state distribution discussed in Sect. 3.2. Knowing the net charge in localised states in addition to that in ionised donors and acceptors, the overall space charge distribution could be obtained as a function of electron energy. A step-by-step solution of Poisson's equation then led to the spatial profile $\varepsilon_b(x)$ of the barrier.

In undoped or weakly doped material, where the net space charge is mainly determined by the localised state contribution, $\varepsilon_b(x)$ differs considerably from the parabolic shape often considered in the crystalline case, assuming a constant charge distribution $|e|N_D^+$. The barrier on undoped a-Si is characterised by a long, shallow initial part (as seen from the specimen volume), going over into a sharply rising profile near the metal contact. This is a direct consequence of the increase in $g(\varepsilon)$ towards the ε_y-peak (Fig. 3.11). In medium and highly doped material, the contribution of the ionised impurities becomes the dominant factor in the space charge distribution and $\varepsilon_b(x)$ approaches the parabolic profile. The model calculations have also been extended to a-Si p-n junctions for various doping levels [3.70].

The differential capacitance C of the metal/a-Si barrier and its bias and frequency dependence can be calculated from the above space-charge distributions. Measurement of C then provides an interesting check of the model calculations and also of the underlying density-of-state distribution. This approach has been used by *Snell* et al. [3.71] who showed that the observed features of the capacitance measurements on Au/a-Si barriers can be explained satisfactorily in terms of the analysis. A similar investigation of Pt/a-Si barriers has been described by *Beichler* et al. [3.72]. This work confirms the strong frequency dependence of C associated with the contribution to the barrier space charge of deep lying localised states. But unlike Snell et al., the above authors do not find an increase in C in reverse bias at very low frequencies where most of the deep states can reach their steady-state occupation.

3.5.2 Junction Configurations and High Current Diodes

The work during the last few years has shown that the most promising a-Si junction configurations are those in which the main part of the device consists of an undoped or lightly doped n-type layer 0.5–1 μm in thickness. The

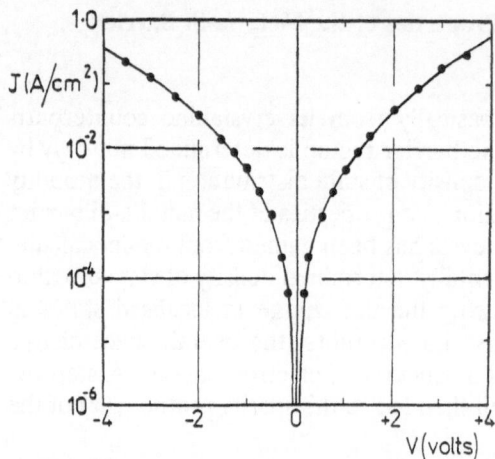

Fig. 3.26. Current density J plotted against the applied voltage V for a symmetric n^+-i-n^+ structure. The graph illustrates the remarkable injecting properties of the n^+-contact [3.51]

junction is formed by thin (~ 100 Å) highly doped n^+ and p^+ films deposited on either side of the central layer to form structures such as n^+-i-p^+, n^+-ν-p^+ or p^+-i-n^+. In this notation "i" and "ν" refer to undoped and lightly phosphorus doped central layers, respectively; the sequence is in the order of deposition of the multilayer film.

The operation of such devices is critically dependent on the lifetime of excess carriers either injected into the low conductivity central region from the highly doped layers or produced by photogeneration. In this section we shall be concerned with the first mechanism. Figure 3.26 for a symmetric n^+-i-n^+ structure, used in the study of SCL current flow (Sect. 3.5) [3.51], illustrates the remarkable injecting properties of thin n^+ contacts on undoped a-Si. It can be seen that the current density J is increased by four orders of magnitude when a few volts are applied across the device. Also the properties of the injecting contacts are accurately reproducible, as shown by the symmetry of the characteristics.

In n^+-i-p^+ or related configurations, strong double injection takes place into the central region under forward bias and the density of excess carriers greatly exceeds the thermal equilibrium densities. We have investigated the common lifetime τ of excess carriers under these conditions [3.73, 74] using a junction recovery method, similar to that often employed for corresponding crystalline devices. The results show that at high injection levels, τ is about 1 μs for the above-mentioned a-Si configurations. At a reduced current level of 10^{-2} A cm^{-2}, τ approaches values between 20 and 30 μs. It must, however, be stressed that τ does not represent the "free" lifetime because the injected carriers interact through trapping and thermal release with the band-tail states. The latter are in quasi-thermal equilibrium with the extended electron states and the measured lifetime includes, therefore, the integrated localisation time in these states. Exactly the same conditions apply to drift mobility measurements on excess carriers so that the two measured quan-

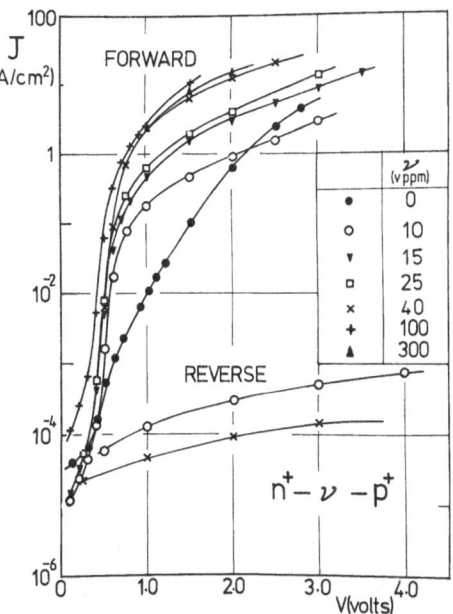

Fig. 3.27. Forward and reverse current density/applied voltage characteristics for n^+-ν-p^+ diodes showing the effect of different gaseous doping ratios [vppm] on the central ν-region [3.75]

tities, τ and μ, should be combined in consistent estimates of drift and diffusion lengths.

One of the attractive features of the amorphous junctions lies in the fact that their profile can be controlled continuously during deposition from the gas phase. This may eventually prove to be an important factor in optimising junction performance for particular applications. The approach has been used in an attempt to improve and optimise the forward current in a-Si junctions [3.75]. Some of the most promising results were obtained from structures of the n^+-ν-p^+ type, about 0.75 μm thick and deposited on stainless steel foil. Figure 3.27 shows a series of characteristics plotted on a semi-logarithmic graph of the current density J against the applied potential V. It can be seen that the current carrying capacity of the n^+-ν-p^+ devices in the forward direction increases systematically with the doping level of the central region; $\nu = 0$ represents a n^+-i-p^+ junction. The dc characteristics terminate at the loading where thermal runaway sets in. For $\nu = 40$ vppm this is over 20 A cm^{-2} and steady current densities of up to 40 A cm^{-2} have been observed. The reverse characteristic for the $\nu = 40$ vppm specimen shows that the rectification ratio at $V = 1$ V (the approximate barrier height) is about 5×10^4. A transition to Zener breakdown is observed at reverse voltages which depend on the doping level. Ideality factors of about 1.5 were found both from the dark characteristics and from photovoltaic experiments. These results are encouraging and suggest that cheap a-Si junction devices may find applications in cases where some relaxation in electrical characteristics from those of crystalline junctions will be acceptable.

3.5.3 The Amorphous Silicon Photovoltaic Junction

At the present time the development of photovoltaic devices for solar energy conversion is a challenging problem. It is likely that a stable thin-film material will ultimately provide the best solution to the large-scale production of cheap, reasonably efficient photovoltaic junctions for use as a supplementary energy source. *Carlson* and *Wronski* [3.76] showed in 1976 that glow discharge a-Si could be successfully applied in photovoltaic cells. Since that time many industrial laboratories have entered the a-Si solar cell field which has become highly competitive. Progress during the last two years has been encouraging and is reviewed in Chaps. 6 and 7.

In the following we shall confine ourselves to two basic material aspects which have important implications for the development of viable p^+-i-n^+ photovoltaic junctions. Figure 3.28 shows a desirable electronic structure through the thickness of a good a-Si p^+-i-n^+ cell. The device is deposited on the stainless steel substrate S and the light enters through the conducting indium-tin oxide (ITO) film (~ 500 Å thick) which also acts as a simple antireflection coating.

In the extraction of the photogenerated carriers, both diffusion and drift in the built-in field play a part. It is therefore relevant to consider the diffusion length L and the drift length δ of the excess carriers. These quantities are given by

$$L = \left(\frac{kT}{|e|} \mu\tau \right)^{1/2} \quad \text{and} \tag{3.4}$$

$$\delta = \mu\tau E . \tag{3.5}$$

Both involve the carrier mobility; with a drift mobility of excess holes lying between 10^{-4} and 10^{-3} cm^2 V^{-1} s^{-1}, about two orders of magnitude smaller than for electrons, it is likely that L_h and δ_h are critical parameters in the performance of photovoltaic devices. Direct measurements of L_h were made

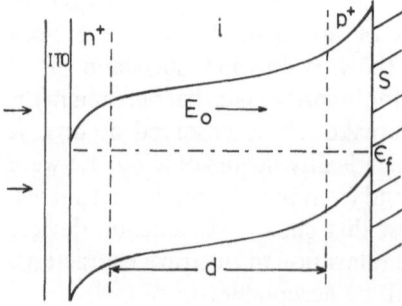

Fig. 3.28. Electron energy profile through an a-Si p^+-i-n^+ photovoltaic junction. (*S*) stainless steel substrate; (E_0) built-in electric field; (ITO) indium tin oxide transparent electrode

by *Staebler* [3.77] who investigated the collection of photogenerated holes as a function of intrinsic layer thickness on a wedge-shaped specimen. He estimated an upper limit to L_h of 350 Å. This appears to be rather low in view of more recent measurements by the PEM effect [3.78] and by the surface voltage technique [3.79] which gave values of about 0.1 μm and 0.4 μm, respectively.

The hole diffusion length can also be obtained from mobility and lifetime measurements using (3.4). For example, if it is assumed that the junction recovery results, mentioned in Sect. 3.5.2, can also be applied to photogenerated carriers in the intrinsic region then, at the appropriate current density (10 mA cm^{-2}), τ lies between 20 and 30 μs. Combining this with an average hole drift mobility of 5×10^{-4} cm^2 V^{-1} s^{-1} gives $\mu_h \tau \simeq 1.3 \times 10^{-8}$ cm^2 V^{-1} and $L_h \simeq 0.2$ μm. Finally, the recent work of *Abeles* et al. [3.80] on a-Si Schottky barriers leads to L_h values between 0.1 and 0.2 μm (for $d < 1$ μm). It appears, therefore, that on the basis of presently available information, hole diffusion lengths are a fraction of the thickness of the intrinsic region in present photovoltaic devices.

As both L_h and δ_h are determined by the $\mu_h \tau$ product, it is important to keep a close check on this quantity in optimising material properties. A useful transient method has recently been described [3.68] whereby $\mu_h \tau$ and E_0 of the p^+-i-n^+ junction can be determined under operating conditions. Figure 3.29 shows the arrangement. A short light flash from an Ar spark source is used to probe the steady state existing in the device without appreciably perturbing it. The metal bottom electrode B of the cell is connected to the load resistor R_L, the ITO top contact T to the potential V_T. The external potential V across the cell can be changed by positive or negative V_T, and/or by steady illumination which increases $|V_B|$. To a reasonable degree of approximation, the average field in the i-region is given by

$$E \simeq E_0 + \frac{V}{d} = E_0 + \frac{\pm |V_T| - |V_B|}{d} . \tag{3.6}$$

In the experiment, highly absorbed blue or green probe flashes are used to generate electron-hole pairs near the electrode T. Electrons are rapidly

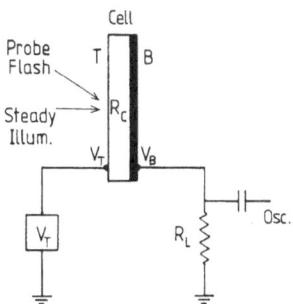

Fig. 3.29. Basic circuit used in the transient experiments [3.68]

extracted and excess holes drift through the *i*-region in the internal field E. By making R_L sufficiently large (>0.5 MΩ), the integrated charge displacement Q of the holes can be measured directly on the oscilloscope connected across R_L and investigated as a function of E. The ratio Q/Q_{sat}, where Q_{sat} represents the total extracted charge as $E \rightarrow \infty$, is given by the well-known Hecht formula [3.81].

$$\frac{Q}{Q_{sat}} = \frac{\mu_h \tau E}{d} \left[1 - \exp\left(-\frac{d}{\mu_h \tau E} \right) \right].$$ (3.7)

This relation is based on a simple model in which the range of the generated holes is limited by deep trapping. It should be applicable to the primary transient photocurrent observed here; the small excess carrier densities used and the low pulse repetition rate ensure that any trapped space charge does not perturb E to any extent.

The experimental points in Fig. 3.30 represent measurements of the integrated pulse height as a function of V/d, see (3.6); the solid line is a computer fit of (3.7) to the experimental data which determines the Q/Q_{sat} scale and the $\mu_h \tau$ value, 1.0×10^{-8} cm^2 V^{-1} in this case. Also, from the intercept of the fitted curve with the V/d axis we obtain a value of $E_0 \simeq 13$ kV cm^{-1}. As a check μ_h, τ and E_0 were also determined from time-resolved transient measurements [3.68] on the same series of solar cells. For the above junction, $\mu_h \tau$ was found to be 1.1×10^{-8} cm^2 V^{-1} and $E_0 \simeq 11$ kV cm^{-1}, in good agreement with the experimentally simpler pulse height experiments.

The results suggest that in a reasonably efficient p^+-i-n^+ photovoltaic cell with a fill factor approaching 0.6, the field E_0 lies between 10 and 20 kV cm^{-1}. Under short-circuit conditions the drift distance $\mu_h \tau E_0$ is therefore 1–2 μm, appreciably longer than the diffusion length. However, as the

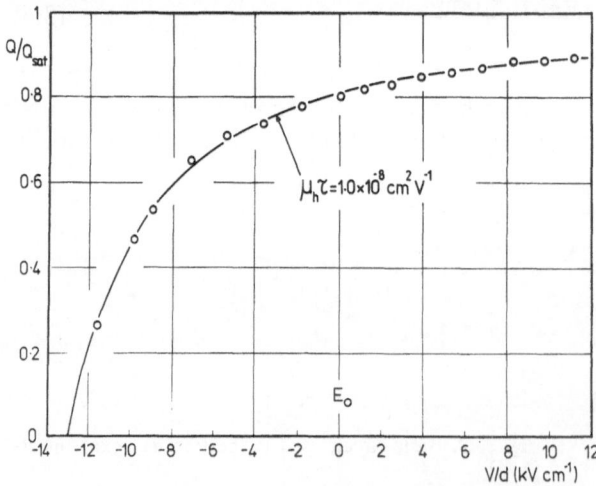

Fig. 3.30. Results of transient integrated pulse height measurements. The computer fit to (3.7) gives the $\mu_h \tau$ value of 1.0×10^{-8} cm^2 V^{-1} [3.68]

external load connected across the illuminated cell is increased, the internal field is reduced below its maximum value E_0 and range limitation for holes will set in as δ approaches the specimen thickness d. It is important, particularly for a satisfactory fill factor, that the limiting condition $\delta \simeq d$ should correspond to as low a field as possible. When the conversion efficiency for a batch of solar cells is correlated with the device thickness, it has been found by several groups that with the presently available material there is an optimum value at $d \sim 0.5$ μm. We believe that this is a direct consequence of the range limitation for holes and suggests that material improvements in this direction should be of importance in the a-Si solar cell development.

The second topic in this section concerns the problem of geminate recombination in a-Si and its possible fundamental limitation of solar cell performance. In a geminate recombination process, the photogenerated electron thermalises and then recombines with the hole from which it was excited. Such pairs do not contribute to photoconduction and therefore reduce the overall generation efficiency per absorbed photon, which will be denoted by η. The most detailed theory of geminate recombination and its dependence

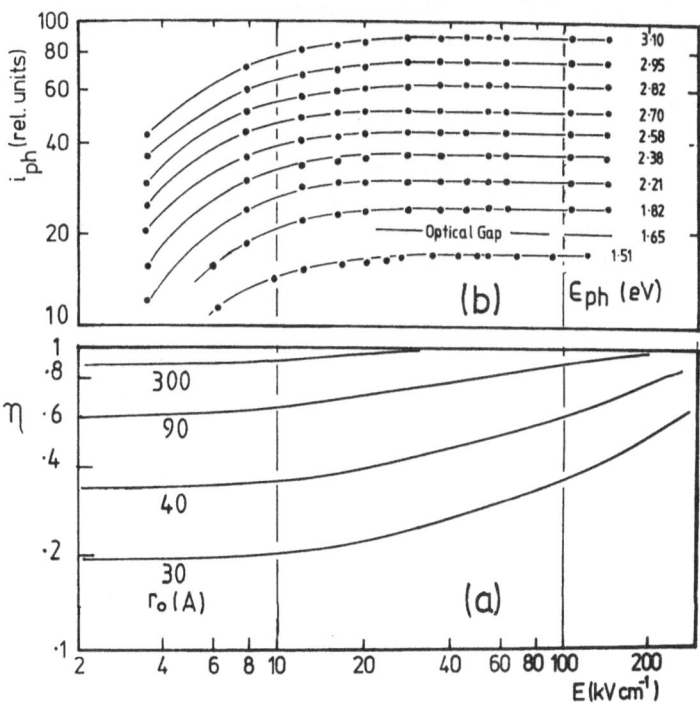

Fig. 3.31. (a) Field dependence of the generation efficiency η calculated from the Onsager theory [3.82] for the thermalisation lengths r_0 given on the diagram. **(b)** Observed field dependence of the short-circuit photocurrent in a p^+-i-n^+ junction. i_{ph} is plotted in arbitrary units and the curves have been separated to show the absence of any field dependence above about 20 kV cm^{-1} [3.91]

on applied field was originally given by *Onsager* [3.82] and the results were used to interpret photogeneration in a-Se [3.83], anthracene [3.84] and other organic photoconductors. The theory considers the photoexcited electron in the Coulomb field of the hole. The spatial extent of the interaction between the charges is defined in terms of the Coulomb radius r_c at which the potential energy in the well equals kT. In a-Si, $r_c \simeq 50$ Å. The excited electron thermalises within a distance r_0 and the relative values of r_c and r_0, as well as E, determine whether separation or geminate recombination of the pair is most likely to occur. Figure 3.31a shows the characteristic field dependence of η predicted by the Onsager theory for a-Si with $r_c = 50$ Å. The curves refer to r_0 values of 30, 40, 90 and 300 Å which give low-field efficiencies between 0.2 and 0.9.

Recently, Mort and his collaborators have investigated geminate and nongeminate recombination in a-Si by combining xerographic discharge measurements and delayed-collection field techniques [3.85–87]. The results show that, depending on specimen preparation, η ranges from 0.44 to 0.55 and that the field dependence of η fits the Onsager curves, such as those in Fig. 3.31a, suggesting r_0 values from 45 to 80 Å. In a subsequent paper, *Chen* and *Mort* [3.88] reached the somewhat controversial conclusions that (i) geminate recombination presents a fundamental limitation to photogeneration in a-Si and (ii) that carrier lifetime limitations are unimportant in present photovoltaic cells.

The first conclusion is not supported by results on photovoltaic cells obtained in several laboratories working on such devices (e.g., [3.89, 90]). Collection efficiencies of 0.8 or above are commonly observed in the green part of the spectrum, substantially larger than the maximum value of about 0.65 calculated by Chen and Mort from $\eta = 0.55$. This discrepancy implies that the generation efficiency in a-Si should generally be much closer to unity than is suggested by the Xerox experiments.

Further evidence, pointing in the same direction, is provided in Fig. 3.31b from some of our recent work [3.91]. The curves show the shape of the field dependence of the primary photocurrent in p^+-i-n^+ junctions under steady illumination with photon energies close to and above the optical gap of a-Si. i_{ph} was measured under short-circuit conditions and the built-in field E_0 was increased or decreased by an applied potential as described above (3.6). The i_{ph} scale is in arbitrary units and the curves have been separated vertically to show conclusively the absence of any measurable field dependence for $E > 20$ kV cm^{-1}. At smaller fields, the lifetime limitation of the holes (discussed earlier) is the most likely cause for the decrease in i_{ph}. The quantum efficiency measured independently on these devices lies between 0.9 and 1 and is therefore consistent with the observed saturation of i_{ph}. These and other results strongly suggest that there is little or no evidence for geminate recombination in a-Si devices prepared in several laboratories. It appears, therefore, unlikely that this mechanism should present a serious limitation in the photovoltaic efficiencies ultimately achievable.

3.5.4 Application to Electrophotography and Image Tubes

The high photosensitivity of a-Si throughout most of the visible spectral region has stimulated exploratory work, particularly in Japan, on its application in electrophotography and in image tubes of the Vidicon type. An essential requirement for the operation of both these devices is good photoconductivity combined with a high dark resistance which should normally be larger than the 10^{12} Ω cm bulk resistivity measured for "intrinsic" a-Si (Fig. 3.24). Two approaches have so far been investigated to overcome this problem. In the first, sufficiently insulating layers have been produced by the deliberate addition of small amounts of oxygen to the gas used in reactive sputtering or in glow discharge deposition [3.92]. This increased the specimen resistivity, presumably due to the introduced defects, but also tended to reduce the photosensitivity. Nevertheless, viable electrophotographic images have been obtained by this technique [3.93]. In the second approach, which ultimately may prove to be the most promising, the high photosensitivity of the intrinsic glow discharge material is maintained but suitable blocking layers are deposited near the specimen surfaces to increase the resistivity to that determined by thermal generation in the material.

Shimizu and his collaborators [3.94–96] have investigated the use of reverse biased *p-i* (or *n-i*) junctions for this purpose as well as thin insulating dielectric blocking layers. Some of the main results will now briefly be summarised. Starting with the work on a photoreceptor for electrophotography, the upper diagram in Fig. 3.32 illustrates the specimen configuration for use with positive corona charging. A *p*-type a-Si layer about 0.2 μm thick is deposited from 350 vppm of diborane in silane on to the grounded Ni–Cr film. For the given electric field, this doped layer prevents the injection of

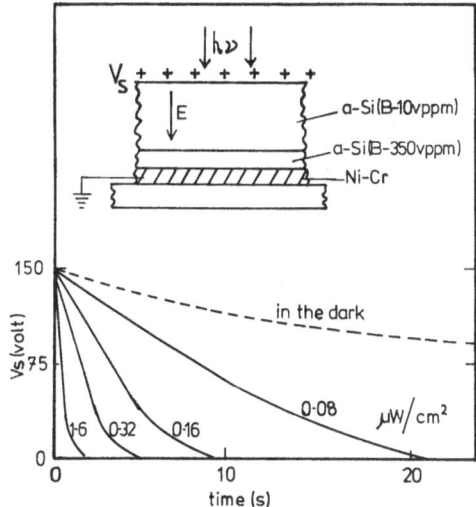

Fig. 3.32. a-Si photoreceptor for electrophotography. The diagram shows the structure of the layer and the time dependence of the surface potential V_s in the dark and at the light levels indicated [3.94]

excess electrons from the substrate. It is followed by 6 μm of "intrinsic" a-Si which is deposited with a small diborane content (10 vppm) to move ε_F to the intrinsic position in the centre of the mobility gap (Fig. 3.24). The dependence of the surface potential V_s on time is also shown in Fig. 3.32. The rate of dark decay is now acceptable for electrophotographic application and the photodischarge curves obtained with 600 nm light at the given intensities show that no residual charge remains after illumination. In recent experiments [3.94] the surface potential at $t = 0$ has been increased to about 400 V in a 10 μm a-Si layer which allows development with a conventional toner.

These results are encouraging and suggest that a-Si may find application as a highly sensitive photoreceptor in electrophotography. Further attractive features in this connection are the mechanical hardness and the nontoxicity of the material.

In a Vidicon image tube the unelectroded surface of the photoconductive layer is scanned by a low energy electron beam which stabilises it to cathode potential. Figure 3.33 illustrates the structure of an a-Si Vidicon target investigated by *Shimizu* et al. [3.94, 96]. The image is focussed on to the glass disc which carries a positive biased transparent ITO film. A n-type blocking layer (doped with 50 vppm of phosphine) prevents the injection of holes from the ITO. As an alternative solution, a blocking layer consisting of 300 Å of glow discharge silicon nitride was found to be successful. The main part of the target consists of lightly B-doped "intrinsic" a-Si up to 4 μm in thickness (to reduce capacitative lag). Finally, the scanned surface is covered with a thin ($\simeq 1000$ Å) "electron blocking" layer of Sb_2S_3 (or other chalcogenide) which tends to improve the resolution of the picture. The a-Si Vidicon performance compares favourably with that of the more conventional tubes: the resolution is better than 800 TV lines for the 1″ a-Si tube and the decay lag in the picture from a 4 μm thick target is about 5% after 50 ms (3 fields).

3.5.5 Application to a Memory Device

In this section we should like to discuss briefly a new a-Si memory device [3.97] developed jointly by the Edinburgh and Dundee groups. It is an elec-

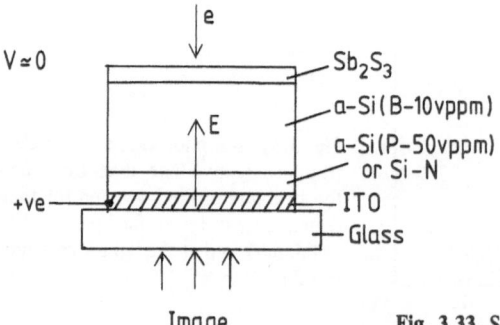

Fig. 3.33. Structure of an a-Si Vidicon target [3.94]

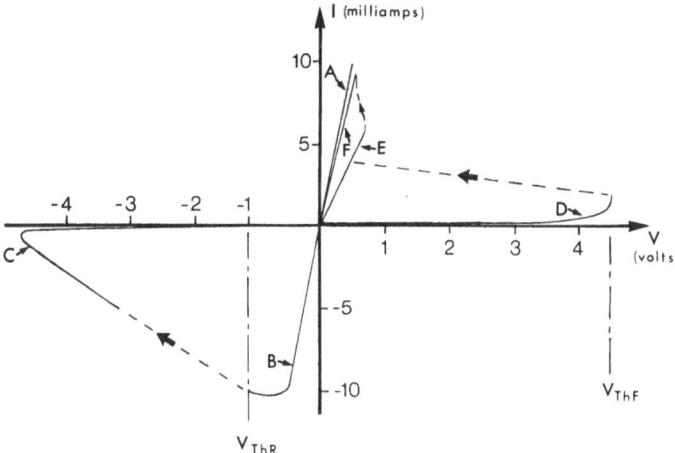

Fig. 3.34. DC characteristics of a nonvolatile, programmable a-Si memory device. V_{ThF} and V_{ThR} are the forward and reverse threshold voltages for switching the device into the on and off-state, respectively [3.97]

trically programmable, nonvolatile memory element which in terms of speed, retention time, operating voltages and stability compares very favourably with crystalline MNOS (metal-nitride-oxide-semiconductor) or FAMOS (floating-gate-avalanche-metal-oxide-semiconductor) devices currently used for nonvolatile programmable storage.

The most promising configuration investigated so far is of the *p-n-i* type. The layers are deposited on a stainless steel substrate giving a total device thickness of between 0.5 and 1.0 μm. A series of Au or Al dots is then evaporated onto the surface to form the top contacts.

After an initial forming step the structure operates permanently as a nonvolatile memory. Figure 3.34 shows a typical set of static (dc) current/ applied voltage characteristics. Immediately after forming, the device is in its on-state so that small positive and negative voltages trace out curve AB; on-state currents of 10 mA or more are generally observed. On increasing the reverse potential (i.e., a negative voltage applied to the *p*-doped region), a reverse threshold voltage V_{ThR} is reached beyond which the device switches to an off-state with a resistance of the order of 1 MΩ, represented by the characteristic CD. The reverse threshold $V_{ThR} \simeq 1$ V and the off-state is stable for voltage swings of ± 4 V. If the forward potential is now increased beyond a value of V_{ThF} (the forward threshold voltage), the device switches back into its high conductivity state AB, in some cases through an intermediate state such as E or F.

The above cycle has been repeated up to 10^5 times without observable changes in characteristics or threshold voltages. Devices set in on and off-states have been monitored for several weeks without detectable change in either characteristic. Operation at temperatures of up to 180 °C shows little change in the threshold voltages.

The dynamic characteristics of the *p-n-i* devices has been investigated with 10 V forward and reverse pulses of 100 ns duration. The experiments demonstrate the important result that both the on-off and off-on transitions are completed within 100 ns and that there appears to be no observable time delay in the switching response. It is estimated that the energy absorbed during either transition is extremely low, typically in the range 10^{-6} to 10^{-8} J.

The mechanisms underlying the switching phenomena described above are unknown at present. However, it will be useful to draw some comparisons with other related devices. Much attention has been given over the past 10–15 years to memory switching devices fabricated from multicomponent chalcogenide glasses in which the reversible memory action is associated with the growth and destruction of a crystalline filament [3.98, 99]. The switching phenomenon in the a-Si structures reported here is clearly very different, at least operationally. The most obvious difference is the completely nonpolar character of switching in chalcogenide glass devices, in contrast to the marked polarity-dependence of the a-Si memory switches. More importantly perhaps, the switching times for the a-Si device are much faster compared with those in chalcogenide devices, which are at least several milliseconds long. In addition, the energy involved in the a-Si switching process is considerably lower.

The closest parallel to the a-Si devices described in this section seems to be the observation of memory switching in heterojunctions of expitaxial *n*-type ZnSe grown on *p*-type (single-crystal) Ge substrates reported by *Hovel* [3.100]. The ZnSe–Ge heterojunction devices are polar and the transition times are both in the region of 100 ns or less. The switching has been tentatively and qualitatively explained by a model involving the filling and emptying of traps in the Zn–Se and the formation of a current filament in the on-state. We believe that on the basis of the present experimental evidence it is likely that the on-state of the a-Si memory devices also involves formation of a current filament.

3.6 Microcrystalline Silicon Produced in the Glow Discharge Plasma

During the last two years the possibility of depositing thin films of microcrystalline (μc) Si directly in the glow discharge has aroused considerable interest. On the fundamental side this material, which we shall denote by gd μc Si, permits an extension in the study of the electronic properties just beyond the a-phase. The films possess very limited long-range order with crystallite sizes of up to 80 Å; the latter can be controlled to some extent by the preparation

conditions. On the applied side, the attractive feature is that simply by turn-
ing a tap during an a-Si deposition run, one can produce a highly conductive
gd µc surface layer which could be most useful, for example, in the photovoltaic development.

In the following sections we shall summarise and discuss some of the
recent work in this new field.

3.6.1 Preparation and Structure of gd µc-Si

The possibility of depositing a microcrystalline semiconductor specimen in a
glow discharge plasma was first recognised by *Veprek* and *Maracek* in 1968
[3.101] using chemical transport in a hydrogen plasma. This method offers a
versatile approach for the study of the kinetics of the silicon-hydrogen system
[3.102] and the conditions for the direct deposition of a µc film. The work of
the Zürich group suggested that stable nuclei for µc growth are likely to be
formed under plasma conditions whereby a chemical equilibrium is
approached at the plasma-solid interface [3.103, 104].

This condition is expressed by the reactions [3.103]

$$\text{SiH}_x(g) \rightleftarrows \text{Si}(s) + x\text{H (plasma)} . \tag{3.8}$$

The forward reaction (left to right) represents the deposition of the gaseous
Si–H fragments to form the a-Si film and to add gaseous hydrogen to the
surrounding plasma. In the reverse direction the interaction of the plasma
with the solid Si film leads to erosion. Normal deposition conditions for a-Si
are such as to make the forward reaction the dominant one so that the system
is far from chemical equilibrium.

Work by several Japanese groups [3.105–107] showed that µc Si specimens can also be produced in the silane glow discharge reactors described in
Sect. 3.1. In this case chemical equilibrium is approached by strong hydrogen
dilution and an increase in the normal level of rf power. Evidently the former
will enhance the reverse reaction in (3.8); the observed deposition rate is
slowed down to 0.1–0.2 Å s^{-1} in spite of the increased rf power, indicating
that the system is fairly close to a dynamic equilibrium.

Under these conditions stable nuclei are formed which stimulate the
growth of a gd µc-Si layer. This is demonstrated convincingly by the electron
diffraction patterns [3.108] shown in Figs. 3.35 a, b. They were taken under
identical conditions in a STEM in the microdiffraction mode. (a) shows the
broad diffraction rings, typical of an amorphous material. The specimen used
for (b) was prepared in the same deposition unit but with strong hydrogen
dilution and somewhat larger rf power. We now see the (110) pattern from a
single microcrystallite and its twin orientation. The crystallite size is about
50 Å.

X-ray and Raman studies on gd μc Si prepared by chemical transport [3.104, 109] and in the silane plasma [3.110–112] have been carried out to provide some insight into the structure of the material. The present data suggest a model in which μc regions, about 50 Å across, are surrounded by "inter-granular" noncrystalline material. A two-phase model of this kind has been widely discussed in the literature for polycrystalline Si and the interpretation of the Hall effect and conductivity measurements in such a material has recently been reviewed by *Orton* and *Powell* [3.113]. Raman spectra are particularly useful in connection with the application of a two-phase model to gd μc. Features near the 520 cm^{-1} crystalline peak can be interpreted in terms of the μc and a-components and lead to estimates of their volume ratio. Recent values for the fraction of the μc component in "good" undoped specimens deposited at 300 °C lie between 0.65 and 0.97; it is not surprising that there appears to be some dependence on the details of the preparation technique.

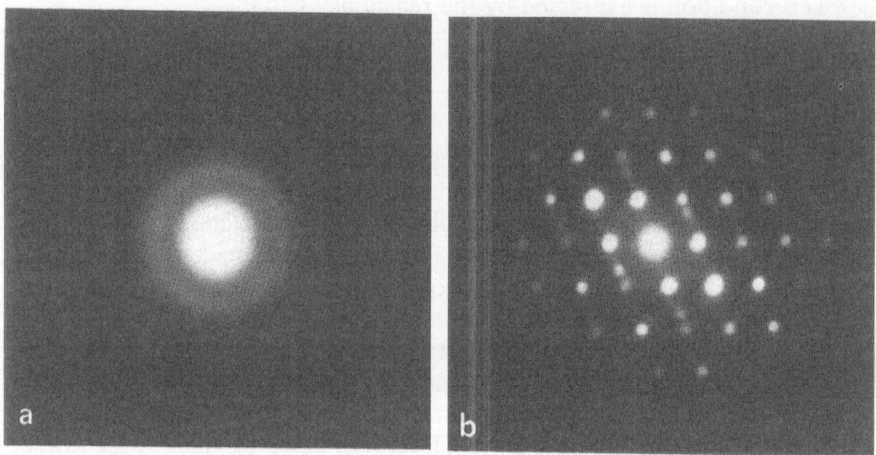

Fig. 3.35 a, b. Electron diffraction results taken under identical conditions in the microdiffraction mode. (a) a-Si, (b) glow discharge microcrystalline Si [3.108]

The deduced range of the μc volume fraction is important in the interpretation of electrical measurements. Estimates based on geometrical models and effective medium theory (see, for instance, [3.113]) indicate that the results of conductivity and Hall effect measurements on "good" gd μc Si specimens should be reasonably representative of the μc phase. Accordingly, we shall regard the information deduced from the electrical measurements in the next sections as average values for the crystallites.

3.6.2 Conductivity

In this and the following sections we shall present the results obtained on a series of undoped and n-type gd μc Si specimens [3.108]. They were deposited in the system shown in Fig. 3.3 with an excitation frequency of 40 MHz and an rf power of 20–30 W. The gas mixture contained 2–3% of silane in hydrogen with the required addition of phosphine for the n-type films. The average crystallite size δ was determined for each deposition run from the halfwidth of the (111) electron diffraction ring by means of the Scherrer formula.

The σ_{RT} versus doping level curve for these specimens has been included in Fig. 3.24 (Sect. 3.5). The curve, denoted by (c), refers to specimens with $\delta \simeq 50$ Å; smaller δ values give conductivities lying below (c) [3.108]. It can be seen that for undoped gd μc Si $\sigma_{RT} \simeq 10^{-2}$ $(\Omega \text{ cm})^{-1}$, although lower values, also corresponding to $\delta \simeq 50$ Å, have been observed in some deposition runs. In phosphorus doped specimens, σ_{RT} lies over two orders of magnitude higher than in a-Si and reaches values around $20(\Omega \text{ cm})^{-1}$ at the higher D.

Boron doped μc specimens are more difficult to prepare in the glow discharge plasma. Curve (d) from the work of *Matsuda* et al. [3.114] refers to p-type μc material prepared at an rf power of 80 W, appreciably higher than for n-type specimens. Curve (e) recently obtained by *Hamasaki* et al. [3.115] increases the σ_{RT} range further, approaching values of $8(\Omega \text{ cm})^{-1}$ on the p-side at the highest doping ratios. It may be significant that these authors used a magnetic field to confine the glow discharge plasma.

Before leaving the subject of conductivity we should like to add a brief comment about the use of SiF$_4$ in the glow discharge, instead of the SiH$_4$ considered so far. The presence of fluorine, a well-known plasma etchant, will enhance the reverse reaction (erosion) in (3.8) and is likely to lead towards chemical equilibrium and a μc structure with little hydrogen dilution and probably at lower power levels. The x-ray and conductivity results of *Matsuda* et al. [3.106] show that this is indeed the case. These results also suggest that the high (apparent) doping efficiency claimed in specimens deposited from SiF$_4$ + 10% H$_2$ [3.116] on the basis of conductivity measurements is mainly due to their microcrystallinity. This throws considerable doubt on the conclusion by *Madan* et al. [3.116] that these results refer to a-Si, but with an appreciably lower density of states than in specimens prepared from a silane glow discharge.

3.6.3 The Hall Effect

The work of *LeComber* et al. [3.117] showed that in a-Si the sign of the Hall effect is anomalous; n-type specimens gave a positive Hall constant and a negative sign was observed for p-type a-Si. Clearly, the crystalline interpretation of the Hall effect can no longer be applied to transport in the random phase region of an a-semiconductor where the mean free path of the charge carriers has been reduced to a few atomic spacings by the disorder. The theories that have been developed take account of the "microscopic" details of the transport and show that the magnitude and sign of the Hall constant will depend on the local geometry in the sites involved [3.118] and also on the nature of the local electronic states [3.119].

In the present work on undoped and *n*-type gd μc Si, it was found without exception that the Hall constant had the negative sign expected from crystalline transport theory, even in specimens with the smallest δ values. This suggests that with the limited long-range order introduced, the mean free path of the electrons becomes sufficiently long for the description in terms of classical electrodynamics to be applicable. We have, therefore, used the simple formula $R_H \simeq (ne)^{-1}$ to obtain an approximate value for the carrier density *n* from the experimental Hall constant R_H.

In Fig. 3.36 *n* has been plotted as a function of the doping ratio *D* for gd μc Si specimens and several thermally crystallised a-Si films (th μc) [3.108]. *U* refers to nominally undoped gd μc specimens. *n* rises almost linearly with the density of PH_3 molecules in the gas phase and is independent of the method of crystallisation and also, within the limited range investigated, of the crystallite size. If one compares the gaseous doping ratio *D* with n/N_{Si} in the solid, one finds about 0.5 *D* for the latter, indicating that doping in gd μc Si must be reasonably efficient. However, it is apparent from the discussion in Sect. 3.4.1 that before a quantitative analysis of doping in

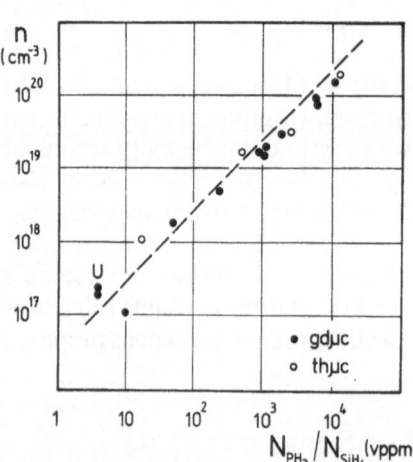
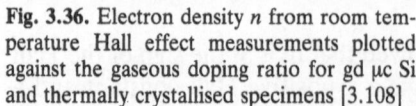
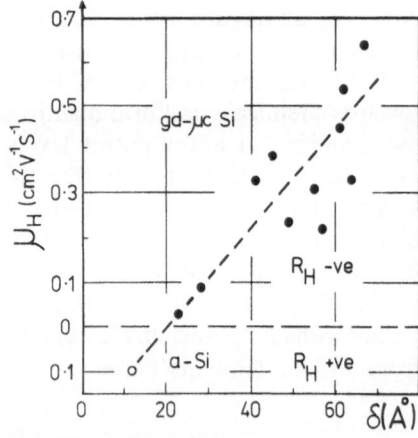

Fig. 3.36. Electron density *n* from room temperature Hall effect measurements plotted against the gaseous doping ratio for gd μc Si and thermally crystallised specimens [3.108]

Fig. 3.37. Hall mobility in gd μc Si specimens plotted against the crystallite size δ. The open circle denotes a typical result for a-Si, showing an anomalous sign of the Hall constant R_H [3.108]

this material can be made, information on the density of states distribution and on the number of incorporated phosphorus atoms will have to be obtained.

From the values of n and the corresponding conductivities the Hall mobility μ_H can be deduced. For the gd μc Si specimens the room temperature values of μ_H generally lie below 1 cm^2 V^{-1} s^{-1} and are largely independent of the doping level. However, μ_H depends on crystallite size, as shown by the upper part of Fig. 3.37 (R_H-negative). In spite of the scatter in the present results, μ_H extrapolates to zero at a critical grain size of about 20 Å. Below $\mu_H = 0$ the open circle denotes a typical result for a lightly phosphorus doped a-Si specimen [3.117], corresponding to an anomalous positive R_H. The magnitude of μ_H for the a-specimen is in agreement with the predictions of the *Friedman* theory [3.118] but no physical significance should be attached to the corresponding crystallite size of about 12 Å. It may perhaps be relevant that δ values between 10 and 12 Å are generally found when the Scherrer correlation graph is extrapolated to the diffraction line width observed with a-Si specimens.

The results given in Fig. 3.37 are of considerable interest: they suggest a limiting size of the ordered regions 20–30 Å for which the Hall effect at the mobility edge of the material can be described in terms of the normal crystalline theory. Once the extent of the ordered regions falls below that limit, then the above microscopic theories become applicable.

The temperature dependence of μ_H has been investigated between 210 K and 440 K and is of the form

$$\mu_H = \mu_0 \exp(-\varepsilon_\mu/kT) \ . \tag{3.9}$$

On the basis of the two-phase model discussed in Sect. 3.6.1, this would imply some form of potential discontinuity of average height ε_μ between the crystallites. The data for the gd μc specimens give values of ε_μ in the range from 40 to 100 meV. The pre-exponential factor μ_0 depends on δ and should be a measure of the Hall mobility in a crystallite of that size. For $\delta \simeq 50$ Å one obtains from (3.9) a typical value of $\mu_0 \simeq 6$ cm^2 V^{-1} s^{-1} at room temperature. Thus, μ_0 is quite comparable to the extended state mobilities in a-Si deduced from drift mobility experiments [3.4] which suggests that the presence of 50 Å μc regions does not yet affect the mobility to any extent. The Hall effect results, therefore, lead to the important conclusion that the appreciable increase in conductivity in the gd μc Si (Fig. 3.24) is almost entirely associated with a larger carrier concentration. The same applies to the th μc specimens; their larger crystallite sizes (80–100 Å) normally lead to higher μ_H values and conductivities above $10^2(\Omega$ cm$)^{-1}$ [3.108].

What is the fundamental reason for the greatly increased carrier density in the μc phase? We originally suggested [3.108] that the order introduced is sufficient to affect the extent of the localised electron tail states. In a-Si this is about 0.18 eV [3.19]; field effect measurements near the mobility edge of gd μc Si suggest that the width of the tail states has been considerably reduced, or that they may have been removed altogether. If the interpretation of these results is correct, the increase in n can be understood. Whereas the rapidly rising tail-state distribution in a-Si limited ε_c-ε_F to about 0.15 eV even at the highest doping levels, the suggested "delocalisation" in the μc material allows ε_F to move close to the current path at ε_c with phosphorus doping. This is clearly shown by the experimental results in Fig. 4.40 and fully explains the observed increase in n and σ.

3.6.4 Thermoelectric Power

A systematic study of the temperature and doping-dependence of the thermoelectric power S in gd μc Si has recently been carried out [3.121]. One of its aims has been to determine whether a consistent interpretation of the results can be given in terms of crystalline semiconductor theory. Figure 3.38 summarises the experimental results on a graph of S versus $10^3/T$. The solid lines represent calculated S-curves which will be discussed below. The numbers refer to the specimens used: 1 is an undoped gd μc Si film and in specimens 2 to 6 the doping ratio is increased from 5×10^{-5} to 10^{-2}. Specimen 7 is a doped thermally crystallised film and 8 an a-Si specimen included for comparison.

The main point brought out by Fig. 3.38 is that all the μc specimens show a similar temperature dependence which varies systematically with the dop-

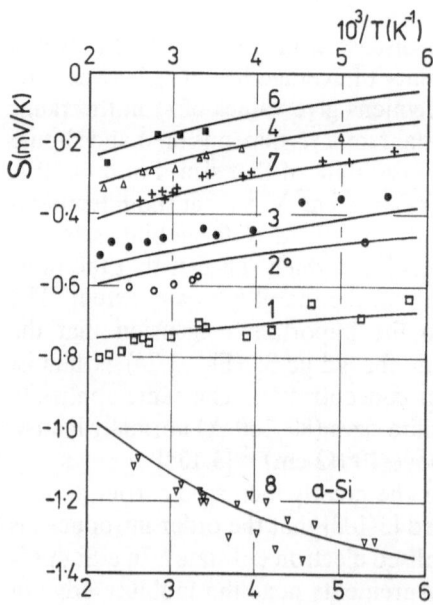

Fig. 3.38. Temperature dependence of the thermoelectric power. Experimental points (*1–6*) gd μc specimens with increasing phosphorus doping; (*7*) thermally crystallised specimen; (*8*) a-Si specimen. Solid lines are calculated curves [3.121]

ing ratio. However, comparison with a typical curve for a-Si (specimen 8) shows that there is a basic difference between the temperature dependence of S in the two cases.

The thermoelectric power for a nondegenerate semiconductor is given by the well-known expression

$$S = -\frac{k}{|e|} \left[\ln \left(\frac{N_c}{n} \right) + A_c \right], \tag{3.10}$$

where N_c is the effective density of states near the bottom of the conduction band and A_c the heat of transport. The latter, normally taken as temperature independent, accounts for the contribution to S of electrons distributed in states above the conduction band edge ε_c and will depend both on state distribution and transport.

As recognised in earlier work on thermally crystallised a-Si [3.120], it is essential in the interpretation of the thermoelectric power data to take account of the temperature dependence of N_c. Clearly, if n does not vary greatly with T, which is the case for the doped μc specimens, then according to (3.10), $S(T)$ should reflect $N_c(T)$. This is not normally evident in the results for a-Si because there n is more strongly activated, obscuring the smaller temperature dependence of N_c.

For crystalline Si, N_c is proportional to $T^{3/2}$. In the present analysis we shall begin with the more general expression

$$N_c = bT^\nu \tag{3.11}$$

and determine the value of ν from the experimental data.

A stringent test of the consistency of the experimental results in terms of (3.10) consists in plotting $-(|e|S/k)$ against $\ln n$ at a given temperature, using the value of n determined from the Hall effect. This has been done at different T throughout the range investigated and Fig. 3.39 shows an example at $T = 300$ K. The points define a line having the expected slope of -1; the intercept on the abscissa gives a value for $\ln(N_c) + A_c$, which according to (3.11) equals $\nu \ln T + \ln b + A_c$. From the temperature dependence of the intercept a value of $\nu \simeq 2$ is deduced.

Unfortunately, the analysis cannot separate the terms $\ln b$ and A_c; for reasons discussed in [3.121], A_c values between 2 and 3 are most likely. At 300 K and with $A_c = 3$, $N_c = 3.5 \times 10^{19}$ cm^{-3}, close to the value of 2.8×10^{19} cm^{-3} for crystalline Si. As a consistency check of the analysis the temperature dependence of S for the μc specimens has been calculated from (3.10, 11) using $A_c = 3$ and the experimentally determined n-values. The results are shown by the solid lines 1 to 7 in Fig. 3.38 which successfully reproduce both the temperature and doping dependence of S. This leads to the conclusion that a consistent interpretation of the Hall and thermoelectric power data can be given by conventional semiconductor theory. The only difference lies in

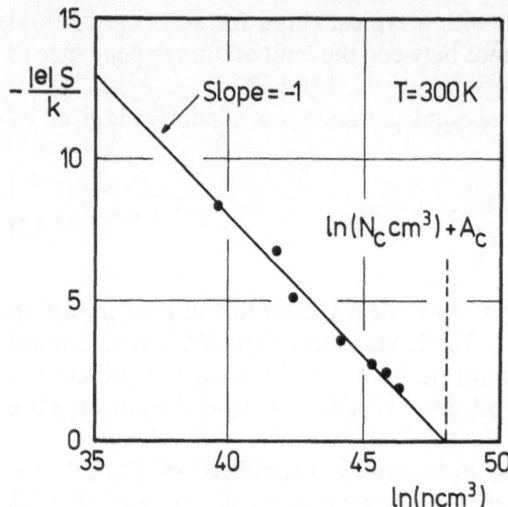

Fig. 3.39. Plot of $-(|e|S/k)$ against ln *n* at $T = 300$ K as a test of (3.10) [3.121]

the temperature dependence of N_c which could be associated with different state distributions above ε_c in μc and single crystal Si.

Finally, the thermoelectric power data provide reliable information on the temperature and doping dependence of the Fermi level position. Using nondegenerate statistics, (3.10) can be written as

$$\varepsilon_c - \varepsilon_F = - kT \left(\frac{|e|S}{k} + A_c \right) . \tag{3.12}$$

Figure 3.40 shows the room temperature value of $\varepsilon_c-\varepsilon_F$ as a function of the doping ratio (expressed in vppm) for $A_c = 3$. $\varepsilon_c-\varepsilon_F$ vanishes at a doping ratio between 1×10^3 and 2×10^3 vppm, corresponding to an estimated donor density of about 3×10^{19} cm^{-3}. This is somewhat higher than the limiting density originally determined by *Pearson* and *Bardeen* [3.122] for phosphorus doped polycrystalline Si.

3.7 Concluding Remarks

In this chapter we have attempted to survey significant developments in the field of glow discharge Si, over as wide a range as possible. The close relationship between the fundamental studies of the material and the applications has been stressed throughout because it is this aspect which has largely contributed to the rapid growth of the a-Si field during recent years. It applies in particular to the two major developments considered in the article, the a-Si FET and the a-Si junction, both of which have opened up a wide

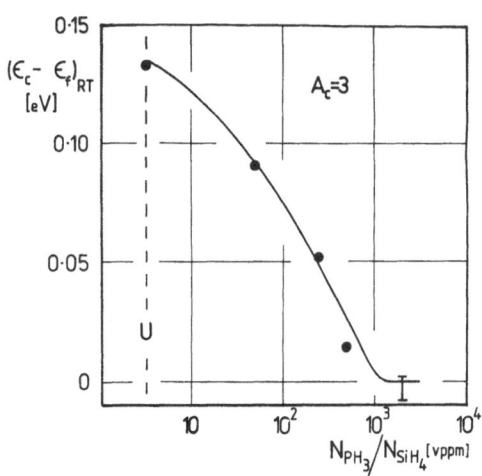

Fig. 3.40. The room temperature value of $\varepsilon_c-\varepsilon_F$ plotted against the gaseous doping ratio for $A_c = 3$ [3.121]

range of new and exciting possibilities for application. The common features underlying these developments are the unique properties of glow discharge Si, particularly the low overall density of gap states which at the present time makes it into the most viable amorphous semiconductor for electronic applications.

As to the future, there is little doubt that with the electronic control achievable, glow discharge Si will remain a useful model material in studies of the fundamental properties of amorphous semiconductors. The applied developments discussed here are very promising and raise the hope that ultimately, a-Si may approach the industrial importance of its crystalline counterpart. However, the history of electronic device developments during the last two decades teaches the need for considerable caution in speculating about the future extent and range of the a-Si development. We believe that one of the decisive factors will be the continued progress in solving the problems of industrialising the glow discharge preparation technique.

References

3.1　H. F. Sterling, R. C. G. Swann: Solid State Electr. **8**, 653–654 (1965)
3.2　R. C. Chittick, J. H. Alexander, H. F. Sterling: J. Electrochem. Soc. **116**, 77–81 (1969)
3.3　P. G. LeComber, W. E. Spear: Phys. Rev. Lett. **25**, 509–511 (1970)
3.4　P. G. LeComber, A. Madan, W. E. Spear: J. Non-Cryst. Solids **11**, 219–234 (1972)
3.5　J. C. Knights: Phil. Mag. **34**, 663–667 (1976)
3.6　J. C. Knights, G. Lucovsky, R. J. Nemanich: J. Non-Cryst. Solids **32**, 393–403 (1979)
3.7　Y. Kuwano, M. Ohnishi: J. Physique **42**, C4, 1155–1164 (1981)
3.8　B. A. Scott, M. H. Brodsky, D. C. Green, P. B. Kirby, R. M. Plecenik, E. E. Simonyi: Appl. Phys. Lett. **37**, 725–727 (1980)

3.9   B. A. Scott, M. H. Brodsky, D. C. Green, R. M. Plecenik, E. E. Simonyi, R. Serino: AIP Conf. Proc. No. 73, pp. 6–9 (1981) (Carefree, Arizona)

3.10   D. A. Anderson, W. E. Spear: Phil. Mag. **36**, 695–712 (1977)

3.11   K. Ogawa, I. Shimizu, E. Inoue: Japan. J. Appl. Phys. **20**, L 639–L 642 (1981)

3.12   R. W. Griffith: *Solar Materials Science* (Academic, New York 1980) pp. 665–731

3.13   A. Matsuda, K. Tanaka: Thin Solid Films **92**, 171–187 (1982)

3.14   R. W. Griffith, F. J. Kampas, P. E. Vanier, M. D. Hirsch: J. Non-Cryst. Solids **35/36**, 391–396 (1980)

3.15   G. Turban, Y. Catherine, B. Grolleau: Thin Solid Film **60**, 147–155 (1979)

3.16   T. Hamasaki, M. Hirose, Y. Osaka: J. Physique **42**, C 4, 807–810 (1981)

3.17   A. Matsuda, K. Nakagawa, K. Tanaka, M. Matsumura, S. Yamasaki, H. Okushi, S. Iizima: J. Non-Cryst. Solids **35/36**, 183–188 (1980)

3.18   F. J. Kampas, R. W. Griffith: AIP Conf. Proc. No. 73, pp. 1–5 (1981) (Carefree, Arizona)

3.19   W. E. Spear, P. G. LeComber: J. Non-Cryst. Solids **8–10**, 727–738 (1972)

3.20   A. Madan, P. G. LeComber, W. E. Spear: J. Non-Cryst. Solids **20**, 239–257 (1976)

3.21   W. E. Spear: In *Amorphous and Liquid Semiconductors*, ed. by J. Stuke, W. Brenig (Taylor & Francis, London 1974) pp. 1–16

3.22   A. Madan, P. G. LeComber: In *Amorphous and Liquid Semiconductors*, ed. by W. E. Spear (CICL, University of Edinburgh, 1977) pp. 377–381

3.23   P. G. LeComber: In *Fundamental Physics of Amorphous Semiconductors*, ed. by F. Yonezawa, Springer Ser. Solid-State Sci., Vol. 25 (Springer, Berlin, Heidelberg, New York 1981) pp. 46–55

3.24   R. Weisfield, P. Viktorovitch, D. A. Anderson, W. Paul: Appl. Phys. Lett. **39**, 263–265 (1981)

3.25   N. B. Goodman, H. Fritzsche: Phil. Mag. **B 42**, 149–165 (1980)

3.26   M. J. Powell: Phil. Mag. **B 43**, 93–103 (1981)

3.27   M. Grünewald, P. Thomas, D. Würtz: Phys. Stat. Sol. (b) **100**, K 139–K 143 (1980)

3.28   R. L. Weisfield, D. A. Anderson: Phil. Mag. **B 44**, 83–93 (1981)

3.29   M. Grünewald, K. Weber, W. Fuhs, P. Thomas: J. Physique **42**, C 4, 523–526 (1981)

3.30   L. Schweitzer, M. Grünewald, H. Dersch: J. Physique **42**, C 4, 827–830 (1981)

3.31   P. Thomas: private communication (1980)

3.32   R. J. Loveland, W. E. Spear, A. Al-Sharbaty: J. Non-Cryst. Solids **13**, 55–68 (1973–4)

3.33   A. Ghaith: Thesis, University of Dundee (1980) (unpublished)

3.34   D. I. Jones, R. A. Gibson, P. G. LeComber, W. E. Spear: Solar Energy Mat. **2**, 93–106 (1979)

3.35   M. H. Tanielian, N. B. Goodman, H. Fritzsche: J. Physique **42**, C 4, 375–378 (1981)

3.36   M. J. Powell, B. C. Easton, D. H. Nicholls: J. Physique **42**, C 4, 379–382 (1981)

3.37   D. L. Staebler, C. R. Wronski: J. Appl. Phys. **51**, 3262–3268 (1980)

3.38   D. Jousse, R. Basset, S. Delionibus, B. Bowden: Appl. Phys. Lett. **37**, 208–211 (1980)

3.39   S. R. Elliott: Phil. Mag. **B 39**, 349–358 (1979)

3.40   I. Solomon: In *Fundamental Physics of Amorphous Semiconductors*, ed. by F. Yonezawa, Springer Ser. Solid-State Sci., Vol. 25 (Springer, Berlin, Heidelberg, New York 1981) pp. 33–39

3.41   H. Fritzsche: Solar Cells **2**, 289 (1980)

3.42   W. E. Spear, R. J. Loveland, A. Al-Sharbaty: J. Non-Cryst. Solids **15**, 410–422 (1974)

3.43   W. E. Spear, P. G. LeComber: Solid State Commun. **17**, 1193–1196 (1975)

3.44   W. E. Spear, P. G. LeComber: Phil. Mag. **33**, 935–949 1976)

3.45   G. Müller, S. Kalbitzer, W. E. Spear, P. G. LeComber: In *Amorphous and Liquid Semiconductors*, ed. by W. E. Spear (CICL, University of Edinburgh, 1977) pp. 442–446

3.46   S. Kalbitzer, G. Müller, P. G. LeComber, W. E. Spear: Phil. Mag. **B 41**, 439–456 (1980)

3.47   D. V. Lang: In *Thermally Stimulated Relaxation in Solids*, ed. by P. Bräunlich, Topics Appl. Phys., Vol. 37 (Springer, Berlin, Heidelberg, New York 1979) chap. 3

3.48   J. Bourgain, M. Lannoo: *Point Defects in Semiconductors* II, Springer Ser. Solid-State Sci., Vol. 35 (Springer, Berlin, Heidelberg, New York 1983)

3.49 a) J. D. Cohen, D. V. Lang, J. C. Bean, J. P. Harbison: Phys. Rev. Lett. **45**, 197–200 (1980)

3.49 b) J. D. Cohen, D. V. Lang, J. C. Bean, J. Harbison: J. Non-Cryst. Solids **35/36**, 581–586 (1980)

3.49 c) D. V. Lang, J. D. Cohen, J. P. Harbison: Phys. Rev. B **25**, 5285–5320 (1982)

3.50 W. den Boer: J. Physique **42**, C4, 451–454 (1981)

3.51 K. D. Mackenzie, P. G. LeComber, W. E. Spear: Phil. Mag. B **46**, 377–389 (1982)

3.52 T. P. Brody, J. A. Asars, G. D. Dixon: IEEE Trans. ED-**20**, 995–1001 (1973)

3.53 P. G. LeComber, W. E. Spear, A. Ghaith: Electron Lett. **15**, 179–180 (1979)

3.54 P. G. LeComber, A. J. Snell, K. D. Mackenzie, W. E. Spear: J. Physique **42**, C4, 423–432 (1981)

3.55 a) A. J. Snell, K. D. Mackenzie, W. E. Spear, P. G. LeComber, A. J. Hughes: Appl. Phys. **24**, 357–362 (1981)

3.55 b) K. D. Mackenzie, A. J. Snell, I. French, P. G. LeComber, W. E. Spear: Appl. Phys. A **31**, 8–92 (1983)

3.56 M. J. Powell, B. C. Easton, O. F. Hill: Appl. Phys. Lett. **38**, 794 (1981)

3.57 a) H. Hayama, M. Matsumura: Appl. Phys. Lett. **36**, 754–755 (1980)

3.57 b) M. Matsumura, S. Kuno, Y. Uchida: J. Phys. (Paris) **42**, C4, 519–522 (1981)

3.57 c) Y. Okubo, T. Nakagiri, Y. Osada, M. Sugata, N. Kitihara, K. Hatanaka: SID 82 Digest, 40–41

3.57 d) S. Kawai, N. Takagi, T. Kodama, K. Asama, S. Yongisawa: SID 82 Digest, 42–43

3.58 H. C. Tuan, M. J. Thompson, N. M. Johnson, R. A. Lujan: IEEE Trans. EDL (in press) (1983)

3.59 I. French, A. J. Snell, P. G. LeComber, J. Stephens: Appl. Phys. A **31**, 19–22 (1983)

3.60 M. Matsumura, H. Hayama: Proc. IEEE **68**, 1349–1350 (1980)

3.61 M. Matsumura, H. Hayama, Y. Nara, K. Ishibashi: IEEE EDL-1, 182–184 (1980)

3.62 a) A. J. Snell, W. E. Spear, P. G. LeComber, K. D. Mackenzie: Appl. Phys. A **26**, 83–86 (1981)

3.62 b) A. J. Snell, P. G. LeComber, K. D. Mackenzie, W. E. Spear, A. Doghmane: Proc. 10th Intern. Conf. on Amorphous and Liquid Semiconductors, Tokyo 1983. J. Non-Cryst. Solids (in the press)

3.63 P. K. Weimer, W. S. Pike, G. Sadasiv, F. V. Shallcross, L. Meray-Hovarth: IEEE Spectrum **6**, 52–65 (1969)

3.64 W. E. Spear, P. G. LeComber, S. Kinmond, M. H. Brodsky: Appl. Phys. Lett. **28**, 105–107 (1976)

3.65 R. H. Williams, R. R. Varma, W. E. Spear, P. G. LeComber: J. Phys. C **12**, L209–L213 (1979)

3.66 W. E. Spear, P. G. LeComber, S. Kalbitzer, G. Müller: Phil. Mag. B **39**, 159–165 (1979)

3.67 W. Beyer, R. Fischer: Appl. Phys. Lett. **31**, 850 (1978)

3.68 W. E. Spear, R. A. Gibson, D. Yang, P. G. LeComber, G. Müller, S. Kalbitzer: J. Physique **42**, C4, 1143–1153 (1981)

3.69 G. Müller, P. G. LeComber: Phil. Mag. B **43**, 419–431 (1981)

3.70 W. E. Spear, P. G. LeComber, A. J. Snell: Phil. Mag. B **38**, 303–317 (1978)

3.71 A. J. Snell, K. D. Mackenzie, P. G. LeComber, W. E. Spear: Phil. Mag. B **40**, 1–15 (1979)

3.72 J. Beichler, W. Fuhs, M. Mell, H. M. Welsch: J. Non-Cryst. Solids **35/36**, 587–592 (1980)

3.73 A. J. Snell, W. E. Spear, P. G. LeComber: Phil. Mag. B **43**, 407–417 (1981)

3.74 A. J. Snell, W. E. Spear, P. G. LeComber: J. Phys. Soc. Japan **49**, Suppl. A, 1217–1220 (1980)

3.75 R. A. Gibson, P. G. LeComber, W. E. Spear: Appl. Phys. **21**, 307–311 (1980)

3.76 D. E. Carlson, C. R. Wronski: Appl. Phys. Lett. **29**, 602–605 (1976)

3.77 D. L. Staebler: J. Non-Cryst. Solids **35/36**, 387–390 (1980)

3.78 A. R. Moore: Appl. Phys. Lett. **37**, 327–330 (1980)

3.79 J. Dresner, D. J. Szostak, B. Goldstein: AIP Conf. Proc. No. 73, p. 317 (1981) (Carefree, Arizona)

3.80 R. Abeles, C. R. Wronski, Y. Goldstein, H. E. Stasiewski, D. Gutkowicz-Krusin, T. Tiedje, G. D. Cody: AIP Conf. Proc. No. 73, pp. 298–301 (1981) (Carefree, Arizona)

3.81 K. Hecht: Z. Phys. **77**, 235–245 (1932)

3.82 L. Onsager: Phys. Rev. **54**, 554–557 (1938)

3.83 D. M. Pai, R. C. Enck: Phys. Rev. **B 11**, 5163–5174 (1975)

3.84 R. R. Chance, C. L. Braun: J. Chem. Phys. **59**, 2269–2272 (1973)

3.85 J. Mort, A. Troup, M. Morgan, S. Grammatica, J. C. Knights, R. Lujan: Appl. Phys. Lett. **38**, 277–279 (1981)

3.86 J. Mort, I. Chen, A. Troup, M. Morgan, J. C. Knights, R. Lujan: Phys. Rev. Lett. **45**, 1348–1351 (1980)

3.87 J. Mort, S. Grammatica, J. C. Knights, R. Lujan: Solar Cells **2**, 451 (1980)

3.88 I. Chen, J. Mort: Appl. Phys. Lett. **37**, 952–955 (1981

3.89 M. Silver, A. Madan, D. Adler, W. Czubatyi: Paper presented at the 14th IEEE Photovoltaic Conference, San Diego (1980)

3.90 Y. Hamakawa: J. Physique **42**, C 4, 1131–1142 (1981)

3.91 F. Carasco, W. E. Spear: Phil. Mag. **B 47**, 495–507 (1983)

3.92 N. Yamamoto, Y. Nakayama, K. Wakita, M. Nakano, T. Kawamura: Japan J. Appl. Phys. Suppl. **20–1**, 305 (1981)

3.93 N. Yamamoto, K. Wakita, Y. Nakayama, T. Kawamura: J. Physique **42**, C 4, 495–498 (1981)

3.94 I. Shimizu, S. Oda, K. Saito, H. Tomita, E. Inoue: J. Physique **42**, C 4, 1123–1130 (1981)

3.95 ibid.: AIP Conference No. 73, pp. 288–292 (1981) (Carefree, Arizona)

3.96 I. Shimizu, T. Komatsu, K. Saito, E. Inoue: J. Non-Cryst. Solids **35/36**, 773–778 (1980)

3.97 A. E. Owen, P. G. LeComber, G. Sarrabayrouse, W. E. Spear: IEE Proc. **129**, 51–54 (1982)

3.98 S. R. Ovshinski: Phys. Rev. Lett. **21**, 1450–1453 (1968)

3.99 A. E. Owen, J. M. Robertson: IEEE Trans. **ED-20**, 105–122 (1973)

3.100 H. J. Hovel: Appl. Phys. Lett. **17**, 141–143 (1970)

3.101 S. Veprek, V. Maracek: Solid State Electr. **11**, 683–684 (1968)

3.102 A. P. Webb, S. Veprek: Chem. Phys. Lett. **62**, 173–177 (1979)

3.103 S. Veprek: Chimia **34**, 489–501 (1980)

3.104 S. Veprek, Z. Iqbal, H. R. Oswald, A. P. Webb: J. Phys. C **14**, 295–308 (1981)

3.105 S. Usui, M. Kikuchi: J. Non-Cryst. Solids **34**, 1–11 (1979)

3.106 A. Matsuda, S. Yamasaki, K. Nakagawa, H. Okushi, K. Tanaka, S. Iizima, M. Matsumara, H. Yamamoto: Japan J. Appl. Phys. **19**, L 305–L 308 (1980)

3.107 T. Hamasaki, H. Kurata, M. Hirose, Y. Osaka: Appl. Phys. Lett. **37**, 1084–1086 (1980)

3.108 W. E. Spear, G. Willeke, P. G. LeComber, A. G. Fitzgerald: J. Physique **42**, C 4, 257–260 (1981)

3.109 Z. Iqbal, S. Veprek: J. Phys. C **15**, 377–392 (1982)

3.110 Y. Hishima, T. Hamasaki, H. Kurata, M. Hirose, Y. Osaka: Japan J. Appl. Phys. **20**, L 121–L 123 (1981)

3.111 A. Matsuda, T. Yoshida, S. Yamasaki, K. Tanaka: Japan J. Appl. Phys. **20**, L 439–L 442 (1981)

3.112 Y. Mishima, S. Miyazaki, M. Hirose, Y. Osaka: Phil. Mag. **B 46**, 1–12 (1982)

3.113 J. W. Orton, M. J. Powell: Repts. Progr. Phys. **43**, 1263–1307 (1980)

3.114 A. Matsuda, H. Matsumura, S. Yamasaki, H. Yamamoto, T. Imura, H. Okushi, S. Iizima, K. Tanaka: Japan J. Appl. Phys. **20**, L 183–L 186 (1981)

3.115 T. Hamasaki, H. Kurata, M. Hirose, Y. Osaka: Japan J. Appl. Phys. **20**, L 84–L 86 (1981)

3.116 A. Madan, S. R. Ovshinsky, E. Benn: Phil. Mag. **40**, 259–277 (1979)

3.117 P. G. LeComber, D. I. Jones, W. E. Spear: Phil. Mag. **35**, 1173–1187 (1977)

3.118 L. Friedman: J. Non-Cryst. Solids **6**, 329–341 (1971)

3.119 D. Emin: Phil. Mag. **35**, 1189–1198 (1977)

3.120 O. Reilly, W. E. Spear: Phil. Mag. **B 38**, 295–302 (1978)

3.121 G. Willeke, W. E. Spear, D. I. Jones, P. G. LeComber: Phil. Mag. **B 46**, 177–190 (1982)

3.122 G. L. Pearson, J. Bardeen: Phys. Rev. **75**, 865–883 (1949)

4. Sputtered Material

Malcolm J. Thompson

With 50 Figures

Sputtering is now established as a very versatile, useful technique for depositing a wide range of elemental and compound thin films. There are many commercially available systems capable of sputtering films on large area substrates fully automated for continuous or batch processing. However, although such sophisticated equipment was not available in the 1960s, many materials scientists started to use sputtering particularly for the deposition of metal and insulator compounds and high-melting-point materials. For instance, thin films of multicomponent amorphous chalcogenide glasses were deposited this way as it was the most reliable and convenient way of obtaining thin films with the required stoichiometry. In the early work on elemental Si and Ge thin films, it was noted that both the sputtered and evaporated material contained many structural defects. Such a large number of defects was not always present in material produced by glow-discharge decomposition of silane. Indeed it was noted that the transport and optical properties of Si produced by this third technique were often quite different from those produced by sputtering and evaporation. It was the work of the Harvard University group firstly on a-Ge and then a-Si that led the way to showing that by reactively sputtering Ge or Si in an Ar and H atmosphere, a hydrogenated film was produced which had markedly different properties from the pure elemental films but similar properties to Si produced from a silane glow discharge. The explosive growth of research on a-Si:H followed the earlier discovery that glow discharge material can be doped and thus devices can be fabricated in a-Si:H. Thus the majority of the research and development has been done on glow discharge produced material following the early successes of this method. However, there are now a number of other techniques available for producing a-Si:H, sputtering being the most widely used of these over the last ten years. Throughout this period continual advances have been made to produce device quality material with low defect density but the volume of research is far smaller than that devoted to silane glow discharge deposited Si. Thus the optimization of sputtered a-Si:H has not progressed as fast as glow discharge Si. However, the attractions of using this technique are considerable both from a material science and physics standpoint; in addition, commercially available equipment is already developed to support large-scale mass production of sputtered metal insulator and semiconductor films.

Most of the research in sputtered a-Si : H has been devoted to the study of optical and electronic properties with some recent microscopy studies. Throughout researchers have attempted to relate film properties to controlled deposition parameters; however, despite a considerable volume of research on sputtering elsewhere, there has been very little *detailed* analysis of the relationship between the plasma and film properties in a-Si : H. Although this is an ambitious task, the first and substantial part of this review attempts to analyze in some detail the plasma processes that can affect film properties as this has not been done in other reviews of sputtered Si. Such a comprehensive analysis is difficult and therefore sometimes qualitative as the information regarding Ar/H plasmas for sputtering is sparse. However, it is hoped that such a treatment will reveal new areas for research which will provide some solutions to present material problems. rf sputtering has been most widely used for Si deposition but as dc sputtering and magnetron sputtering have been used recently, all these techniques are reviewed. Indeed it is suggested that it maybe that the adoption of a number of techniques like magnetron and bias sputtering as well as conventional sputtering may help to determine the plasma species that most affect the properties of the growing film. That is the approach we established at Sheffield having all these techniques available in the same deposition system, so that the effects of geometrical variations and background vacuum evironment may be eliminated. There are many correlations that have been reported to exist between preparation conditions, deposition processes and film properties; however, the issue of cause and effect determining the species in the plasma and growth which are most influential in controlling the physical properties of a-Si : H remains unresolved. Inevitably in the discussion of plasma species, the electron and atomic bombardment effects on the growing film are highlighted. This subject has still not been evaluated to any extent in glow discharge SiH_4 plasmas. Although a few studies of ion beam sputtered Si are now being reported, these will not be reviewed here.

An attempt has been made to evaluate the properties of sputtered a-Si : H taking special care to clarify the sometimes misleading issues arising when comparing results from different laboratories, let alone comparing material produced by different methods. Inevitably it has been necessary to comment on the limitations of the different measuring techniques which is more fully discussed in other chapters. Regretfully it is only on a few occasions that it has been possible for one group of researchers to make measurements on materials from different laboratories to make absolute comparisons. The luminescence studies at Sheffield have been fortunate in being able to do this. However, despite all these issues, some consensus is emerging on the basic optical and structural properties of sputtered a-Si : H. Inevitably the quest has been to fabricate device quality material similar to glow discharge Si. This has sometimes had a negative impact on the understanding of the physics of sputtered a-Si : H. In addition, the assessment of the material and device quality is much dependent on the reliability of the measurement technique and the ability to fabricate devices, as well as the intrinsic materials properties. Despite this, considerable progress has been made in the understanding of the properties of sputtered material along with an improvement in corresponding device properties. Further advances will undoubtedly be made; however, the slope of the "progress" curve is inevitably related to the number of researchers dedicated to sputtered a-Si : H.

4.1 Sputtering Processes and Techniques Applied to Hydrogenated Amorphous Silicon

4.1.1 Sputtering Mechanisms and Plasma

Sputtering is the phenomenon whereby material is removed predominantly in atomic form due to bombardment of energetic gas atoms. One of the most convenient ways to obtain energetic atoms is to accelerate in an electric field the ions formed in a glow discharge plasma. However, these ions are likely to be neutralized as they are accelerated towards the target (the material to be sputtered) due to Auger emission by electrons from the target. When the ion impacts the target, it can set up a series of collisions with the target atoms which will lead to an ejection of one of these atoms. This process is known as sputtering. In a planar diode sputtering system the plasma extends between the parallel target and the substrate which are separated by 4–10 cm. As the electrons have a larger velocity than ions, they will diffuse faster out of the plasma than the ions leaving the plasma with a net positive potential. A sheath or positive space charge region exists between the target and the plasma across which most of the target voltage is dropped (Fig. 4.1). The ions from the plasma are accelerated across the sheath to the sputtering target.

Thin films of a-Si : H have been deposited by both dc and rf sputtering. rf sputtering can be used to deposit insulators in addition to conducting materials as the alternating voltage avoids charge buildup at the target which occurs in dc systems with an insulating target. Electron impact ionization is more efficient in rf than dc discharges and thus the former tends to be operated at lower pressures. Therefore, the difference in operating pressure of these two types of discharge means that certain pressure-dependent mechanisms may play a more or less dominant role. For instance, charge exchange and scattering of sputtered species is more important at higher pressures. The breakdown voltage of dc plasmas is influenced by electron impact ionization in the plasma and secondary electron emission at the electrodes, with the Paschen curve showing a minimum breakdown voltage for a particular pd (pressure × electrode spacing). Loss of charge from the plasma occurs by diffusion and recombination at the walls. The determination of the breakdown voltage of an rf discharge is more complex because the reversing

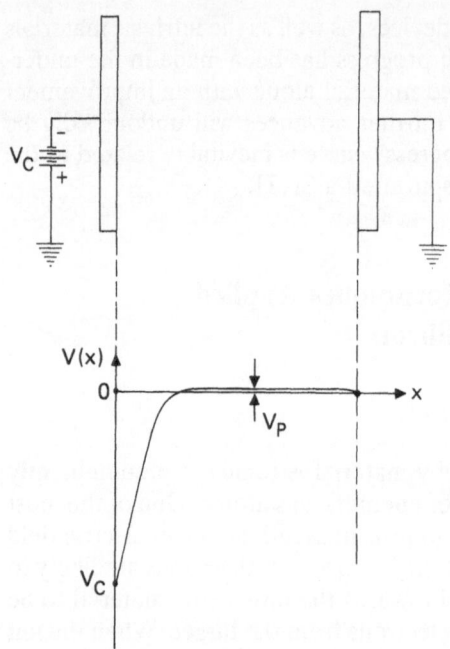

Fig. 4.1. Approximate potential versus distance plot for an unobstructed dc glow discharge [4.1a]

field can cause reduced losses to the walls and increased secondary electron emission by electrons oscillating between the electrodes creating the so-called multipacting phenomenon [4.1]. More details of sputtering plasmas can be found in the literature [4.2].

In reactive sputtering a reactive gas included in the plasma is incorporated in the growing film. Hydrogen and argon are used almost exclusively as the sputtering gases when depositing a-Si:H.

a) Target Processes

There are a number of phenomena that occur at the target, some of which can have an important effect on the properties and growth of thin films of a-Si:H on the substrate situated several centimeters away from the target. These interactions are illustrated in Fig. 4.2 but a precise calculation of the collision processes or target kinetics is extremely difficult. However, the collisions are sufficiently short range to the first approximation that they can be considered as a series of binary elastic collisions [4.3]. There is a minimum energy (threshold yield) required to emit an atom, being around 15–20 eV for Si; it is a function of the binding energy of the target atoms. Between 100 eV and 1 keV the simple elastic collision theory agrees well with experiment showing a near linear dependence of the yield (number of emitted target atoms per incident ion) on the energy of the bombarding ions. At higher energies the dissipated energy is distributed throughout a large vol-

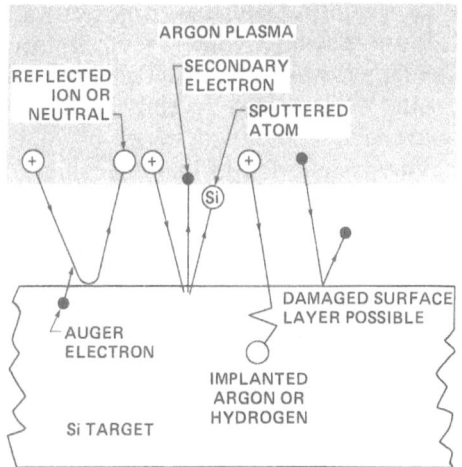

Fig. 4.2. Interaction of species from an A–H plasma with a Si target: (a) A and H can be reflected and may be neutralized with an auger electron; (b) Si atoms will be ejected with a secondary electron; (c) plasma electrons may be refleted at the target; (d) high energy A and H atoms can be implanted in Si target and may cause structural changes at the surface

ume of target material so the energy transferred to the surface layers remains virtually constant. At very high energies the yield can reduce as ion implantation becomes dominant. The sputtering yield for Si up to 5 keV is shown in Fig. 4.3; the saturation at high energies is apparent from the experimental points. Less than 1% of the incident power goes into the ejection of sputtered material; a considerable amount of the energy is lost in collisions and is dissipated thermally, hence the need to water cool the target so that undesirable high temperatures are not generated. The data in Fig. 4.3 was obtained on ⟨111⟩ single crystal Si. The target kinetics are such that the material sputtered from single crystals leaves the surface of the crystal in a number of principal crystallographic directions giving an orientation-dependent yield. Spotted deposition patterns showing clearly enhanced sputter ejections along certain crystal directions have been observed when sputtering single crystal Si [4.4, 5]; however, such patterns have not been observed by others [4.6].

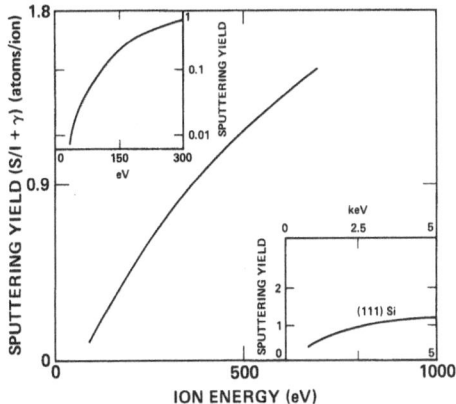

Fig. 4.3. Sputtering yield for ⟨111⟩ Si bombarded by 0–5 keV A ions [4.6b]. Lower insert from [4.6a]; top is insert from [4.6c]

These differences can be explained by scattering and thermalization of Si atoms in the plasma (Sect. 4.1.1 b). More recently, workers studying sputtered a-Si : H have used polycrystalline targets which will presumably give no crystal preferential yield. However, some workers have recently used single crystal Si targets as they are readily available with a wide range of doping (Sect. 4.2.5), but no gross nonuniformity is reported in film deposition by these researchers.

Assuming the sputter ejection of atoms from a target is a quasi-random process, the number of molecules leaving the surface in a solid angle dw at an angle θ to the surface normal is proportional to $dw \cos \theta$; but for low energy bombardment the emission tends to be undercosine. In addition to this cosine emission there is an area of high erosion rate at the edge of the target due to the high electric fields between target and shield causing excessive high energy bombardment here; thus the target geometry and electric field configuration has an influence on the uniformity of sputtering. The larger the sputtering target the greater the uniformity of emission, and thus sputtering is well suited for the large area production of thin films.

It is desirable to identify the species which emanate from the target because under certain conditions these electrons, atoms and ions can bombard the substrate. The target atoms are sputtered and leave the target with a relatively low average energy of 5–10 eV. However, the distribution in energy is highly asymmetric resulting in some sputtered atoms having a very large energy up to several hundred eV. The sputtered Si atoms will travel towards the substrate through the plasma as discussed in Sect. 4.1.1 b.

In order to describe the reflected ions, secondary electrons and yield at the target, the energy of the bombarding ion species must be known. Frequently the dc target voltage is taken as the ion accelerating voltage across the dark space adjacent to the target. In a dc discharge the voltage across the sheath is the target voltage V_t minus the plasma potential V_p; this is the accelerating potential for the ions. However, the energy of the bombarding Ar ions is less than $(V_t - V_p) = V_c$ because of symmetric charge exchange collisions of the type

$$Ar + Ar^+ \rightarrow Ar^+ + Ar . \tag{4.1}$$

The energy distribution for Ar ions striking the target is shown in Fig. 4.4 which shows excellent argreement between experiment and theory giving an effective cross section for charge exchange of 5×10^{-15} cm^2 [4.7]. An increase in the target voltage will result in a larger number of high energy ions reaching the target but the effect of pressure is small. Ar^{2+} ions are present in the plasma but the collision cross section is smaller (7×10^{-16} cm^2) and thus more of these ions reach the target with high energy twice the maximum energy for Ar$^+$. The cross section for charge transfer for H$^+$ is considerably smaller.

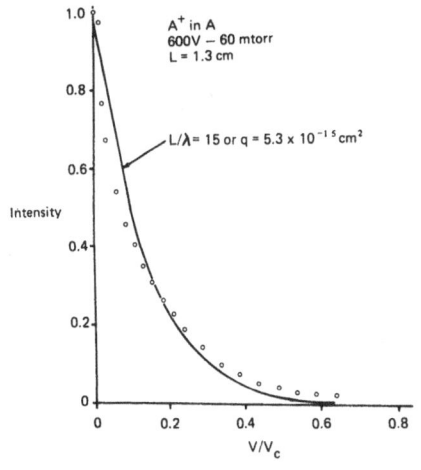

Fig. 4.4. Energy distribution for Ar$^+$ from an argon discharge [4.7]

(——) Theoretical distribution
(ooo) Experimental points

The determination of the energy of bombarding ions in the rf glow discharge commonly used for sputtering a-Si:H is more difficult. The equivalent circuit of an rf planar discharge system is shown in Fig. 4.5. The capacitances represent the sheaths formed at the discharge surface boundaries. The grounded surfaces can be the substrate electrode if grounded and the chamber walls if they are stainless steel or conducting. The problem is that

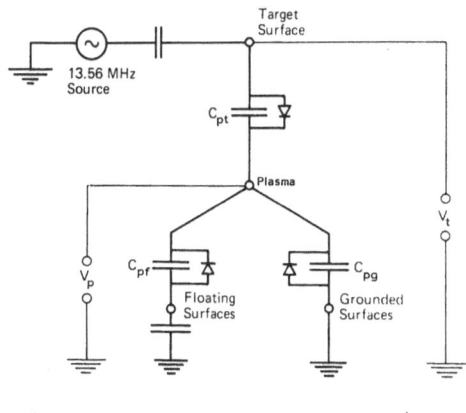

◄ Fig. 4.5. Approximate equivalent circuit of rf planar discharge system [4.8]

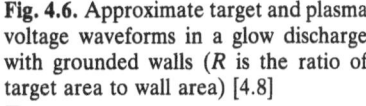

Fig. 4.6. Approximate target and plasma voltage waveforms in a glow discharge with grounded walls (R is the ratio of target area to wall area) [4.8]
▼

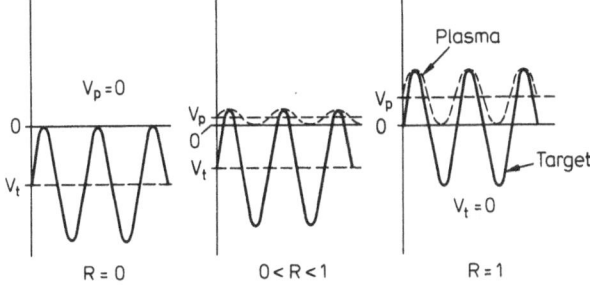

the potentials at the target, plasma and substrate, are a strong function of the system geometry, in particular the relative size of the target, substrate and walls. For instance, if the target area is equal to the area of all other surfaces in contact with the discharge, then all surfaces receive the same ion bombardment and there is no net sputter deposition; thus the target self-bias voltage would be zero. However, as shown in Fig. 4.6, if the target area is small compared with all other surfaces, the dc target voltage is negative and equal to half the peak to peak rf voltage (V_{pp}). In many situations where the ratio of target area to wall area is >0 but <1, the plasma and target potentials are represented by the second case in Fig. 4.6. Thus, *Coburn* et al. [4.8] deduced that the plasma potential can be estimated by measuring V_{pp} and the dc target voltage:

$$2V_p = \frac{(V_{pp}) - |V_{dc}|}{2} \, . \tag{4.2}$$

Bruce [4.9] has recently further clarified the discussion of plasma potentials and shows by direct measurements that the plasma potential oscillates coherently with the applied voltage, and at frequencies below the ion cut-off

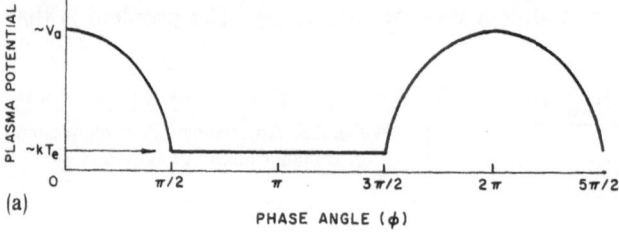

(a)

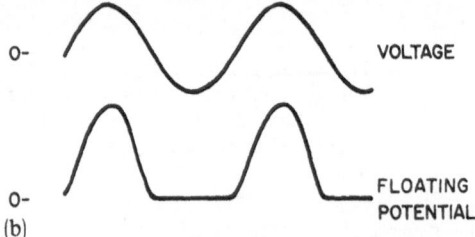

(b)

Fig. 4.7. (a) Dependence of plasma potential on phase of applied fields from model; **(b)** oscilloscope recording of the floating potential in a 40-Pa A discharge with 0.6 W cm^{-2}, 100 kHz excitation (from [4.9])

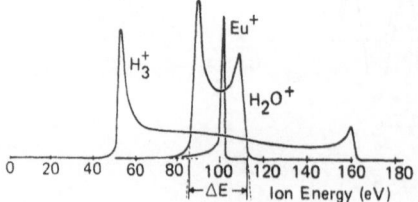

Fig. 4.8. Energy distribution of H$_3^+$, H$_2$O$^+$ and E$^+$ at the substrate plane in a confined discharge. 13.56 MHz rf power = 100 W, argon pressure = 75 mtorr, target = 5 cm diameter [4.8]

frequency, the floating target potential should follow the plasma potential since both ions and electrons can cross the sheath in times short compared to the oscillation period (Fig. 4.7). However, at frequencies of 13.56 MHz used in rf sputtering, the argons ions will respond to the average potential which is Va/Π. However, the ion cut-off frequency for hydrogen is much higher and the energy distribution for ions will be larger because the H^+ will respond to the varying plasma potential as shown by *Coburn* and *Kay* [4.8] in Fig. 4.8. In fact, in hydrogen plasmas at pressures $> 4 \times 10^{-4}$ torr the dominant hydrogen ions in the plasma are H_3^+ [4.10]. Thus hydrogen ion bombardment of the target and substrate occurs with an energy range corresponding to the difference in voltage excursion of the plasma and target or substrate, respectively.

The reflection of the ion species bombarding the target is given from a Rutherford backscattering relationship. As the atomic mass of A is larger than Si, the back reflection of A from the target is expected to be small. From *Eckstein* et al. [4.11] a H^+ reflection coefficient from the target is expected to be 0.25–0.50 for energies in the range 500–1000 eV. The Si and H ions that impinge on and are reflected from the target combine with auger electrons and are neutralized.

Secondary electrons are produced at the target due to electron and ion bombardment. The corresponding secondary electron emission from Si is 1.1 for electron bombardment and between 0.024 to 0.039 for 10–1000 eV argon ion bombardment [4.2]. It is the electrons emitted from the target that early workers found was responsible for the larger heating on the substrate. The electrons are accelerated across the sheath in response to the instantaneous voltage difference between target and plasma. Negative ions can be formed at the target but they have not been found in an a-Si sputtering system.

The presence of hydrogen in the target has been found by *Ross* and *Messier* [4.12] and *Lemperière* et al. [4.13] as discussed in the next section.

b) The Plasma

Only a few studies have been made of Ar–H plasmas formed in Si sputtering systems. The determination and characterization of species contained in the sputtering plasma provides a lifelong challenge. We have already seen that species are injected into the plasma from the target and the secondary electrons from the target provide an ionizing source for the atoms in the plasma. Although a detailed explanation of all the processes and species in the sputtering plasma is not possible, a broad review of the predominant reactions and processes is desirable in order that the ions and electrons bombarding the growing film may be evaluated. Figure 4.9 illustrates the cross sections for electrons in argon gas. The thermal electron energy distribution peaks between 2–10 eV in sputtering plasma but has a large tail and a peak at high energies due to reflected primaries or secondaries which are accelerated in the dark space. Thus the tail of thermal Maxwell-Boltzman distributed elec-

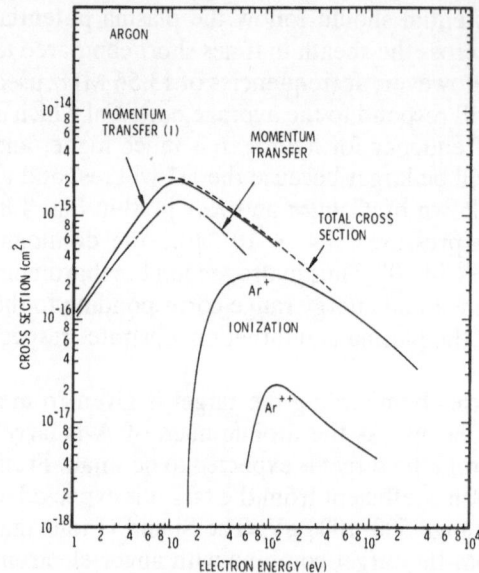

Fig. 4.9. Collision cross section for electrons in argon gas. The data are: total cross section [4.14]; momentum transfer (*1*) [4.15]; momentum transfer [4.16]; ionization [4.17 a]. After [4.17 b]

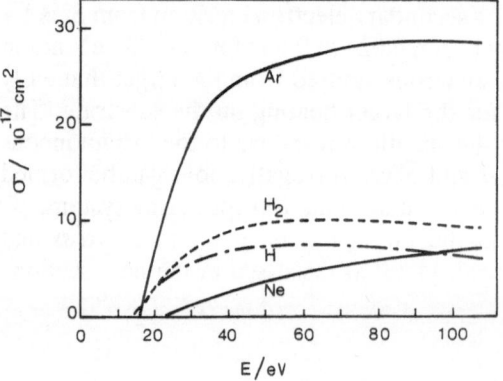

Fig. 4.10. The ionization cross section for Ar, H_2, H and Ne [4.17 c]

trons and those from the dark space have sufficient energy (> 15.7 eV) to ionize the argon atoms. The ionization cross section for A and H_2 are compared in Fig. 4.10. The argon-excitation potential is 11.6 and the excitation cross section reaches a maximum of 4×10^{-17} cm^2 for an electron energy of 21 eV. H and H_2 have a larger excitation cross section. Mass spectrometry analyses of an A and an A/H plasma contained in a dc sputtering system are shown in Fig. 4.11. The most notable features are the H_3^+ and AH$^+$ ions as well as the SiH$_x$ species in the A/H plasma. In addition, the relative intensity of the A^{2+} and A$^+$ changes when H is added to the plasma. *Paesler* et al. [4.18] and *Matsuda* et al. [4.19] using emission spectroscopy disagree as to the presence of SiH in the plasma; however, the results of *Tardy* et al. [4.20]

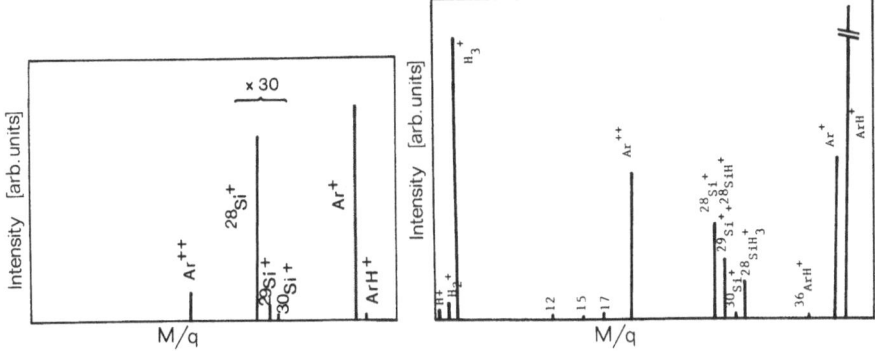

Fig. 4.11. (**a**) Glow discharge mass spectrum in pure argon. $P = 70$ mtorr, power = 1.6 W cm^{-2}. (**b**) Spectrum in argon-hydrogen. $P = 70$ mtorr, $Rg = 0.2$ power = 1.6 W cm^{-2}. The ratio of heights of ArH$^+$/Ar and H$_3^+$/Ar$^+$ are, respectively, 3.6 and 4.1 [4.20]

in Fig. 4.11 confirm the assertions of Matsuda et al. *Coburn* and *Kay* [4.21] first showed that the ionization of the sputtered species can be caused by penning ionization if the metastable A atoms have an excitation energy higher than the ionization energy of the sputtered atom:

$$A^* + Si \rightarrow Si^+ + A + e .\tag{4.3}$$

The ionization energy of Si and SiH$_x$ ($x = 1$ to 3) is between 7.4 and 9.5 eV [4.22], much lower than the two argon metastable levels at 11.55 and 11.72 eV which have a lifetime exceeding 1.35 s. This explains why the Si$^+$ and SiH$_x^+$ peak intensities are only 30 times less than that of A$^+$ ions, whereas the Si/A partial pressure ratio is 10^{-5}–10^{-6}. In addition it is interesting to note that under typical operating conditions the residence time for A atoms in the sputtering systems can be several seconds (see Sect. 4.2.5).

As the power density exceeds 1 W cm^{-2}, *Tardy* et al. [4.20] have shown that Si$^+$ species are larger than the SiH$^+$ and SiH$_3^+$. This occurs over a large range of pressure and Ar/H ratio; these measurements are for dc sputtering extending down to an argon pressure of 5×10^{-2} torr and a hydrogen partial pressure of 5×10^{-4} torr. They also present evidence for the sputtering of hydrogenated species from the target [4.13]. Figure 4.12 shows the species existing in an Ar plasma after the target was submitted to a hydrogen discharge. The large number of SiH species relative to Si is evident; the fall off in the SiH indicates an average hydrogen concentration of 5% in a 500 Å surface layer of the target. This deduction is similar to *Ross* and *Messier* [4.12] who found up to 40% H in Si targets to a depth of several hundred angstroms after sputtering a-Si:H under typical conditions. The density of the Si species in an A/H dc sputtering plasma as a function of hydrogen partial pressure is shown in Fig. 4.13. The normal hydrogen partial pressures used in rf puttering are 0.1–2 mtorr which are unfortunately at the extreme of

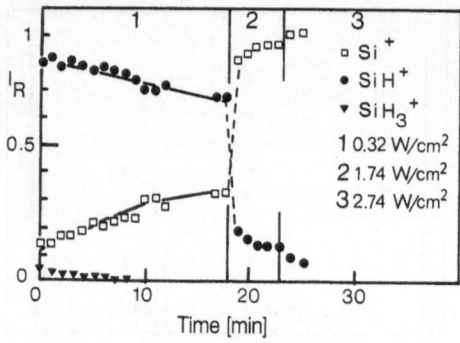

Fig. 4.12. The emission from a Si target sputtered in an argon plasma following the target's exposure to a hydrogen plasma [4.13]

this data in Fig. 4.13; however, these data indicate a significant Si/SiH ratio of sputtered species of ~ 2 at P_H = 1 mtorr. Figure 4.13 also shows that the Si/SiH ratio increases as the A pressure is increased. *Lemperière* [4.13] has suggested that SiH$_3$ is due to gap phase reaction of $SiH^+ + H_2 \rightarrow SiH_3^+$. However, there is a concentration of H_3^+ in plasma which presumably could give rise to

$$Si + H_3 \rightarrow SiH_3^+ . \tag{4.4}$$

Having discussed the species and processes present in the plasma, the next stage is to examine the reactions of the neutrals and electrons from the target as they pass through the plasma towards the substrate. *Anderson* et al. [4.23] have applied *Westwood*'s analysis [4.24] to sputtered Si and calculated the thermalization distance of Si from the target. Starting from Westwood's expression

$$D = \tfrac{1}{2} \left[\ln(E_0/E_g)\ln E\right]\lambda(1 + \cos\langle\theta\rangle) , \tag{4.5}$$

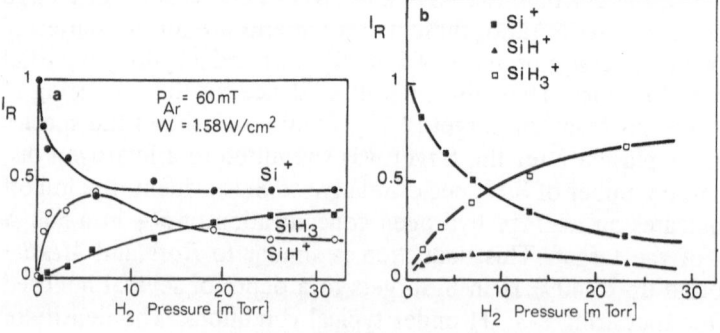

Fig. 4.13a, b. The mass spectrometric analysis of Si species in a A/H plasma as a function of hydrogen partial pressure (P_{H_2}). (a) P_A = 60 mtorr, power 1.58 W cm^{-2}, (b) P_A = 101 mtorr, power = 4 W^1 cm^{-2} [4.17c]

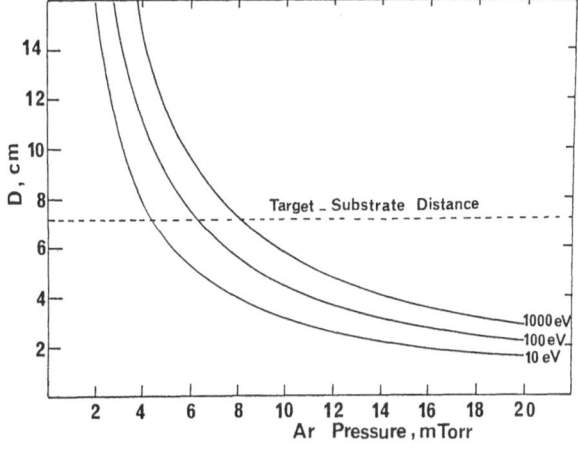

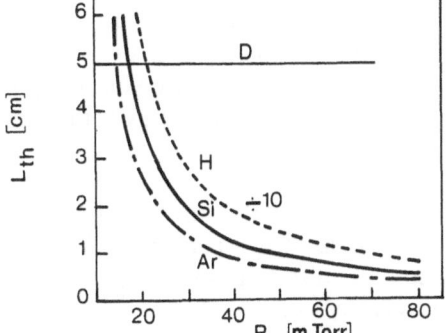

Fig. 4.14. (a) Distance required for thermalization of Si atoms leaving sputtering target with energies of 10, 100 and 1000 eV, plotted as a function of the argon sputtering pressure. After [4.23]; (b) Thermalization distance versus sputtering pressure [4.17c]

where D is the thermalization distance, E_0 is the initial Si energy, E_g is the gas temperature, E is the average fractional energy loss per collision, λ is the mean free path of the sputtered atom and $\cos\langle\theta\rangle$ takes into account the angular distribution of the ejected sputtered Si, they derive the expression for Si in an argon plasma

$$D = Par^{-1}\, 5.7 \ln(E_0/E_g) \ . \tag{4.6}$$

The thermalization distance for Si as a function of pressure is shown in Fig. 4.14a; Fig. 4.14b shows the calculation of Tardy for H, Si and A over a higher pressure range with a A/H ratio of 0.25. These results indicate that under the operating conditions frequently used in an rf sputtering system where the argon partial pressure is 6–10 mtorr, the high energy sputtered Si atoms in the tail of the Maxwell-Boltzman distribution may not be thermalized depending on the target-substrate separation. The higher the energy of the electrons emanating from the target sheath, the lower the cross section for interacting with the species in the plasma. Although the fast electrons do cause some ionization in the plasma, many will pass through unattenuated. The A thermalization distance is less than that of Si so may be less likely to cause significant high energy bombardment of the substrate; however, there are many more A ions than high energy Si ions. Note from Fig. 4.14b that the hydrogen ions are not thermalized within 5–10 cm, the target-substrate distances normally used.

4.1.2 Substrate-Plasma Interactions

The most crucial materials issues in understanding the growth and properties of a-Si : H is the influence of the plasma on the film nucleation and growth. Life on the substrate in an rf plasma is complex in that it can be influenced by numerous phenomena which can be difficult to identify let alone control; this is no less true of silane discharges in which plasma-deposited Si is grown.

At a pressure of 5 mtorr there are 2×10^{18} Ar atoms cm^{-2} striking the substrate which are about four orders larger than the number of Si atoms reaching the substrate. Most of these bombarding A atoms are not incorporated in the film but they can dissipate kinetic and potential energy at the film surface which can influence nucleation and growth. The excited A species arrive at the growing film and can release the excitation energy of 11.72 eV. There are also a number of neutral species reaching the substrate from the target; the energy of these ions can be thermalized by operating at large A partial pressures. The thermalized sputtered Si diffuse towards the substrate where they condense. There is little direct control over the energy of the neutral species arriving at the substrate. However, the energy of the charged

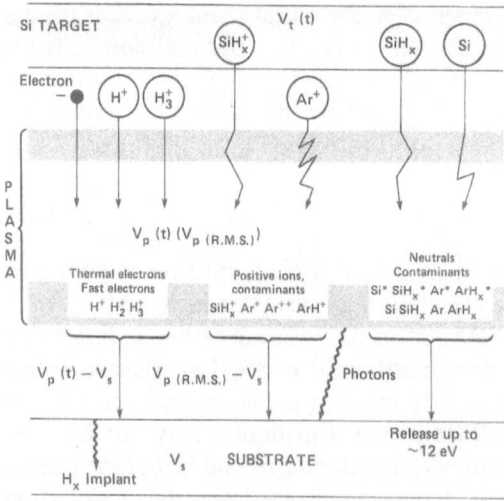

Fig. 4.15. The species from the target and plasma bombarding the substrate

species bombarding the substrate is influenced by the difference in potential between the substrate and plasma. In the simplest case of a dc discharge where the substrates are conducting and grounded, there exists a sheath around the substrate across which the plasma potential is dropped (Fig. 4.1). Thus, positive ions will be accelerated and electrons slowed across the sheath; fast electrons from the target will have sufficient energy to overcome this potential. If the substrates or growing films are insulating, then a negative potential can be induced on the surface due to electron diffusion from the plasma being faster than that of the ions, similar to the wall potential established on any insulating surface in the plasma. In an rf plasma the situation is more complex but can be compared to that at the target. The A ions are influenced by some rms plasma potential whereas the electrons and lighter H ions can follow excursions in the rf voltage and thus will be accelerated across the sheath with a time varying field. As has been previously noted, the potential of the plasma, target and substrate is greatly influenced by the geometry of the systems. The species striking the substrate are summarized in Fig. 4.15. The Si and Ar species from the target have partly or completely thermalized. These ions thermalized from the target and the ions generated in the plasma obtain an energy $e[V_p(\text{rms}) - V_s]$ as they cross the substrate sheath whereas the neutral species will be unaffected by the sheath potential. Some high energy ions and neutrals of H^+, H_2^+ and H_3^+ and electrons that are emitted from the target are relatively unattenuated in the plasma and bombard the substrates; the ions reflected from the target are influenced by the oscillating plasma and sheath voltages. *Brodie* et al. [4.25 a] measured a power dissipation on the substrate from fast electrons of 5.5 kW m^{-2} from an input power of 1.2 kW to a copper target in an A plasma. The electron and positive ion current at the substrate and target have been measured in an rf sputtering system as illustrated in Fig. 4.16. Note that the positive ions at the substrate are relatively low energy compared to the electrons. However, in sputtering a-Si : H most researchers have used very low power densities (1 W cm^{-2}), but still high ac potentials exist on the target

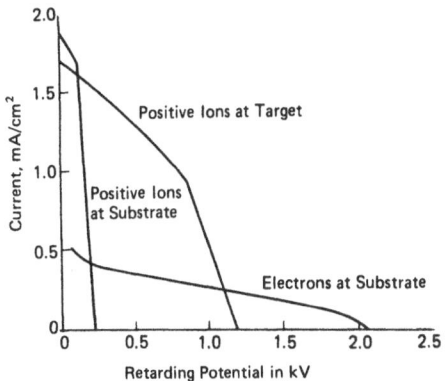

Fig. 4.16. Electron and positive ion currents at substrate and target in an rf sputtering system. Pressure ~ 5 mtorr. Maximum ion energy $\sim$ equivalent dc sheath energy [4.25 b]

(1–2 kV) and thus fast electrons will be created. Radiation damage can be caused by soft x-rays or photons from a plasma [5.26]. No information exists on the effects in sputtered Si, although neutral charge traps have been created in SiO_2.

An important issue is to determine how the energy from the bombarding species is dissipated in the film and how this affects the nucleation and growth. Clearly high energy Si or Ar ions can give effective momentum transfer to the Si lattice to create significant structural rearrangement. Si and H atoms could be sputtered off the surface if the ions are energetic enough, as discussed later, or the impinging species can be implanted into the film. Hydrogen bombardment could result in rearrangement of the H in the lattice or implantation. A commonly held view is that hydrogen ions bombarding the film are included in the growing surface as the Si structure forms. This kinetic model is discussed later in the light of the evidence for deposition of SiH species discussed earlier. The continuity and coordination of the film structure are markedly influenced by the ability of the arriving atoms to diffuse or move around on the substrate. The substrate temperature will affect the atomic mobility but so also can the bombarding species give increased mobility to the Si and H atoms at the growing surface. *Eltoukhy* and *Greene* [4.27] have measured the effect of low energy bombardment on the interdiffusion at InSb/GaSb interfaces. They show a depth-dependent enhanced interdiffusion coefficient

$$D^*(x) = D^*(0)\exp(-x/L_d) , \tag{4.7}$$

where $D^*(0)$ was five orders of magnitude greater than thermal diffusion coefficients. For higher bombardment energies (1000 V), *Strack* [5.28] reported an enhancement of the diffusion of B and P in Si by a factor of 10^5 due to ion bombardment sustained in either $H_2 + B_2H_6$ or $H_2 + Ph_3$ discharge giving diffusion lengths of 0.3 μm for P in Si.

4.1.3 Bias Sputtering

In order to control more directly the energy of the plasma species bombarding the film, a bias voltage can be applied to the substrate. Such control of the substrate bias has a strong influence on the properties of the film. The most common configuration is to use an rf bias to the substrate with the power from a single source split to the target and substrate electrodes (Fig. 4.17) via matching units. Different phase shifts can be introduced in these matching units causing the rf voltage on the substrate and target to be out of phase. This phase biasing can produce various effects on the electron trajectories and bombardment; thus by adjusting the substrate cable length, *Logan* et al. [4.29 a] showed that phase biasing could control and minimize the electron bombardment energy on the substrate. DC biasing can be applied to electri-

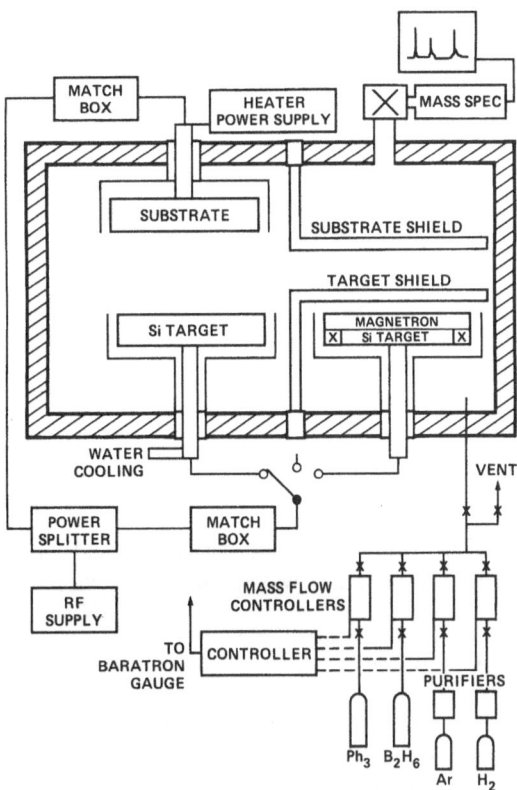

Fig. 4.17. An a-Si:H rf sputtering system with magnetron and planar diode target source as well as bias sputtering capability

cally conducting substrates. Bias sputtering of a-Si:H has been used by the Sheffield and Exxon groups and the results will be discussed later.

4.1.4 Magnetron Sputtering

Two configurations can be used for magnetron sputtering; both have been used to sputter a-Si:H. The planar magnetron sputtering system is like a conventional diode system but permanent magnets are placed behind the target (Fig. 4.18). The other configuration is a cylindrical magnetron sputtering system which has a central permanent magnet cathode with a surrounding cylindrical anode on which the substrates are placed. The main feature of both these systems is that high deposition rates can be obtained with low cathode dc potentials and low electron substrate bombardment. The electrons are trapped in the magnetic field giving high ionization efficiency. DC bias voltages on Si targets have been at least halved using magnetron compared with conventional sputtering and the deposition rate is increased by at least 4. Only 3 laboratories have reported the use of magnetron systems for sputtering a-Si:H, one group of researchers used a cylindrical configuration

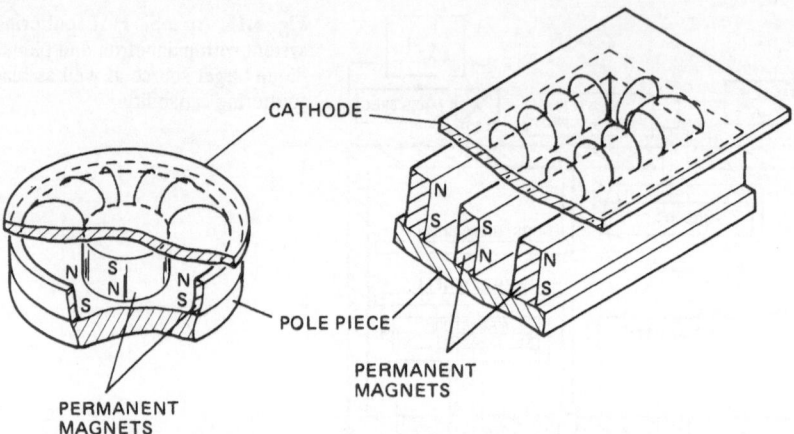

Fig. 4.18. Circular and rectangular planar magnetron sputtering sources. Curved lines represent magnetic field lines [4.29 b]

target while the other groups used a planar configuration. The results indicate that high quality a-Si : H can be produced this way as will be shown later.

4.1.5 Deposition Parameters and Conditions

The main feature of a rf sputtering system is shown in Fig. 4.17. The main parameters that have been adjusted to control Si film growth are temperature, power, gas pressure (P_A, P_H) of A and H, substrate-target separation and substrate bias. The monitoring of the substrate temperature is not trivial; with a large substrate electrode and many substrates it is difficult to obtain uniform substrate heating. *Ross* [4.30] has described the problem and shown how large the errors in temperature determination can be; in addition, the plasma potential was measured using a Langmuir probe. Many researchers have recorded the rf power to the target but more useful parameters are the rf target voltage and current and the dc voltage from which an approximate calculation of the plasma potential may be made. There is no need in theory to pump a sputtering system but because of the desorption of gases from the walls and the relatively low deposition rate, a reasonably good pumping station is required. In many systems the pumping speed is not high so that in order to obtain the appropriate A pressure, the pump must be throttled; however, the disadvantage of doing this is that the contaminants are then less effectively removed from the chamber. The inclusion of a mass spectrometer in the chamber allows residual gas analysis before or even during each run. It is clear that a-Si : H is much more sensitive to impurities than first thought; however, there is little information on the effect of Si target purity and gas purity on the film properties. The target is generally pre-sputtered before each run to remove any contamination on the surface of the Si target, with the substrate being shielded.

4.2 Properties of Sputtered Silicon and Their Relationship to Preparation Conditions

4.2.1 Structure and Morphology

A discussion of this subject brings to mind a round-table meeting of experts at a summer school in Rhode Island a decade ago. There was considerable emphasis on attempting to relate electrical properties to structural properties of evaporated, sputtered and glow discharge Si. The presence and size of voids in films produced by different techniques was highlighted as a cause of the large variation in electronic properties. No mention was made of the inclusion and role of hydrogen in glow discharge material; sputtered and evaporated films were not prepared in a hydrogen environment at that time. These experts and many others had ignored a volume of literature on silicon hydride formation in films prepared in silane plasmas stretching back to 1880. However, it was soon realized by most that the inclusion of hydrogen had a profound effect on the electronic and optical properties of a-Si. Thus, since 1976 there has been a considerable volume of work on the investigation of silicon-hydrogen bonding and quantifying the hydrogen content of a-Si thin films. Following this the structural studies attracted much less attention than the early days; however, more recent studies of homogeneity of the material has been stimulated by the first evidence for columnar growth in glow discharge a-Si:H.

Meanwhile, from a totally different perspective, much interest has been generated in the growth mechanisms and morphology in a wide range of films

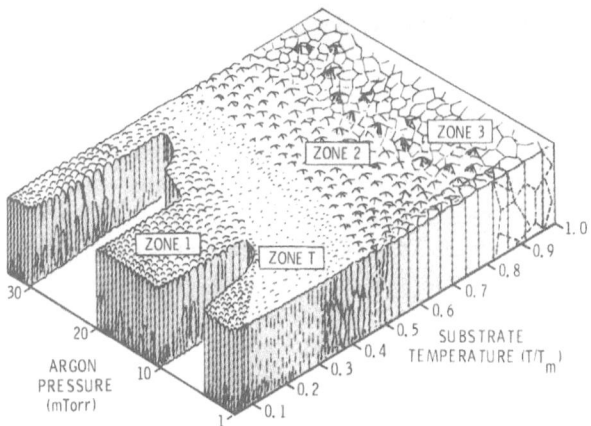

Fig. 4.19. Schematic representation of the influence of substrate temperature and argon pressure on the structure of metal coatings deposited by sputtering using cylindrical magnetron sources. T is the substrate temperature and T_m is the melting point of the coating material in absolute degrees [4.31]

produced by sputtering and more recently magnetron sputtering. In 1974 *Thornton* [4.31] presented the schematic representation in Fig. 4.19 on the dependence of film structure on deposition parameters for magnetron sources. This model is based on the control of adatom mobility over the microstructure. The first atoms arrive at the substrate surface and due to a limited mobility will tend to stick at some low potential point in close proximity to where they arrive. These atoms will preferentially condense on the surface protrusions on the substrate and thus subsequent adatoms, presuming they have sufficient mobility, will migrate to these nucleation sites and initiate growth of the film in 3 dimensions. The film grows until the clusters nucleated at various points touch, forming the boundaries to the columns. The substrate surface roughness and protrusions, along with a relatively low mobility of the adatom species, creates the preferential nucleation and growth which promotes columnar morphology. Despite this evidence it is only recently that columnar growth and structural inhomogeneity have been studied in sputtered a-Si : H. *Ross* and *Messier* [4.32] have provided the most extensive study showing that columnar growth can be eliminated by bombardment of the growing surface.

To return to the issue of hydrogen incorporation in sputtered material, copious evidence exists primarily from ir studies that monohydride and dihydride Si can exist in a-Si : H films. NMR studies appear to indicate that films are inhomogeneous with fluctuations of composition occurring even in films containing only monohydride species.

The emphasis of this section will be to try to relate the structure and morphology of sputtered a-Si : H films to preparation conditions. The aim is to firstly illustrate the correlation of growth conditions and film property but secondly and more significantly, to try to evaluate whether these results can be justifiably interpreted as cause and effect. A more detailed discussion of structural characterization in a-Si : H is given in Chap. 2.

a) Microstructure

It is evident that void, microstructure and microcrystalline formation in sputtered a-Si : H is dependent on the deposition conditions. The most influential deposition parameters are substrate temperature, gas partial pressure and deposition rate. The first two parameters have been related to adatom diffusion on the substrate. However, even when diffusion is significant, if the deposition rate is extremely high, then atomic motion on the substrate is limited as atoms become locked in a fast growing network.

One of the main problems in comparing structural data is that the techniques which have been used for these studies are sensitive to different sizes of microstructure and usually require films to be prepared on special substrates. In addition, it is easy to obtain a null result and detect no microstructure by not performing the experiment correctly. *Ross* and *Messier* [4.32] have shown that at low argon partial pressure, no SEM resolvable micro-

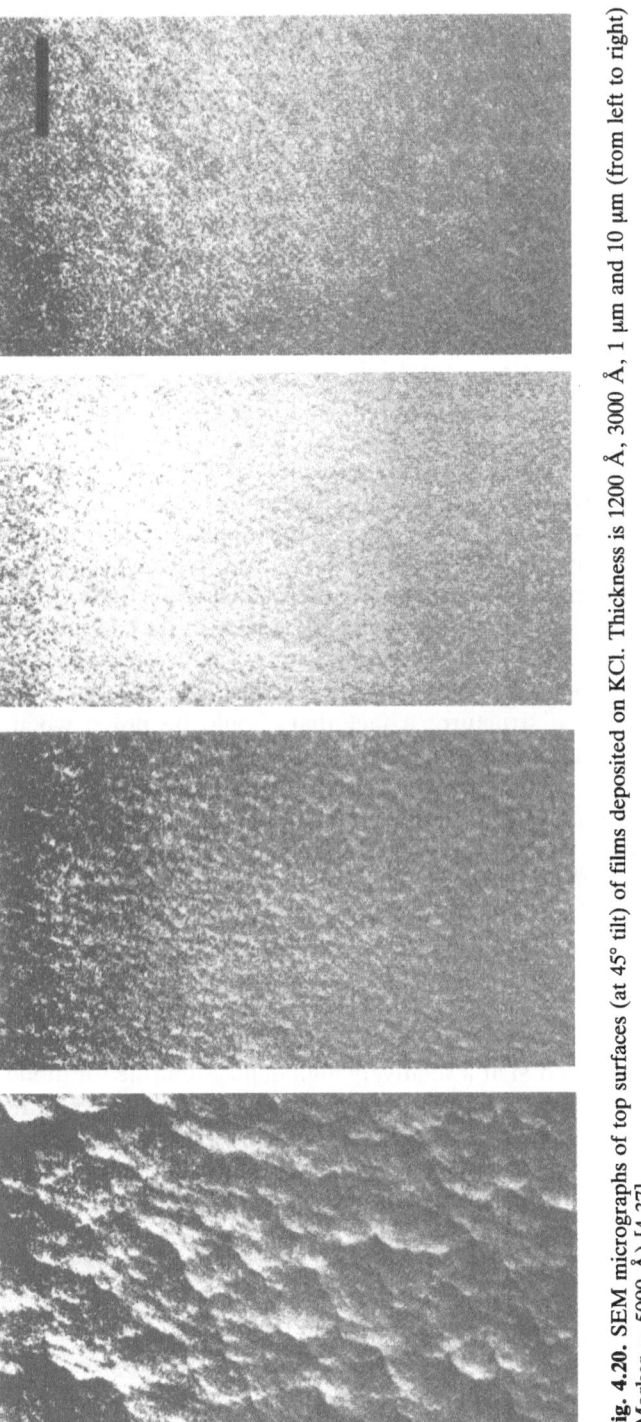

Fig. 4.20. SEM micrographs of top surfaces (at 45° tilt) of films deposited on KCl. Thickness is 1200 Å, 3000 Å, 1 μm and 10 μm (from left to right) (Marker = 5000 Å) [4.37]

structure is present in sputtered a-Si : H films; these films are dense, compressively stressed and high in A content ($\sim 5\%$). Small-angle scattering has shown that no detectable voids exists for films deposited near optimum preparation conditions for device quality material but voids can exist which are rich in hydrogen in material deposited at very high rates using magnetron sputtering. *Bellisent* et al. [4.33] found that small-angle neutron-scattered intensity from sputtered Si was lower than previous workers but here they used a relatively high deposition temperature of 400 °C. These data on small-angle scattering are measured on relatively thick films similar to those used for NMR which will be discussed later. *Messier* and *Ross* [4.34] have shown a link between the small scale nanostructure of 100 Å feature size observed in TEM and the large structural inhomogeneities viewed on fracture surface and top surfaces by SEM. They in fact identified island structures of different dimensions on films of thicknesses 1200 Å, 3000 Å, 1 μm and 10 μm (Fig. 4.20). The islands in the thinnest film were between 50–200 Å across in a uniform honeycomb type structure but as the film thickness increased, the network of voids became nonuniform with a mixture of larger voids and the remnants of the small void network found in the thinnest film. They identified four different superstructures, the smallest ranging from 50–200 Å and the largest 2000–5000 Å. This appears to indicate a continual evolution of void and superstructure as the film thickness increases with the consequence that measurements on thick films are more likely to reveal macroscopic nonuniformity in structure, a fact that should be noted when evaluating NMR and small-angle scattering data where mostly only thick films of ≥ 5 μm have been measured. A further interesting observation made by Ross and Messier is that the microstructure was uninfluenced by the surface roughness of their KCl substrates, indicating a low epitaxy effect. *Shirafuji* et al. [4.35] have reported that microcrystalline sputtered Si of 200 Å grain size has been produced when sputtering under high hydrogen partial pressure. They noted a correlated change in the electrical transport properties giving evidence for nondispersive transport in the microcrystalline phase and dispersive transport in the amorphous phase. Microstructure observations by *Noda* and *Ishida* [4.36] were closely correlated to hydrogen bonding environments but they deposited Si at a relatively high deposition rate for diode sputtering of 500 Å s^{-1}; thus a comparison with other data may prove difficult.

Before moving on to discuss in detail hydrogen incorporation and the possible correlation between void structures and different hydrogen bonding environments, it is essential to study the conditions under which films with columnar morphology are produced. As *Knights* and *Lujan* [4.39] reported that films with columnar morphology also have poor electronic properties, there has been little incentive and some reluctance to reveal detailed structure studies by those who are striving to demonstrate they can produce good quality material. Again we have to rely on the work of Messier and Ross to give us some understanding of how the growth conditions of sputtered a-Si : H are influential on the microstructure. The most critical parameter in

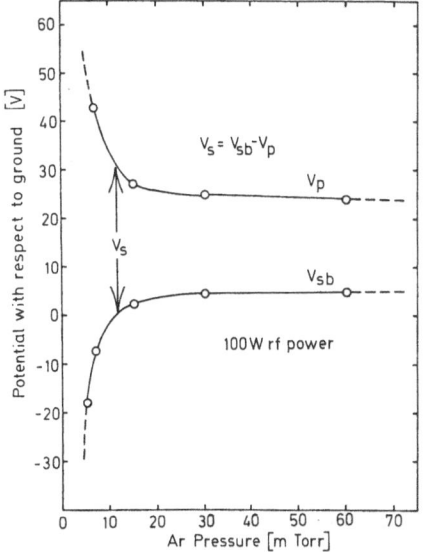

Fig. 4.21. The substrate and plasma potential as a function of argon pressure for 100 W rf power [4.30]

influencing microstructure is argon partial pressure; they found hydrogen partial pressure and deposition temperature (from 0 to 200 °C) had no discernable effect on film microstructure. They attributed the lack of microstructure in films grown at low argon pressure (< 30 mtorr) as due to low energy A^+ ion bombardment of the growing film resulting from the large induced substrate voltage at low pressure. These thermal ions derived from the plasma are accelerated towards the substrate by the potential difference between the plasma (V_p) and the substrate surface (V_{sb}). V_{sb} is a strong function of argon partial pressure becoming increasingly negative at low pressure because of increased electron current to the substrate (Fig. 4.21). The reduction in microstructure when $V_s < -20$ V is related to the threshold sputtering yield of -20 V for Si. Thus, it was concluded that bombardment of the growing surface causes removal of loosely bound species or enhanced surface diffusion of atoms. The films without columnar microstructure are dense with no accompanying post depositional oxidation and are compressively stressed with a high argon content of 5%. In order to investigate their model further, *Messier* and *Ross* [4.37] sputtered Si in Ne and Kr as well as A. These additional data confirmed that the lack of microstructure is related to large substrate bias-plasma potential and thus bombardment effects of the substrate. The evidence they recognized was not totally definitive in inferring that the bombardment was by plasma ions in that at low pressures, because of the decreased mean free path, the bombardment could be from high energy species emitted from the target. The only significant additional information is that with Ne, for which the largest thermalization of Si is predicted, the 2090 cm^{-1} ir mode predominates; they claim a positive correlation between ion bombardment from target species and the presence of 2090 cm^{-1} absorption. These results cannot be compared with glow discharge samples produced in SiH diluted in Ar, Ke, He as the deposition mechanism is quite different.

In conclusion, the evidence that microstructure in sputtered Si can be eliminated by sputtering at low inert gas pressures is in no doubt. The correlation of lack of microstructure due to bombardment is quite convincing. However, the actual effect of the bombardment on the growing surface is less clear. The preferential condensation at protrusions leading to columnar growth can be eliminated by resputtering or by enhanced adatom diffusion. The evidence for the former is that the elimination of microstructure occurs for substrate-plasma potentials < -20 V which corresponds well with the exponential rise of sputtering yield at low energy. However, in plasma research it is well known that Langmuir probe measurements of rf plasma potentials are difficult and not highly accurate. Thus, sticking coefficient measurements are required to give more direct evidence of resputtering. Evidence from sputter deposition of other materials would support that the major effect of bombardment is increased adatom diffusion. The species that are responsible for the bombardment-induced structure change is probably A, as indicated by the large A content of the films. However, the best way to differentiate between ion bombardment from the plasma and species coming from the target is to use a combination of bias and magnetron sputtering techniques at high A pressure; under these conditions all atomic species from the target should be thermalized. *Moustakas* [4.53] has recently used bias sputtering to determine the role of electron bombardment on the Si–H bonding as will be discussed in the next section. Hydrogen does not appear to play a major role in microstructure formation except at high hydrogen pressure > 30 mtorr where microcrystalline Si forms, presumably due to hydrogen etching effects, creating chemical equilibrium as discussed by *Veprek* et al. [4.38] for glow discharge Si. Of course, for much of this study we have had to rely on the excellent study of Messier's group, but as they have not done such comprehensive photoelectronic measurements, it is not possible to know the exact correlation of structure with electronic properties. As small angle x-ray and neutron scattering measurements on material prepared over a wide range of conditions would be valuable, particular attention should be paid to whether there is a material thickness dependence of these results as the microstructure size may be dependent on film thickness as in highly columnar material.

b) Hydrogen Content

It is often assumed that the hydrogen content of a sputtered film can be controlled by simply regulating the hydrogen partial pressure. However, this is an oversimplification in that it has been found that the hydrogen content of the film is influenced by other parameters such as rf power, in addition to substrate temperature and bias. The relationship between the hydrogen content and hydrogen partial pressure is not a simple one. Reports of hydrogen content in sputtered Si up to 40% have been made. The primary techniques which have been used to evaluate hydrogen incorporation are hydrogen evolution, ir absorption, nuclear reaction, SIMS and nuclear magnetic resonance. There is some considerable discussion in the literature as to the accuracy of these various techniques. It is not appropriate that any detailed discussion is made here as these techniques are discussed in Chap. 2 and [Ref. 4.39, Chap. 7]. An intensely debated topic is also the distribution and location of hydrogen in the Si film which has been evaluated by ir absorption and more recently, nuclear magnetic resonance has been used to address this issue. Much of the ir data analysis has been preoccupied with the discussion of the 2090 cm^{-1} absorption attributed by many to SiH$_2$ vibrations. The reason is that in sputtered material this is frequently observed and there has been a longstanding belief from glow discharge a-Si:H studies that material with good photoelectronic properties does not contain SiH$_2$ [4.40a]. The intense arguments of the 2090 cm^{-1} assignments have thus been somewhat loaded, which results, for example, in the publishing of a paper entitled "Inferior electronic properties of rf-sputtered a-Si:H films with only the 2000 cm^{-1} ir absorption band" [4.40b]. Thus, from all these rather emotive discussions can we hope to extract any useful data? The best that can be probably done is to identify any overall trends in the dependence of hydrogen incorporation on deposition conditions and to create an overall picture which at this stage can only be somewhat qualitative. Before reviewing this data it is appropriate to summarize the limitations of the various techniques which are more extensively reviewed elsewhere.

Hydrogen evolution often gives a low estimate of hydrogen content due to hydrogen absorption in the substrate and other parts of the vacuum system. A direct measure of hydrogen partial pressure using mass spectrometry is more desirable than measuring total pressure increase.

IR absorption is a very useful technique for analyzing Si–H configurations. Apart from the arguments about whether the 2090 cm^{-1} absorption can be uniquely assigned to the SiH$_2$ stretching mode, there is evidence that the calculation of total hydrogen content from the integrated intensity under the Si–H stretching mode can be unreliable due to the uncertainty in the value of oscillator strengths.

The nuclear reaction technique appears to be a reliable technique for evaluating hydrogen content. Unfortunately it is not universally used as it is not always available; in addition, *Knights* points out in Chap. 2 that care has to be taken that beam heating does not produce hydrogen diffusion.

SIMS has been used to profile hydrogen in sputtered a-Si:H. It is well known that matrix effects can often provide problems in SIMS analysis. *Ross* et al. [4.41] have found that matrix effects are found when analyzing material incorporating SiH and SiH$_2$ which leads to the necessity of having standards close to the composition of the unknown.

Nuclear magnetic resonance measurements have provided an interesting new perspective on the homogeneity of a-Si:H. Two superimposed NMR lines are present even in material containing only SiH and no dihydride [4.42, 43]. The narrower line has been attributed to distributed monohydride species and the broader line to densely clustered monohydride; these observations have been made on material which contains no visible microstructure. Although the interpretation of the data is still to be scrutinized, the evidence for a two-phase heterogeneity is compelling. As the hydrogen content of the film is increased from 0 to 10%, the H in the broad line increases from one-half that of the narrow line to approximately the same. Further increases in the hydrogen content go into the broad line [4.43]. Unfortunately, a relatively large volume of material is required for NMR analysis, thus generally only thick films have been evaluated.

As shown in Fig. 4.22, it appears that there is good agreement between hydrogen concentration determined by the nuclear reaction technique and ir absorption for [H] < 17%. *Ross* et al. [4.41] have pointed out that anomalies can occur in comparing ir and nuclear reaction data expressed as number of H atoms if the density of the films is unknown. The data in Fig. 4.22 assumes a Si atomic density of 5×10^{22} atoms cm^{-3}; thus at high hydrogen concentrations if the density varies, the nuclear and ir absorption techniques could be in better agreement. The dependence of hydrogen concentration on hydrogen partial pressure for planar diode sputtered films is shown in Fig. 4.23; the results of *Moustakas* et al. [4.44], *Ross* et al. [4.41], *Martin* and *Paulewicz* [4.45] and *Freeman* and *Paul* [4.46] are in reasonable agreement, all using similar deposition conditions. The hydrogen incorporation in the film reduces with increasing power at all argon pressures.

The next issue is whether the wealth of data on ir absorption gives us any substantial information on the location of hydrogen in the Si matrix. The

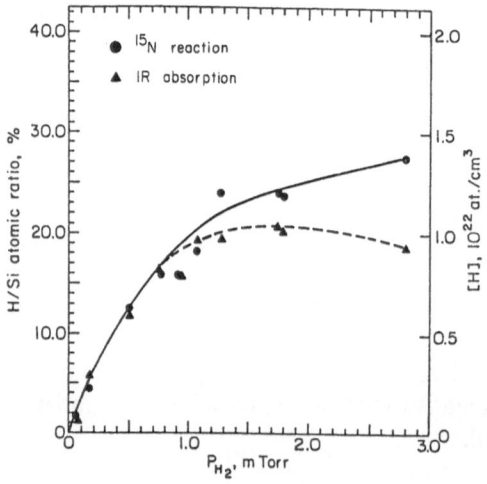

Fig. 4.22. Hydrogen concentration of reactively sputtered a-SiH$_x$ films as a function of hydrogen partial pressure, P_{H_2}. The left-hand y-axis is for ^{15}N nuclear reaction and the right-hand y-axis for ir absorption spectrometry. The two axes agree for the hypothetical Si atomic density of 5×10^{22} atoms cm^{-3} [4.41]

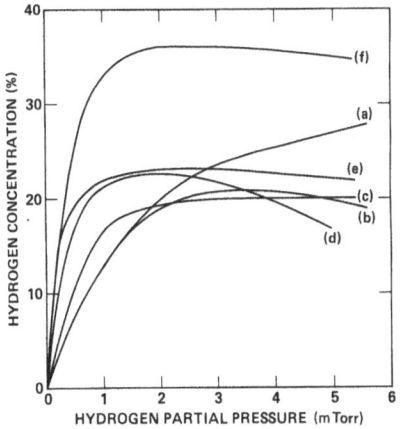

Fig. 4.23. The hydrogen concentration versus hydrogen partial pressure (*a*) ir data, (*b*) nuclear data (after [4.41]), (*c*) ir data [4.44], (*d*) ir data [4.46], (*e, f*) ir data for rf power 13.1 W cm^{-2} and 2.2 W cm^{-2}, respectively [4.45]

detailed assignments of the stretching, bending and wagging vibration of SiH, SiH$_2$ and SiH$_3$ or (SiH$_2$)$_n$ complexes were discussed in the early work of *Brodsky* [4.47]. However, most of the controversy in sputtered materials is the assignment of the 2090 cm^{-1} absorption to an SiH$_2$ stretching mode and the SiH stretching mode at 2000 cm^{-1} is often compared with 2090 cm^{-1} peak (or shoulder). The occurrence of the appropriate SiH$_2$ bending mode is often taken as confirmation for the SiH$_2$ 2090 cm^{-1} peak assignment. The SiH bond type (2000 cm^{-1}) is always dominant at low hydrogen concentrations [4.48, 49] (Fig. 4.24) and at a high deposition rate. The Sheffield group, along with *Jeffrey* et al. [4.50] have noted the increase in SiH/SiH$_2$ ratio with increasing deposition rate or rf power. Jeffrey attributed this effect to increased substrate bombardment at higher rf powers; however, this evidence is not conclusive as at higher rf powers less hydrogen is incorporated in the film. Similar results for magnetron sputtered Si are shown in Fig. 4.25;

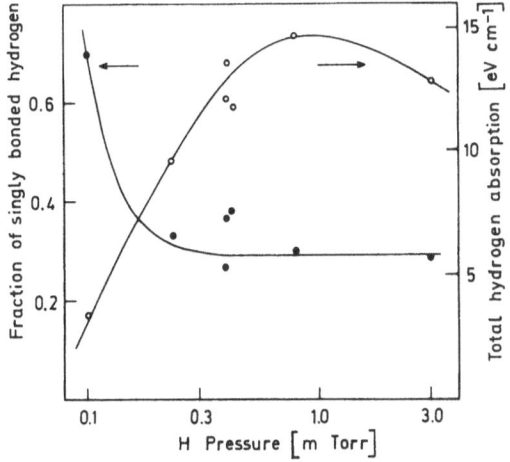

Fig. 4.24. Total area under hydrogen stretch bands (right-hand scale) as a function of hydrogen partial pressure. Ratio of area under 2000 cm^{-1} band to total (left-hand scale)

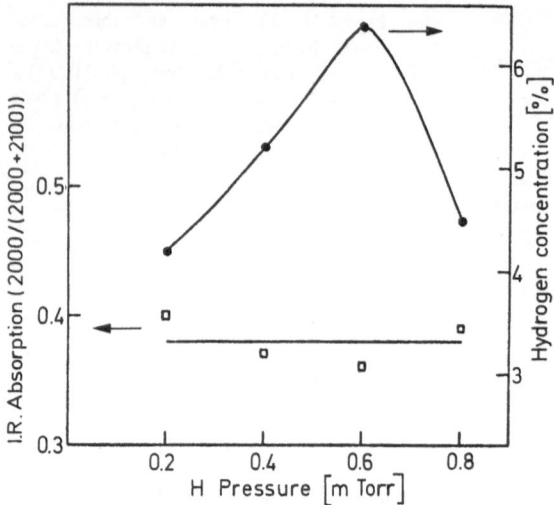

Fig. 4.25. Fraction of singly-bonded hydrogen and hydrogen content versus varying hydrogen partial pressures for magnetron sputtered a-Si:H

here the monohydride content does not vary strongly with P_H but is proportionally higher than in diode sputtered material but the total H content is lower [4.51]. The substrate bombardment effects from target electrons are very small and the ions from the target have lower energy in a magnetron system. The effect of substrate bombardment on hydrogen bonding was demonstrated by *Turner* et al. [4.52] using rf bias sputtering where the SiH concentration is shown to be a strong function of substrate bias (Fig. 4.26). *Moustakas* [4.53] has recently shown that under positive substrate bias conditions which favor electron bombardment, the 2000 cm^{-1} stretching mode is dominant; they have indicated that under negative bias the 2100 cm^{-1} mode is favored. They have also claimed that secondary electrons from the target which bombard the growing film are responsible for the structural

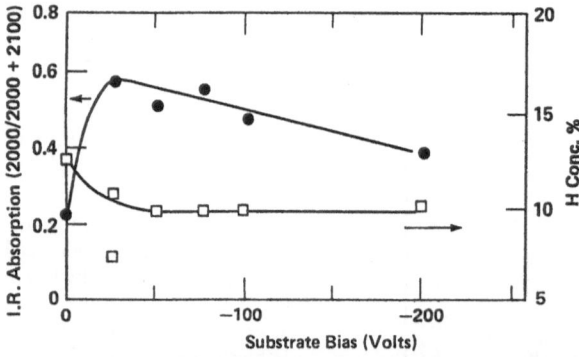

Fig. 4.26. The fraction of singly-bonded hydrogen (2000/2000 + 2100) and hydrogen content for samples grown at various bias voltages and $P_H = 4 \times 10^{-4}$ torr

homogeneity of the film, not the A bombardment as discussed by Messier. However, their latter assertion is more difficult to understand because in magnetron sputtering these electrons are trapped at the target and yet structurally homogeneous magnetron sputtered a-Si:H has been produced. Moustakas made an important observation that the electronic properties of the films were not well correlated with the Si:H bonding but are related to the total amount of bonded hydrogen. This contradicts the Harvard data, recent results on magnetron sputtered Si which contain low hydrogen content, and evidence from glow discharge material. Clearly these results on material produced with positive substrate bias are interesting and warrant further investigation which takes into account the ac potentials present at substrate and plasma, as it is these which control the electron and hydrogen bombardment.

Oguz et al. [4.54] reported changes in oscillator strength of the vibrational modes due to high energy bombardment of 100 and 200 keV He ions. However, although there is no doubt that the sole use of the vibrational modes to calculate hydrogen content and designation is risky, the ir results in the literature on sputtered Si are reasonably consistent providing the P_H and deposition rate dependence are similar. *Shanks* et al. [4.55] have looked at the bonding in sputtered a-Si:H using both ir and NMR studies. They identified distributed and clustered phases of SiH and associated the latter with the stretching mode at 2090 cm^{-1}.

The deposition rate dependence of hydrogen incorporation has led *Moustakas* et al. [4.44] to present a kinetic model similar to the ideas of *Connell* and *Pawlik* [4.56]. It was assumed that H incorporation arises from surface reactions and the amount of hydrogen incorporated in the Si is strongly dependent on that present in the gas at the growing surface. Their model fitted their data well for the range $P_H = 0$–1.5 mtorr that they used. This data would also fit to a model where the SiH$_x$ species from the hydrogenated target were incorporated in the film. Unfortunately, they did not discuss the ir absorption spectra in any detail.

In conclusion, the belief that a-Si:H contains an inhomogeneous distribution of hydrogen is at present popular because of the NMR data. The significance of the appearance of SiH$_2$ is not established although there is sufficient evidence to indicate that H can diffuse to a new site in the lattice following ion or electron bombardment. There is insufficient evidence to indicate deep H implantation in the film but tis may be expected from the plasma analyses. It is not hard to believe that "excess" hydrogen is taken up in a clustered phase giving SiH or SiH$_2$ locally. Does this clustered phase correspond to the point of coalescence of the growing columns? Does the boundary of the column in the strongly columnar material relate to the clustered phase in material where no microstructure is observed? It is tempting to deduce that the cluster sites may act as recombination centers which would affect the transport properties but this remains unproven. Whether SiH$_2$ or SiH sits at these sites may be unimportant but the density of clusters may be the critical parameter. More NMR, ir data needs to be generated on films with and without columnar growth and the thickness dependence of these data is important. In particular with larger sputtering systems now being used, it should be possible to examine whether the NMR data is thickness dependent which would provide information on noncolumnar film growth like the increase in microstructure size with thickness observed by *Ross* et al. [4.32] on their columnar material. Clearly the electron and ion bombardment issue can be resolved by using bias and magnetron sputtering with proper considerations of the rf as well as dc potentials present in the system.

4.2.2 Optical Properties and Defects

a) Optical Absorption

The shift in absorption edge with increasing H content of the film has been widely reported. The first attempt to comprehensively analyze optical data on sputtered a-Si : H was reported by *Freeman* and *Paul* [4.46, 57]. The band gap of the a-Si : H is often deduced from the absorption data plotted as in Fig. 4.27:

$$(\alpha h\nu)^{1/2} = \text{const}(E_0 - h\nu) , \qquad (4.8)$$

where α is the absorption coefficient.

This region corresponds to the absorption coefficients above 10^3 cm^{-1}. The constant contains an average matrix element where the equation is only valid if the bands are both parabolic. The shift in the absorption edge to

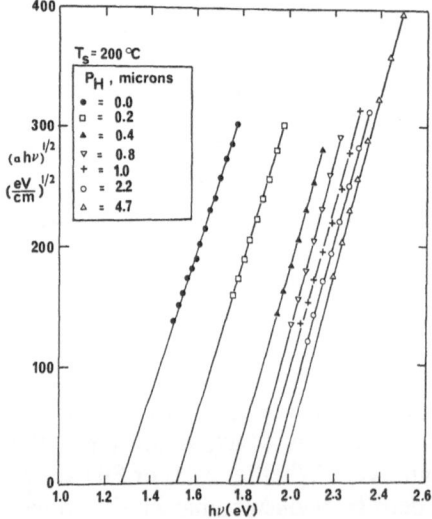

Fig. 4.27. Shift with P_H of linear portion of plot of $(\alpha h\nu)^{1/2}$ as a function of $h\nu$ [4.46]

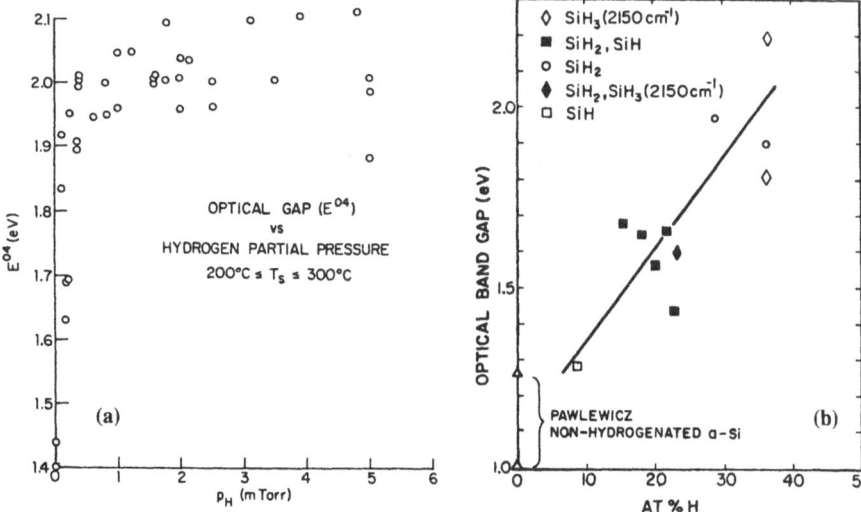

Fig. 4.28. (a) The optical energy gap E^{04} (energy at $\alpha = 10^4$ cm^{-1}) as a function of P_H [4.63]. (b) Influence of hydrogen concentration on optical band gap for dominant Si–H type(s) [4.49]

higher energies with increasing hydrogen content is due to removal of tail states near the band edges. *Freeman* and *Paul* [4.46] have shown that there tends to be a saturation in energy gap at high hydrogen content.. However, there was considerable scatter in data which indicates that samples containing similar hydrogen content had different optical gaps. They have suggested that this is due to the different bonding configurations which give different band gap densities. The sensitivity of the optical gap to preparation conditions is shown by the scatter in the data in Fig. 4.28; note that this is the case even for unhydrogenated material. *Cody* et al. [4.58] have suggested that the disorder is the fundamental determining factor of the optical gap and the H affects the bandgap indirectly through its ability to relieve strain in the network. There is evidence to support Cody's proposal that the substrate temperature is more directly influential on the optical gap than the total hydrogen content [4.59]. At high hydrogen contents $> 20\%$, alloying could produce the increase in band gap observed by some. The data for material with high hydrogen contents and/or prepared in Ar pressures > 20 mtorr should be treated with considerable caution because without structural or micrography measurements, it is possible that the material is inhomogeneous leading to considerable analysis problems. Although the Urbach edge which is exponential between $\alpha \sim 10$ to 10^3 cm^3 is observed in all materials, the slope of the edge can vary around 0.7 eV depending on deposition conditions.

Extrinsic absorption measurements have been made for $\alpha < 10^2$ cm^{-1}. However, the various conventional optical transmission measurements are somewhat inaccurate and unreliable. Photothermal deflection spectroscopy (PDS) measurements which measure absorbed power have clarified the con-

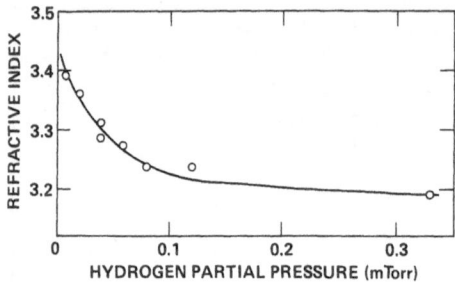

Fig. 4.29. The variation of refractive index with the hydrogen partial pressure in which a-Si:H films were grown

troversy associated with defect-related extrinsic absorption [4.60]. However, PDS measurements have not been made on sputtered material; thus it is concluded that little useful information can be deduced from extrinsic absorption measurements in sputtered material. Photoconductivity measurements have been used to deduce α but in many cases the energy dependence of the generation efficiency has not been measured directly, which invalidates the technique. In addition, internal photoemission effects can interfere with low energy photoconductivity data in Schottky barriers and band bending grossly distorts low energy adsorption in gap cells [4.61].

There is general agreement [4.46] that the index of refraction in sputtered a-Si:H decreases with increasing hydrogen concentration (Fig. 4.29), similar to that observed for a-Ge by *Connell* and *Pawlik* [4.56].

b) Steady State and Transient Luminescence

Luminescence measurements provide an interesting technique for the study of defects. The most extensive measurements on sputtered Si have been made by the Sheffield and Harvard groups; the former have also made a direct comparison of sputtered material with glow discharge samples from the Dundee group.

A single featureless photoluminescence peak is observed at low temperatures in the range of 1.25–1.40 eV with a width of 0.3 eV. A second peak at 0.9 eV is observed in most undoped sputtered and doped material. The first peak is generally agreed due to a transition between tail states while the second peak, it is suggested, is due to a tunnelled electron from the conduction band in a dangling bond recombining with a self-trapped band-tail hole. The detailed interpretation of the photoluminescence data is not totally without controversy; however, this will not be discussed here. We shall attempt to evaluate the sensitivity of the luminescence data to sputtering conditions and how this might reveal defect creation related to growth condition.

The photoluminescence efficiency of rf diode and magnetron sputtered Si for samples grown in different P_H is shown in Fig. 4.30. The Sheffield diode sputtered films exhibited a saturation in the photoluminescence intensity with P_H [4.62], whereas the Harvard samples [4.63] showed a reduction in efficiency at high P_H like the Sheffied magnetron sputtered samples. This

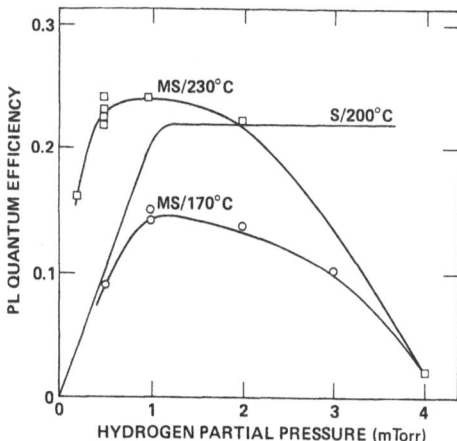

Fig. 4.30. The photoluminescence quantum efficiency of diode sputtered (*S*) and magnetron sputtered (*MS*) Si prepared at various substrate temperatures as a function of hydrogen partial pressure [4.59, 62]

implies that nonradiative recombination occured because the combined strength of the main and defect luminescent band is lower at high P_H for the magnetron samples. *Paul* and *Anderson* [4.63] have suggested that the decrease in photoluminescence at high P_H may be due to microstructure. In fact, some recent data from the Sheffield group on photoluminescence in microcrystalline Si is very revealing in this context [4.64, 65]. *Anderson* [4.66] has reported that the defect band width (fwhm) is 0.2 eV in the Harvard material, whereas in the Sheffield planar and magnetron material and in glow discharge Si, the fwhm is close to 0.3 eV. Figure 4.31 gives the width of the 0.9 eV defect band as a function of temperature for various Si samples and shows that in the a-Si : H produced by glow discharge and sputtering, the width of the defect band remains virtually constant with temperature up to 100 K. The microcrystalline Si shows a large variation in band width up to 200 K. A significant result is that the Harvard sputtered sample (point H) sits on the microcrystalline curve, indicating that the sputtered Si sample probably contained a significant volume of microcrystalline material. It is also interesting to note the presence of the 0.9 eV band in microcrystalline material implies the defect is common to the amorphous and crystalline phase with a photoluminescence line width which is sensitive to the degree of disorder. The evidence from ODMR confirms that the defect level responsible for the 0.9 eV is related to a dangling bond and is present in all material [4.67, 68]. *Rhodes* et al. [4.59] have shown there is a correlation between the strength of the 0.9 eV luminescence with film deposition rate as shown in Fig. 4.32. At high deposition rates the strength of the 0.9 eV band is high and the low temperature ϱ_L efficiency is low. As the deposition rate decreases at higher P_H (1–2 mtorr), the ϱ_L efficiency increases and the relative strength of the 0.9 eV band decreases. The Harvard data shows an increase in the 0.9 eV with P_H, but this different result is probably due to their defect band (which in fact is at 0.95 eV) being associated with a microcrystalline phase. Figure 4.33 shows a comparison of the luminescence spectra from diode

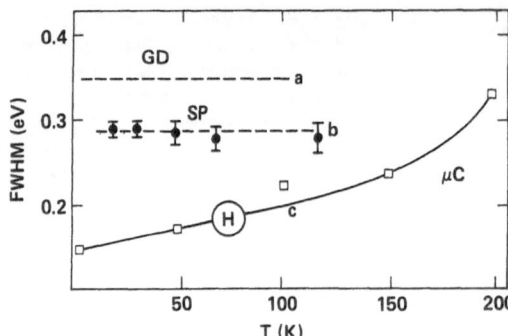

Fig. 4.31. Width of the 0.9 eV peak. (*a*) gd a-Si (Xerox, Dundee, Marburg); (*b*) Sputtered a-Si (Sheffield); (*c*) Microcrystalline Si (Dundee). H on the lower curve corresponds to a sample sputtered in a hydrogen partial pressure

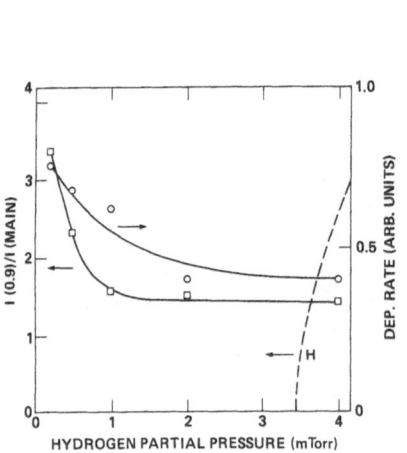

Fig. 4.32. The ratio of the intensity of the 0.9 eV defect band transition to the main band transition and the deposition rate at a function of hydrogen partial pressure P_H

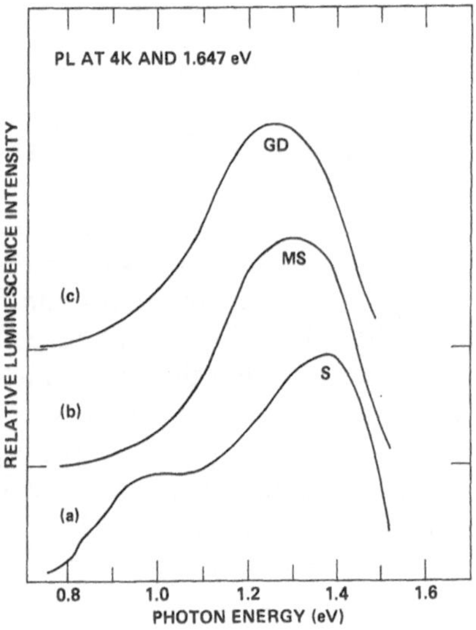

Fig. 4.33. Photoluminescence spectra for Sheffield diode sputtered (*S*) and magnetron sputtered (*MS*) samples and Dundee glow discharge a-Si : H (gd)

sputtered and magnetron Si from Sheffield and Dundee glow discharge Si under low energy excitation as subband gap excitation enhances the 0.9 eV band. The magnetron sputtered and glow discharge Si show very similar spectra but the planar sputtered Si has a relatively large 0.9 eV band even though the higher energy band luminescence efficiency is the same as the glow discharge and magnetron sputtered Si. The magnetron sputtered Si contains around 4–8% H which is lower than the planar sputtered optimum samples. *Bhat* et al. [4.65] have suggested that the lack of a defect band in the magnetron sample is due to the reduction in low energy bombardment from ions in magnetron sputtering compared with diode sputtering. To support their suggestion they note that *Street* et al. [4.67] have shown that the defect band luminescence can be enhanced by electron and ion bombardment. They concluded that as the magnetron samples have an integrated photoluminescence efficiency similar to the best glow discharge films, the defect density is comparably low.

The large shift in the main photoluminescence band in diode sputtered material with increasing P_H is correlated with a shift in the band edge with P_H. However, in magnetron sputtered Si the photoluminescence peak wavelength and the absorption edge do not change significantly up to 2 mtorr with increasing P_H whereas the hydrogen content does (Fig. 4.25). The infrared data for these films shown in Fig. 4.34 suggests that the 890 cm^{-1} and the 2090 cm^{-1} originate from the same SiH configuration and that the ratio of $(SiH_2)_n/SiH_2$ increases with P_H. The photoluminescence peak energy shifts to lower energy as the deposition temperature decreases and the hydrogen content increases with decreasing temperature, but the fraction of singly-bonded hydrogen remains constant up to 4 mT. This data is consistent with the deduction that the major influence on the position of the band edge and the luminescence peak is the deposition temperature which influences the structural order and the distribution of band-tail states [4.58a]. As the deposition temperature is increased, the structure will be more ordered while the existence of large amounts of $(SiH_2)_n$ appears to correlate with a reduction in rigidity of the lattice towards a more inhomogeneous material. The photoluminescence is a maximum for samples prepared at 200° and 230 °C; even though those prepared at 230 °C contain only 4% hydrogen, this must be sufficient to compensate for a large fraction of the dangling bonds.

From the transient photoluminescent decay measurements shown in Fig. 4.35, it can be seen that sputtered and glow discharge material have similar recombination kinetics. A fast, roughly exponential decay with $\tau \approx 15$ ns is followed by a nonexponential decay extending to ~ 1 ms. The decay curve of the sputtered material with the highest photoluminescence

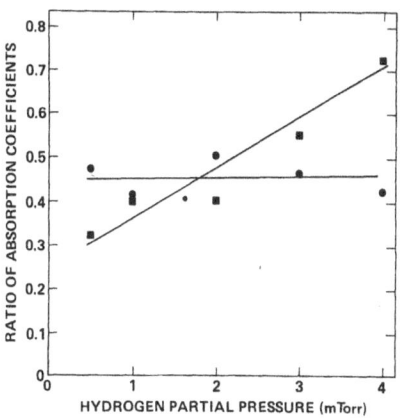

Fig. 4.34. The ratios of the absorption constants $\alpha_{890}/\alpha_{2090}$ (●) and $\alpha_{845}/\alpha_{890}$ (■) as a function of hydrogen partial pressure

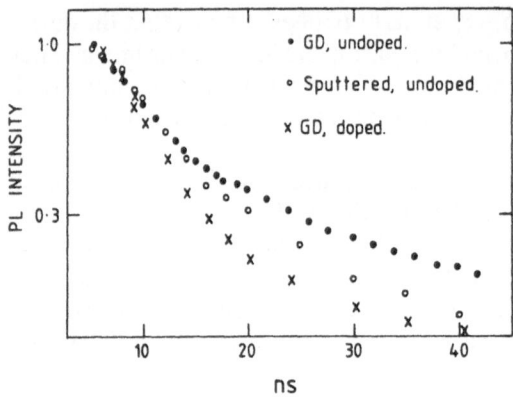

Fig. 4.35. The photoluminescence decay at 1.44 eV for glow
discharge (gd) and sputtered
samples [4.58 b]

efficiency is very close to that for high efficiency glow discharge films. The
time resolved shift down in energy of the luminescence band, it is argued, is
no direct manifestation of the width of the conduction band. Both glow
discharge and sputtered Si show a similar shift in the luminescent band. The
ODMR studies reveal similar transitions for the Dundee glow discharge and
Sheffield diode sputtered material but the strength of each transition is different in the two types of material. Unfortunately there are no decay measurements available on the magnetron sputtered material. The ODMR data confirms the dangling bond contribution to the nonradiative recombination.
Depinna et al. [4.68] have reported the spectral dependence of the ODMR in
which they show the resonances that are present. The interpretation of the
recombination mechanisms deduced by ODMR is not without controversy.
However, it is not clear whether some of the problems are not due to the
semantics of the definitions for such things as geminate and distant pairs.

As an analytical tool, luminescence provides some useful information
which can, with optical absorption and hydrogen measurements, be used to
correlate material properties with preparation conditions. However, the
transient measurements have been most valuable in identifying mechanisms
and they show different features; depending on film deposition conditions it
is very difficult to deduce systematic trends which indicate such parameters as
the width variation of band tails between different samples.

4.2.3 Gap State Density

To date there is no technique which gives a reliable comprehensive description of the density of states of a-Si : H; as will be shown, this measurement is
most difficult in undoped material. This section will review briefly the techniques for evaluation of gap state density and the corresponding data published on sputtered material.

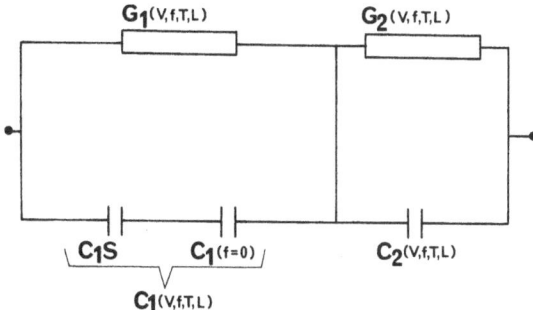

Fig. 4.36. Equivalent circuit of the complete Schottky barrier device. G_2 and C_2 are the conductance and the capacitance of the bulk, G_1 and C_1 are the conductance and the capacitance of the depletion region, where C_1 is the sum of the capacitance due to electrons in energy levels that are able to respond to the signal frequency [$C_1(f = 0)$] and the geometrical capacitance of the part that behaves as a dielectric within the depletion region [C_{1s}]

a) Capacitance Measurements

The standard formulae for junction capacitance of metal-crystalline Si barriers cannot be used for amorphous semiconductors due to the relatively large density of localized states which contribute to the space charge in the depletion region. Using a measured or assumed density of states [4.69–72], the capacitance variation with voltage and frequency and the barrier profile can be calculated; however, the interpretation of the density of states measurements is not unique. The most popular approach is to use an equivalent circuit similar to that in Fig. 4.36. Often the capacitance and resistance of the injecting contact is ignored which is appropriate in some situations but may not be when very low defect a-Si is used (Sect. 4.2.4 a) or the sample is strongly forward biased. The capacitance and conductance terms of the depletion region are a fnction of a number of variables. Poisson's equation can be written:

$$d(d\psi/dx)^2 = - (2/\varepsilon)\varrho(\psi)d\psi , \tag{4.9}$$

where ψ is the band bending function, ε is the permittivity and $\varrho(\psi)$ is the charge density at a point $\psi(x)$. Integrating the capacitance per unit area for a band bending ψ_s can be written as

$$C(V, f, T, L) = \frac{\varepsilon^{1/2} p\{\psi_s'(V, f, T, L)\}}{\left[2 \int\limits_{0}^{\psi_s(V, F, T, L)} p[\psi(V, f, T, L)d\psi(V, f, T, L)] \right]^{1/2}} , \tag{4.10}$$

where $\psi_s'(V, f, T, L)$ is primed to account for the fact that ψ_s is not necessary at the surface but occurs at the limit where the charge can contribute to C.

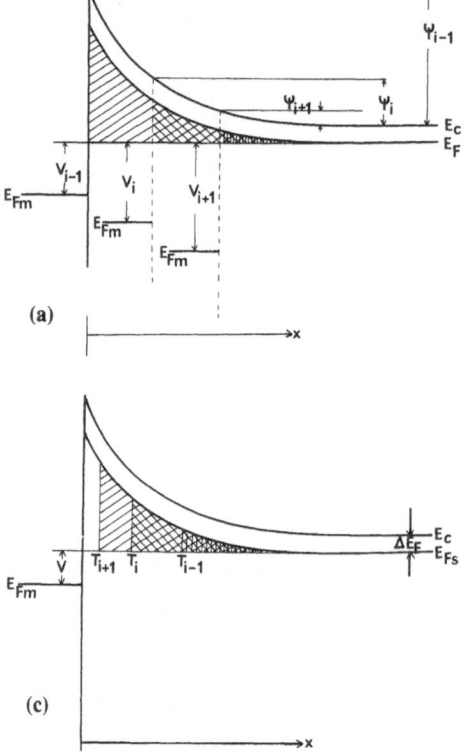

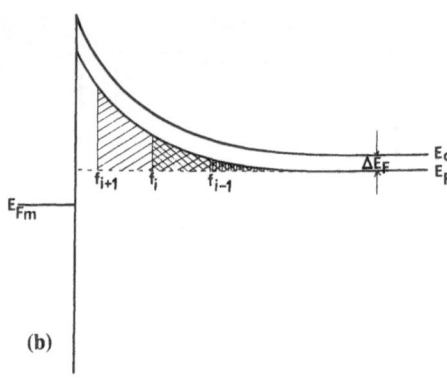

(a)

(b)

(c)

Fig. 4.37. (a) Schematic diagram of energy bands in the barrier region at different applied voltages. E_{fm} is the fermi level in the metal. **(b)** Regions of energy and space occupied by states that are able to respond to an applied signal at three different frequencies $f_{i-1} > f_i > f_{i+1}$. **(c)** Regions of energy and space occupied by states that have been pulled above E_f and are able to respond for three different temperatures

The variables are voltage V, frequency f, temperature T and illumination L. Expressions for ψ_s' can be obtained under conditions where some variables are kept constant. The voltage and frequency dependence have been studied mostly. The three cases that can be solved are illustrated in Fig. 4.37, with the first being the zero-frequency unilluminated low temperature measurement. Here all the charge in the depletion region contributes to the capacitance, the electron energy diagram for three different voltages is shown and ψ_s is a function of V only:

$$\psi_s(v) = eV_D - eV \tag{4.11}$$

where V_D is the flat band voltage. Assuming zero temperature the density of states can be evaluated in a number of ways to render the expression [4.73, 74]

$$N(E_{F_s}) = 1/\varepsilon e \left[C_s^2 + \int_0^{\psi_s} C(\psi)d\psi \; dC_2/d\psi_2 \right]. \tag{4.12}$$

The barrier profile and depletion width can be found from the expression

$$X_p = X(\psi_p) = \int_{\psi_p}^{\psi_s} \frac{d\psi_1}{1/\varepsilon \int_0^{\psi_L} C(\psi)d\psi} \, . \qquad (4.13)$$

The zero-frequency case has to be approximated by using a very low frequency. The low frequency capacitance voltage relationship for a Pt/sputtered a-Si : H Schottky barrier shown in Fig. 4.38 was derived experimentally by integrating over 1000 s the current due to a potential step for reverse bias, and a lock-in technique at 0.2 Hz for forward bias. The density of states derived from this diode is shown in Fig. 4.39 with the main source of error in this calculation being the uncertainty in the value of V_D, the built-in potential. This can be obtained in a number of ways from far forward IV characteristics, the difference between barrier height and activation energy and the $1/C^2$-V curve extrapolated to zero for various frequencies. All of these techniques for evaluating V_D have a number of problems, especially for very low defect-density material where conductivity measurements are uncertain, as will be discussed in Sect. 4.3.4 a. For the data in Fig. 4.39, the error in calculating the density of states is $< 50\%$; note this is a sample which does not have a particularly low defect density. *Viktorovitch* [4.75] found a density of states ranging from $> 10^{18}$ to 10^{15} eV^{-1} cm^{-3} depending on the sputtering conditions. *Tiedje* et al. [4.72, 76] measured density of states down to 10^{15} cm^{-3} eV^{-1} for samples prepared at $P_H = 1.5$ mtorr and showed an inverse linear relationship between the log density of states versus P_H. However, the assumptions regarding the potential distribution and built-in voltage may be

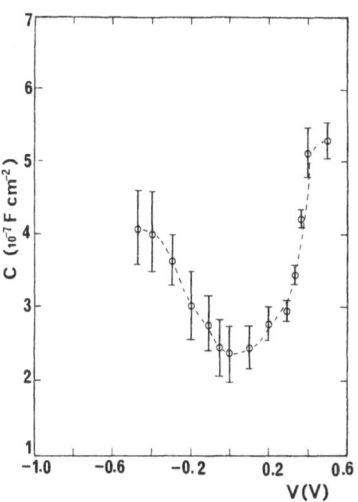

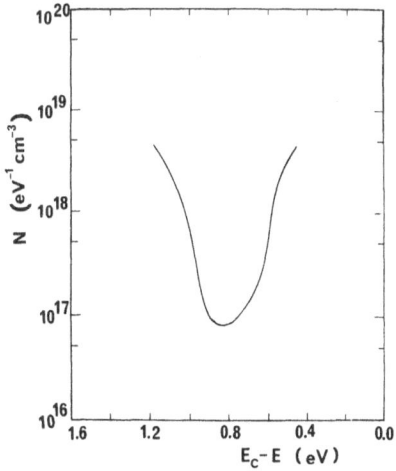

Fig. 4.38. The capacitance voltage curve for a Pt–a-Si : H Schottky barrier measured at a low frequency

Fig. 4.39. The density of states distribution $N(E)$ in a-Si : H; $(E_c - E)$ is the energy below the conduction band

incorrect along with the assumption that the density of states between E_F and $E_F - V_0$ is the same for all samples. All the calculations of potential distribution show a sharp drop in potential near the junction and a long tail. Many of the analyses techniques, in addition, assume that the quasi Fermi level is flat throughout the depletion region which will not be true under biased conditions if diffusion limited transport is dominating; that is why Viktorovitch used zero applied voltage data. It is clear that for samples having a wide range of defect density, the dominant transport mechanism in the samples may be different.

The second case which has been studied is the frequency dependence of the capacitance under constant voltage, low temperature and unilluminated conditions. Figure 4.36b represents the positive charge due to band bending that can respond at a determined frequency. The response time for an electron in a localized state is taken as the average thermal release time into the conduction band

$$\tau = \tau_0 \exp[\Delta E_F + e\psi(x)] \; ; \tag{4.14}$$

then the limiting condition for a particular state to contribute to the depletion capacitance is $f\tau = 1$.

The dielectric capacitance C_{1s} in the depletion region is given by

$$C_{1s} = X(\psi_{sf})/\varepsilon = \int_{\psi_{sf}}^{\psi_s(X=0)} \frac{d\psi_1}{\int_0^{\psi_1} C(\psi)d\psi} \; . \tag{4.15}$$

The conduction and capacitance variation with frequency is shown in Fig. 4.40; the experimental results (data points) are in good agreement with the theory (solid line). The third case is where the temperature is varied and the frequency and voltage remain fixed. Figure 4.36c shows the positive charge introduced due to band bending that is able to contribute to C at three different temperatures. Figure 4.41 shows that as T is increased, the deeper states respond faster and will eventually contribute to the depletion capacitance; therefore the capacitance at this frequency increases.

In conclusion, the capacitance variation of Schottky barriers with a number of parameters have been measured and modeled showing good agreement between theory and experiments. However, despite some interesting overall trends which indicate that the minimum density of states decreases with P_H, there remains such a significant number of assumptions and errors in calculating certain parameters that capacitance data cannot be relied upon by itself to provide useful data to determine the correlation between density of states and barrier profile information with general material properties and preparation conditions. Perhaps the most dangerous practice that has been adopted in the past is to attempt to apply a particular

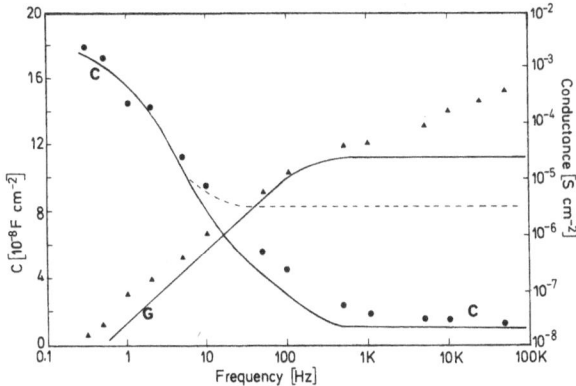

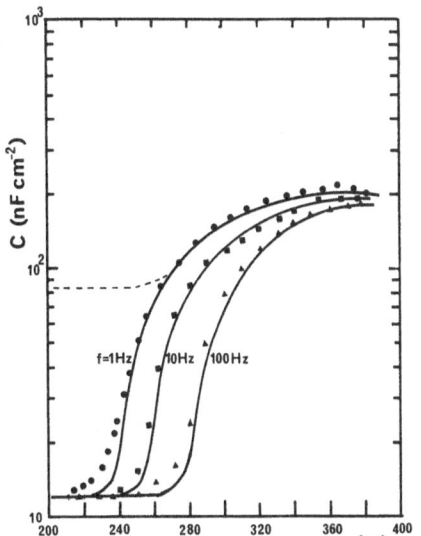

▲
Fig. 4.40. Calculated capacitance and conductance as a function of frequency with no applied voltage, at room temperature, for a metal a-Si:H device. The points are the experimental data and the dashed line indicates the calculated capacitance of the barrier

◄ **Fig. 4.41.** Calculated capacitance as a function of temperature for various frequencies for a metal/a-Si:H device. The points are the experimental data measured in the dark at zero-applied voltage and the dashed line indicates the calculated capacitance of the barrier

capacitance theory with its accompanying assumptions to a-Si:H materials which have a wide range of properties. It is probable that various mechanisms and analysis problems will be appropriate only for a narrow range of material. One advantage of capacitance measurements is that surface state effects can be separated from bulk effects; however, in thin films of 0.5–1 μm which have been commonly used in the past there may be significant nonuniformities and gradation of film properties. Measurements on much thicker films are also required because now much larger depletion widths can be achieved and in thin samples the depletion region will extend to the n^+ ohmic contact.

b) Field Effect Measurements

This technique pioneered by *Spear* and *LeComber* [4.77] was the earliest method used to measure the density of states in a-Si : H. It has been much scrutinized over the years, various new procedures [4.78, 79] for data analysis have been used, but the most controversy associated with measurement is concerned with differentiating surface and bulk properties. *Goodman* and *Fritzsche* [4.80, 81] demonstrated that there is no unique solution to the field effect data and the detailed differences in derived density of states should be treated with caution, particularly near the density of states minimum. *Weisfield* and *Anderson* [4.79], in their field-effect analysis, addressed the issue of determining the flat band voltage. The density of states derived from the field effect data for a sputtered sample is shown in Fig. 4.42 which gives a value of $N(E_F) = 10^{17}$ cm^{-3} eV^{-1} which is in satisfactory agreement with the bulk density 6×10^{16} cm^{-3} eV^{-1} deduced from analysis of capacitance measurements. *Weisfield* et al. [4.82] deposited an SiO$_x$ layer at the interface of an a-Si : H-insulator interface and reduced the surface states; this was an excellent illustration of the problem with field effect measurements.

In conclusion, because of the problem of separating surface states from bulk states, field effect measurements only reveal a maximum for the density of bulk states.

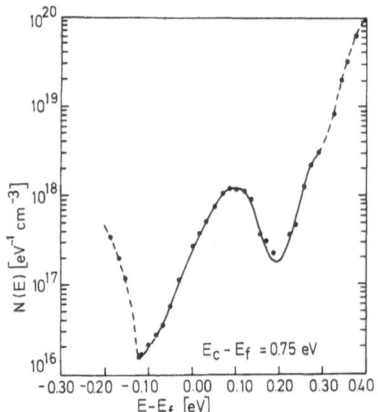

Fig. 4.42. The density of states derived from field-effect data [4.79]

c) Deep-Level Transient Spectroscopy (DLTS)

DLTS is an extremely powerful technique [4.83] which has been used extensively to analyze defects in crystalline semiconductors. Both current and capacitance DLTS have been used to study a-Si : H. The current or capacitance transients are measured often as a function of temperature following a trap filling event in a Schottky barrier or *p-n* junction. Optical excitation or injection forward-biasing is used to fill traps at low temperatures and the current or capacitance transient due to trap emptying is measured at fixed time intervals after the diode is returned to a reverse bias condition. The temperature is scanned and the difference in current at times t_1 and t_2 of the transient are recorded and plotted against temperature revealing peaks in the spectrum. The energy depth and concentration of traps can then be calculated but an important assumption made in DLTS is that the motion of the charge is emission rate limited. Thus, any long retrapping event (long compared with emission time) would lead to an underestimation of the trap density. The application of DLTS to materials containing a continuous distribution of states is very difficult. *Crandall* [4.84 a] was the first to apply the technique to a-Si : H using current transients and he revealed electron traps due to air contamination in samples were associated with the Staebler-Wronski effect. Density of states measurements have been evaluated by *Cohen* et al. [4.84 b] using numerical analysis of capacitance data. However, capacitance DLTS cannot be made on low conductivity undoped material, thus their studies are restricted to doped material where the detailed density of states will probably be different from undoped material. In addition, in order for the detailed spectrum to reflect accurately a density of states, the signal should be saturated over the entire spectrum in order to ensure the traps are completely full. This is sometimes not easily observed at the minima of the DLTS signal, which corresponds to the much quoted minimum in the density of states. In addition, a detailed understanding of the Schottky or junction devices is necessary in order for the DLTS to provide useful information. It is clear that in the future this technique may provide some very useful data; however, no useful systematic data on sputtered a-Si : H has emerged.

d) Electron-Spin Resonance (ESR)

ESR is a most reliable method [4.85] for revealing defect densities in a-Si : H but it only gives a lower limit on density of states per unit volume near the Fermi level by measuring the total number of unpaired spins. The reduction of spin density due to the inclusion of hydrogen in sputtered a-Ge was observed by *Lewis* et al. [4.86 a] and *Connell* and *Pawlik* [4.56]. Several groups have measured spin densities of 10^{16} cm^{-3} in sputtered Si, first reported by *Pawlick* and *Paul* [4.86 b]. However, little data has been published recently on relating spin density to preparation conditions in sputtered a-Si : H. One disadvantage of the technique is that the most accurate data can only be obtained on thick films (> 2 μm).

4.2.4 Electrical Transport

Conductivity measurements on a-Si : H have been extensively reported in the literature as these measurements are relatively easy to perform; however, there are a number of problems with the interpretation of the data. An irony is that as material issues are solved and the "quality" of the a-Si : H improves,

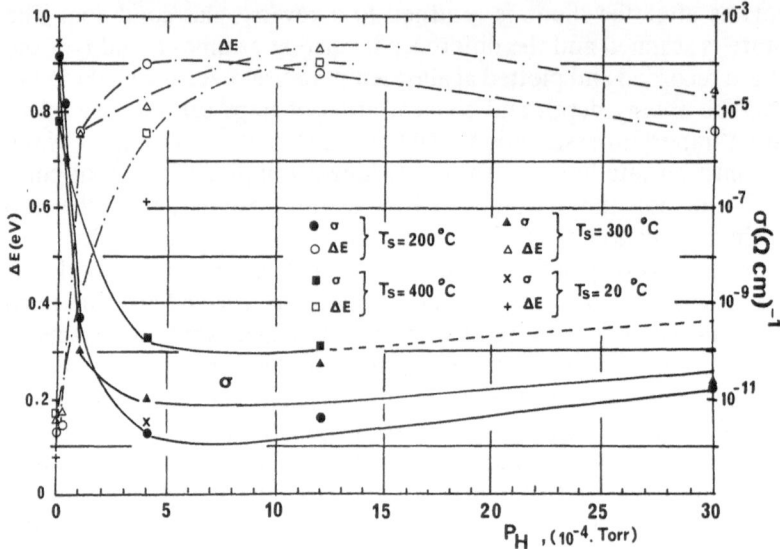

Fig. 4.43. The room temperature conductivity and activation energy as a function of hydrogen partial pressure and substrate temperature

the interpretation of transport phenomenon becomes more difficult, as will be discussed later. A further complication discussed by *Paul* and *Anderson* [4.87] is that of inhomogeneous material (i.e., samples containing columnar morphology) where the transport path through the solid is unknown.

a) Dark Conductivity and Activation Energy

The conductivity and activation energy shows marked changes when the hydrogen content of the film is increased, as shown for Sheffield material in Fig. 4.43; this was first demonstrated by the Harvard group. As the hydrogen is added to the material, the dominant transport mechanism changes from localized hopping in the band gap to extended state transport, the hydrogen compensating for deep defect states. The conductivity-temperature charac- teristics for the Harvard and Sheffield material is shown in Figs. 4.44, 45. Though there is agreement in the general trends on the influence of prepara- tion conditions on film conductivity for samples from many laboratories, there is considerable scatter in the data. A large distribution of results can be obtained for samples prepared in the same sputtering system under what appeared to be the same conditions. It is clear that the transport measure- ments are very sensitive to various material parameters, some of which are difficult to identify and control. It can be seen from Fig. 4.45 that two distinct regions in the activation energy plots exist for samples prepared below or above an optimum partial pressure of 4×10^{-4} torr. At this "optimum" there only appears to be extended state conduction over the entire temperature

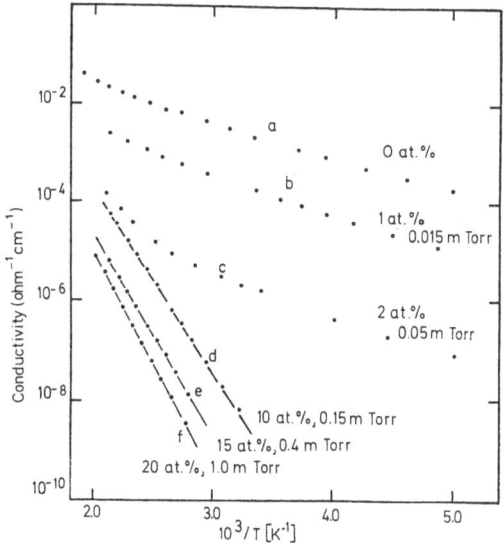

range of the measurements. In samples prepared at $P_H > 1 \times 10^{-4}$ torr the conductivity can be described as

$$\sigma = \sigma_{01}\exp[-(E_C - E_F/kT)] + \sigma_{02}\exp[-(E - E_F + W/kT)] , \qquad (4.16)$$

where the first term dominates at temperatures above 300 K with $(E_C - E_F)$ between $(0.76–0.90)$ eV and the second term at lower pressures with $(E - E_F + \Delta W)$ between $(0.2–0.2$ eV$)$. This therefore implies that even above the optimum hydrogen pressure where there is $>10\%$ hydrogen, localized conduction can be observed at low temperature and thus the inference is that in some way the "extra" hydrogen is creating a defect hopping conduction path. As we have seen, the material contains more SiH_2 complexes when prepared at higher P_H. *Anderson* and *Paul* [4.88] have shown that at high P_H $(>3$ mtorr$)$, the activation energy for the conductivity between 200 K and 400 K is singly activated with a low activation energy. Their explanation for these variations in temperature-dependent conductivity characteristics with P_H was made on the basis of a model of transport through mixed phases of material with different properties. The difference in the properties of the islands and tissues is that the latter is a highly defective state with a large H content. In addition, they went further and explained anomalous σ_0 values and the high temperature kink in the activation energy plot, often observed in their material prepared in high P_H, in terms of a change in the dominant conduction path from the island to the tissue. *Messier's* [4.34] observation of microstructure in films prepared at high P_H provides support for their model. Further evidence is provided from their own photoluminescence data which they did not relate to in this context. As

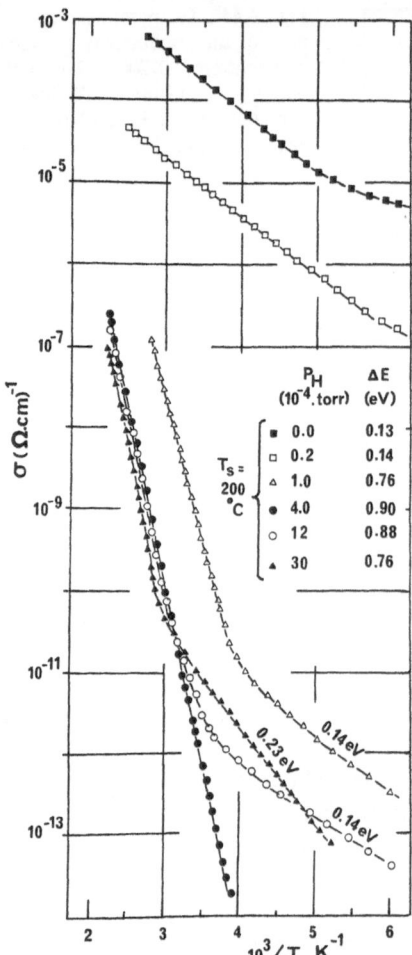

Fig. 4.45. Conductivity temperature dependence for sample deposited at 200 °C in a range of P_H

shown in Sect. 4.2.2c, the narrow defect band is similar to that obtained in microcrystalline material. Thus, the high P_H films which exhibit high conductivity and low activation energy probably contain a microcrystalline phase which dominates the transport. However, it is difficult to see how the existence of this microcrystalline phase is responsible for the kinks in conductivity and thermopower activation energy plots obtained in material with lower hydrogen content. However, *Anderson* et al. and *Paul* [4.88] claim a good agreement between their model of an inhomogeneous material and the conductivity data. The existence of the two phases of H from the NMR and the ir data indicates the possibility of heterogeneous material even in the lowest defect samples; however, the link between the two phases and parallel conduction paths in all material is not proven. However, the presence of microstructure and microcrystalline material samples prepared in high P_H makes the Harvard model plausible for these samples.

The two major difficulties in the measurement of conductivity relate to band bending effects. These effects can only be large in materials with a low defect density so it is easy to see why many of these phenomena are not observed in all material. *Tanielan* et al. [4.89] have demonstrated the effect of surface absorbate on conductivity on glow discharge samples. *Jackson* and *Thompson* [4.61] have recently shown with a thin film transistor structure that band bending has an enormous effect on gap type conductivity and photoconductivity measurements. Thus, measurements on gap cell structures should be treated with considerable suspicion, unless the band bending effects are analyzed. The second major issue is that in high mobility, low defect density samples, large depletion and diffusion lengths in excess of 1 μm have been measured; thus, depending on the nature of the contacts, a flat band condition may not exist in the sample unless it is thick enough; this can apply to samples with n^+ ohmic contacts as well as to those with Schottky contacts. In addition, with large $\mu\tau$ products, carriers can drift large distances (> 10 μm) without recombining, thus can pass completely through the sample at relatively low fields. Under these conditions the carrier concentration in the conduction band can be dominated by injected charge, not the thermally excited carriers. Space-charge effects will also complicate conductivity and activation energy measurements and can dominate transport processes. There is no evidence in the literature that any published conductivity measurements on sputtered material is influenced by these problems.

The conduction in sputtered Si is also sensitive to other preparation conditions such as substrate temperature and substrate bombardment. The activation energy and conductivity varies with rf substrate bias conditions during growth, as shown in Fig. 4.46. The photoconductivity is also a maximum at a bias voltage of 25 V. At high bias level, bombardment effects produce defects in the film.

b) Carrier Mobility

Only a few measurements of carrier mobility in sputtered Si have been published. *Tiedje* et al. [4.90] noted that their sputtered samples exhibited dispersive transport with an electron mobility of 0.1 cm^2(V s)$^{-1}$ which was invariant of hydrogen content from 14 to 19.5%. The multiple trapping model is given by

$$\mu_D = \mu_0(1 - a)(vt_0/1 - a)^{(a-1)/a} \tag{4.17}$$

for $a < 1$, where L is the sample thickness, V is the applied bias and $t_0 = L^2/\mu_0 V$ is the free carrier transit time; v is the thermal emission rate from localized states. They found $\mu_0 = 20$–24 cm^2(V s)$^{-1}$, $v = 1$–2×10^{13} s^{-1}. From their analysis, the band-tail states have a width of 411 or 419 K which is higher than 312 K obtained for glow discharge samples. The band tail is independent of hydrogen concentration between 14–19.5% even though the

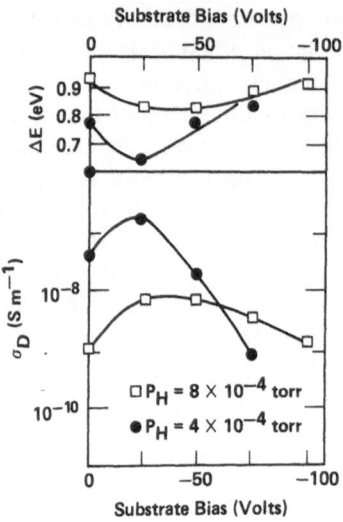

Fig. 4.46. Activation energy and dark conductivity for samples grown at various bias voltages and P_H

mid gap defect density varies in this range. This is consistent with the model of *Cody* et al. [4.58] which relates the band-tail shape with degree of disorder in the lattice.

c) Photoconductivity

Early data on photoconductivity in a-Si : H showed that the photocurrent is a strong function of the hydrogen content on the film. The photoconductivity is given by $\Delta\sigma = eG\tau\mu$, where G is the generation rate and τ the free electron lifetime, and it is τ that is the functionally dependent variable with hydrogen content. The magnitude of the photoconductivity is influenced by the position of the Fermi level in that the latter indicates whether a recombination site in the gap is occupied. A large scatter exists in photoconductivity data [4.87], even in samples produced in the same laboratory; this is probably due to band bending effects at the surface. Analysis of the intensity dependence of the photocurrent in terms of monomolecular or bimolecular recombination must be treated with some caution because of such effects [4.61]. Transient photoconductivity measurements by *Paul* and *Anderson* [4.87] revealed electron mobilities of around 0.2 cm^2(V s)$^{-1}$ and mobility activation energies of about 0.2 eV. Photoconductivity measurements with subband gap illumination revealed a shoulder in the photoconductivity response in the lower energy region indicative of defect impurity photoconductivity. *Moustakas* [4.53] found that this shoulder was particularly prominent in P and B-doped samples but, unfortunately, photothermal deflection spectroscopy measurements have not been made on sputtered samples; this would aid the interpretation of the photoconductivity data. *Moustakas* et al. [4.91] found that electron and hole recombination occurred through states in the middle of the gap and obtained a linear dependence of $\mu\tau$ on midgap defect density. They also

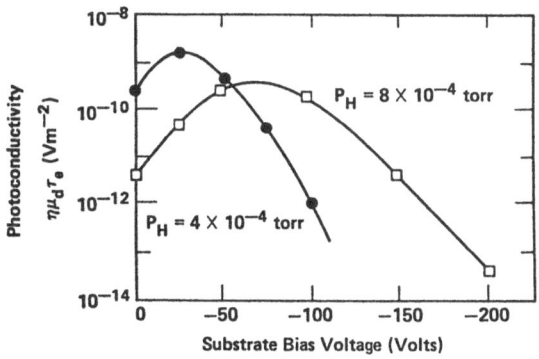

Fig. 4.47. The photoconductivity for sample grown at various bias voltages and P_H

found that the photoconductivity did not relate to SiH bonding; the samples they made with highest photoconductivity had both 2000 and 2100 cm^{-1} absorption.

The photoconductivity and thus the recombination center is particularly sensitive to substrate bombardment during growth, as revealed in Fig. 4.47. For samples that underwent low energy bombardment, the photoconductivity is enhanced. However, for those with higher bombardment, the photoconductivity was much reduced presumably due to defect states being created. *Suzuki* et al. [4.92] also saw an enhancement in photoconductivity in unhydrogenated samples prepared under bias conditions.

Further extensive measurements of primary and secondary photoconductivity is required on more recent low defect density samples. However, the band bending effects will always put a limit on this usefulness of dc and transient photoconductivity measurements unless elaborate steps are taken to quantify these effects.

d) Device Measurements

Schottky barriers and more recently *p-i-n* devices have been fabricated in sputtered a-Si:H for solar cell applications. Unlike many other measurements, minority carrier properties are influential in these devices. The first notable success was the achievement by the Sheffield group [4.93] of a Pt–a-Si:H Schottky barrier solar cell of efficiency ~2%. *Alkaisi* and *Thompson* [4.94] reported at length on the temperature and illumination dependence of the Schottky barrier devices. The dark characteristics of the cell were given by the conventional relationship

$$J = J_0 \exp(eV/nkT) , \qquad (4.18)$$

where the ideality factor $n = 1.1$–1.4, J is the current density, V is the voltage, T the temperature, and J_0 the saturation current density is given by

$$J_0 = A^* T^2 \exp(-\varphi_B/kT) , \qquad (4.19)$$

where φ_B is the barrier height and A^* is Richardson's constant.

The open-circuit voltage of the cell is given by

$$V_{oc} = nkT/e[\ln(J_{sc}/J_0 + 1)] \, , \tag{4.20}$$

where J_{sc} is the short-circuit current density.

They showed that with different quality devices under various illumination conditions and temperatures, three different transport mechanisms could be identified (Fig. 4.48) [4.95]. At low temperatures V_{oc} was temperature independent due to the field emission transport across the barrier but at higher temperatures thermionic emission dominated. In nonoptimum samples, localized conduction and tunneling through the barrier became the primary transport mechanism. *Moustakas* et al. [4.96] measured the dependence of V_{oc} and J_{sc} on the P_H the samples were grown in. The V_{oc} changes reflected the band gap variations and the J_{sc} data implied, similar to photoconductivity measurements, that defects appear to be created at higher P_H. These results indicate that it is difficult to tailor the band gap without affecting the defect density.

Thermally stimulated current (TSC) measurements on sputtered Si Schottky barriers have revealed relatively sharp peaks which are due to a high density of deep hole traps [4.97]. Similar measurements revealed that electron traps were created under intense illumination and were the cause of the Staebler-Wronski effect. Clearly these measurements provide some interesting information but as TSC is known not to be a reliable quantitative technique, further measurements are required using other techniques like DLTS.

Recently the Exxon group [4.98] has had some success in fabricating *p-i-n* structures and have achieved a solar cell efficiency of 4%, which is a milestone for sputtered a-Si:H. The current-voltage characteristics for this cell

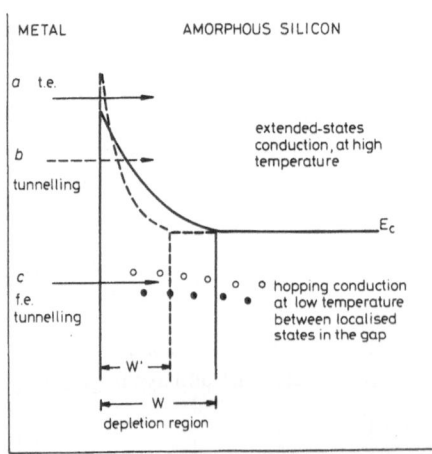

Fig. 4.48. Model of transport mechanism in an a-Si:H Schottky barrier

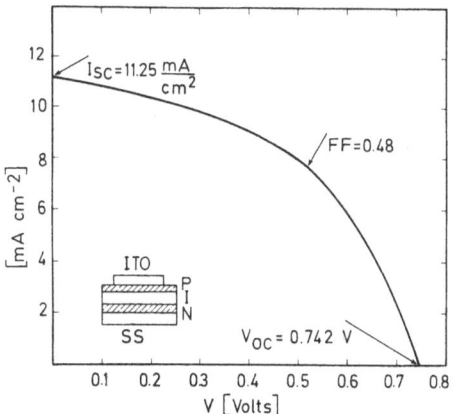

Fig. 4.49. Current voltage characteristics of a metal/*n-i-p*/1 TO solar cell structure. The engineering efficiency of the device is 4% [4.98]

are shown in Fig. 4.49. They have also measured a hole diffusion length of 3000 Å in their material. The fill factor was quite low and should be improved; thus a significantly improved cell efficiency can be expected in the near future.

4.2.5 Doping

The Harvard group were first to demonstrate that P and B doping of a-Si : H could be achieved by reactive sputtering in Ar–H atmospheres containing PH_3 and B_2H_6 [4.99]. High doping levels can be achieved as illustrated by conductivity activation energy measurements by the values of V_{oc} obtained in solar cell measurements. No detailed analysis has been done of the properties of doped sputtered material, as done on glow discharge Si by *Street* et al. [4.100]. *Moustakas* et al. [4.101] discussed a major problem when constructing an efficient *p-i-n* cell, that being the contamination of the intrinsic layer with dopant gas from previously deposited films. This is a serious issue in both sputtered and glow discharge material. It has not been conclusively proven in the literature exactly where the dopant is coming from, whether it is etching or sputtering from the walls of the chamber or from residual gas in the gas lines. Determination of doping profiles near heavily doped-intrinsic Si is required. Moustakas et al. compensated the contaminated intrinsic layer with B. *Thompson* et al. [4.102] overcame the use of heavily doped contact layers in Schottky barriers by using a thin unhydrogenated layer. Because this layer has a high number of defects, it formed only a very thin depletion layer with the metal through which all the carriers tunnelled.

Some extremely interesting work on nitrogen doping in dc sputtered Si was published in 1978 [4.103]. The authors' measurements indicate that the Fermi level is pushed to within 0.07 eV of the conduction band and the photoconductivity in the sample is strongly enhanced with doping. Unfortunately no comparable study or follow-up has appeared in the literature since

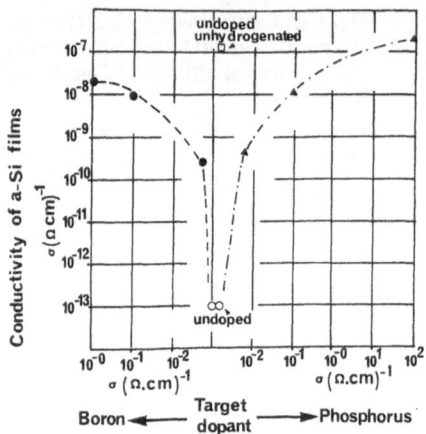

Fig. 4.50. Conductivity of a-Si : H films plotted as a function of target doping

then. *Toulemonde* et al. [4.104] found that the hydrogen and argon content of the film was markedly different in doped material than in the intrinsic film which they suggested indicated a grain microstructure in the doped material. *Anderson* and *Paul* [4.105] concluded from their P-doped Si studies that conduction was dominated by intrinsic band transport above 400 K and conduction in a P-related defect between 200–400 K.

A totally different technique from gas phase doping which could solve contamination problems was reported by the Sheffield group [4.106] in their early work and by *Van Dong* and *Hai* [4.107]. They used doped Si targets and showed that the dopants were sputtered from the target and actively incorporated in the film (Fig. 4.50). The range of doping levels achieved were limited as they used powdered doped Si targets which were oxygen contaminated and, in addition, it has since become clear that they were not operating under optimum sputtering conditions. This technique, although at that time was thought to be inflexible, could be very useful for forming abrupt junctions and limiting dopant contamination issues provided solid targets are used. Oxygen contamination of sputtered films has been reported by a number of researchers. Many of the reports relate to post contamination doping in highly columnar material generally deposited at high Ar pressures. In most homogeneous films, the O level is < 1% unless insufficient care has been taken to get rid of excess water vapor in the chamber.

Ar contents of 7% have been measured by *Messier* and *Ross* [4.34] in films grown at $P_{Ar} = 5$ mtorr and 0.5% for those grown at 70 mtorr. *Tanaka* et al. [4.108] found 8 at.% Ar for about 8 mtorr Ar plasma pressure. As we have seen earlier, it appears that under high substrate bias conditions at low pressure, the Ar content is higher and the film is more homogeneous and highly compressively stressed.

4.3 Present Status and Future Developments

Highly columnar amorphous or microcrystalline material is produced above a critical A (~ 20 mtorr) and H (~ 3 mtorr) pressure, respectively. The creation of noncolumnar uniform material is associated with growth conditions which encourage bombardment of the film by low energy plasma ions. Bombardment of the substrate by high energy Ar and Si species from the target can be avoided by operating at sufficiently high A pressure; this results in lower substrate bombardment conditions for plasma ions but they can be controlled independently by bias sputtering. High energy electrons from the target will bombard the substrate unless a magnetron target is used where the magnetic field traps the electrons near the target. An additional advantage of magnetron sputtering is that the target voltages are much smaller than conventional diode sputtering; therefore, the energy of the bombarding and reflected species is thus reduced. However, the magnetron target cannot contain the neutral species, thus the thermalization distance of the neutrals must be less than the target substrate separation. Under normal operating conditions, the H species (predominately H_2 and H_3 ions and neutrals) will not be thermalized; also they are light enough to respond to the alternating rf electric fields. These species will bombard the growing film, but the ion bombardment can be controlled by bias sputtering including phase biasing. The monohydride and dihydride species in the film can be influenced by substrate bombardment; however, no correlation has been established between electronic properties and the existence of the dihydride phase. In order to produce a relaxed structure, H and/or A is included under certain bias conditions. From NMR it is deduced that H is distributed uniformly in a dilute phase but it appears that the clustered phase which can contain SiH and SiH_2 may be responsible for the relaxation of the strain in the film. Further NMR measurements on a range of samples with different thicknesses and prepared under different bombardment conditions will indicate if there is any change in heterogeneity with thickness and with adatom diffusion. In planar diode sputtering, 15–20% H is included in the film in order to relax the structure and reduce the lowest defect density in the material. There is at present no correlation between the existence of SiH_2 and any electronically active defect. In magnetron sputtered Si, the H content for optimum material is often lower with luminescence measurements indicating that the defect luminescence which is correlated with a Si dangling bond is extremely low, similar to that in the best glow discharge material for a H content of 4–6%. However, the electronic transport properties for magnetron material have not been thoroughly evaluated.

The interpretation of transport measurements in both glow discharge and sputtered material is still highly controversial; ironically, as the material properties are improved and long carrier diffusion and depletion lengths are obtained, band bending and injection effects complicate these measurements. It is probable that measurement of the depletion and diffusion using a number of methods is the best means of evaluating the electronic quality of the material. Both extended state and localized transport are evident in nonoptimum materials. Inhomogeneity of the material provides further complications in the analysis of conductivity data. No gap state spectroscopy measurement has been successfully used to identify defect levels that are responsible for dominating recombination processes and transport [Ref. 4.39, Chap. 5]. It is known that under high electron and ion bombardment (> 30 eV) growth conditions, the photoconductivity and electronic properties deteriorate presumably due to the creation of defects.

In order to improve the control of deposition parameters and to evaluate bombardment species on the films, a mass spectrometer analysis of the substrate species must be undertaken with magnetron and bias sputtering. Doped targets may provide some useful technological advantages in producing doped layers for device structures.

4.4 Conclusions

It is clear that bombardment effects now dominate the discussion of growth in sputtered a-Si : H. With a more quantifiable measurement of the bombarding species, controlled deposition conditions can be used to fabricate better quality material. It must be recognized that the electron and H ion species are controlled by the alternating rf voltages at the target, plasma and substrate. Provided with this improved understanding and control of growth conditions, the precise physical properties of sputtered a-Si : H can be evaluated. Resolving issues of heterogeneity and the relationship between certain defects, hydrogen incorporation sites and recombination processes still provide a considerable challenge.

Acknowledgment. I am indebted to my colleagues at Sheffield University for their collaboration over the years I worked there. I would also like to thank Vi Moffat at PARC for her diligence in preparing this chapter.

References

4.1   a) J. W. Coburn: *Plasma Etching and Reactive Ion Etching* (AVS Monograph, New York 1982)
       b) E. W. B. Gill, A. von Engel: Proc. Roy. Soc. A **192**, 446 (1948)
4.2   See, for example, B. Chapman: *Glow Discharge Processes* (Wiley, New York 1980)
4.3   H. F. Winters: *Radiation Effects on Solid Surfaces*, ed. by M. Kaminsky, Adv. in Chemistry Series No. 158 (American Chemical Society 1976)
4.4   G. S. Anderson, G. K. Wehner: J. Appl. Phys. **31**, 2305 (1960)

4.5   E. B. Henschke, J. Derby: J. Appl. Phys. **34**, 2458 (1963)
4.6   a) A. L. Southern, W. R. Willis, M. T. Robinson: J. Appl. Phys. **34**, 153 (1963)
       b) S. P. Wolsky, E. J. Zdanuk: J. Appl. Phys. **32**, 782 (1982)
       c) J. Comas, E. A. Wolicki: Electrochem. Soc. **117**, 1197 (1970)
4.7   W. D. Davis, T. A. Vanderslice: Phys. Rev. **131**, 219 (1963)
4.8   J. W. Coburn, E. Kay: J. Appl. Phys. **43**, 4965 (1972)
4.9   R. H. Bruce: J. Appl. Phys. **52**, 7064 (1981)
4.10   W. A. Chupka, J. Berkowitz: J. Chem. Phys. **48**, 5769 (1968)
4.11   W. Eckstein, F. E. P. Matschke, H. Verbeck: J. Nuc. Mat. **63**, 199 (1976)
4.12   R. C. Ross, R. Messier: AIP Conf. Proc. No. 73, 53 (1981)
4.13   G. Lemperière, J. M. Poitevin, J. Tardy: Proc. 8th Intern. Vacuum Congress, Cannes (1980)
4.14   J. L. Decroix: *Introduction to Theory of Ionized Gases* (Wiley, New York 1960) p. 128
4.15   L. G. Christopherou: *Atomic and Molecular Radiation Physics* (Wiley, New York 1971) p. 283
4.16   L. G. H. Huxley, R. W. Crompton: *The Diffusion and Drift of Electrons in Gases* (Wiley, New York 1974) p. 611
4.17   a) L. J. Kieffer, G. H. Dunn: Rev. Mod. Phys. **38**, 1 (1966)
       b) J. A. Thornton, A. S. Penfold: In *Thin Film Processes*, ed. by J. L. Vossen, W. Kern (Academic, New York 1978) p. 84
       c) J. Tardy: PhD Thesis, Lyon (1982)
4.18   M. A. Paesler, T. Okumura, W. Paul: J. Vac. Sci. Tech. **17**, 1332 (1980)
4.19   A. Matsuda, K. Nakagawa, K. Tanaka, M. Matsumura, S. Yamasaki, H. Okushi, S. Iizima: J. Non-Cryst. Solids **35/36**, 183 (1980)
4.20   D. J. Tardy, J. M. Poitevin, G. Lemperière: J. Phys. D **14**, 339 (1981)
4.21   J. W. Coburn, E. Kay: Appl. Phys. Lett. **18**, 435 (1971)
4.22   G. Turban, Y. Catherine, B. Grolleau: Thin Solid Films **67**, 309 (1980)
4.23   D. A. Anderson, G. Moddel, M. A. Paesler, W. Paul: J. Vac. Sci. Tech. **16**, 902 (1979)
4.24   W. D. Westwood: J. Vac. Sci. Tech. **15**, 1 (1978)
4.25   a) I. Brodie, L. T. Lamont, D. O. Myers: J. Vac. Sci. Tech. **6**, 124 (1969)
       b) H. R. Koenig, L. I. Maissel: IBM J. Res. Dev. **14**, 168 (1970)
4.26   D. DiMaria, L. M. Ephrath, D. R. Young: J. Appl. Phys. (1979)
4.27   A. H. Eltoukhy, J. E. Greene: J. Appl. Phys. **51**, 4450 (1980)
4.28   H. Strack: J. Appl. Phys. **34**, 2405 (1963)
4.29   a) J. S. Logan, J. H. Keller, R. G. Simmons: J. Vac. Sci. Tech. **14**, 92 (1977)
       b) R. K. Waits: In *Thin Film Processes*, ed. by J. V. Vossen, W. Kern (Academic, New York 1978) p. 131
4.30   R. C. Ross: PhD Thesis, University of Pennsylvania (1981)
4.31   J. A. Thornton: J. Vac. Sci. Tech. **11**, 666 (1974)
4.32   R. C. Ross, R. Messier: J. Appl. Phys. **52**, 5329 (1981)
4.33   R. Bellisent, A. Chenevas-Paule, M. Roth: Physica B 2 C **117** & **118**, 941 (1983)
4.34   R. Messier, R. C. Ross: J. Appl. Phys. **53**, 6220 (1982)
4.35   J. Shirafuji, H. Matsui, A. Narukawa, Y. Inuishi: Appl. Phys. Lett. **41**, 535 (1982)
4.36   M. Noda, H. Ishida: Japan J. Appl. Phys. **21**, L 195 (1982)
4.37   R. Messier, R. C. Ross: J. Appl. Phys., in press
4.38   S. Veprek, Z. Iqbal, H. R. Oswald, F. A. Sarott, J. J. Wagner, J. Physique **42**, C 4–256 (1981)
4.39   J. Jannopolous, G. Lucovsky (eds.): TAP 56
4.40   a) J. C. Knights, R. A. Lujan: Appl. Phys. Lett. **35**, 244 (1979)
       b) S. Oguz, D. K. Paul, J. Blake, R. W. Collins, A. Lachter, B. G. Yacobi, W. Paul: J. Physique **42**, C 4–679 (1981)
4.41   R. C. Ross, I. S. T. Tsang, R. Messier, W. A. Langford, C. Burnam: J. Vac. Sci. Tech. **20**, 406 (1982)
4.42   J. A. Reimer: J. Physique **42**, C 4–715 (1981)

4.43　M. E. Lowry, R. G. Barnes, D. R. Torgeson, F. R. Jeffrey: Solid State Commun. **28**, 113 (1981)

4.44　T. D. Moustakas, T. Tiedje, W. A. Langford: AIP Conf. Proc. **73**, 20 (1981)

4.45　P. M. Martin, W. T. Paulewicz: J. Non-Cryst. Solids **45**, 15 (1981)

4.46　E. C. Freeman, W. Paul: Phys. Rev. B **18**, 4288 (1978)

4.47　M. H. Brodsky: Thin Solid Films **50**, 57 (1978)

4.48　T. S. Nashashibi, T. M. Searle, I. G. Austin, K. Richards, M. J. Thompson, J. Allison: J. Non-Cryst. Solids **35–36**, 675 (1980)

4.49　P. M. Martin, W. T. Paulewicz: Solar Energy Mat. **2**, 143 (1980)

4.50　F. R. Jeffrey, H. R. Shanks, G. C. Danielson: J. Appl. Phys. **50**, 7034 (1979)

4.51　A. R. Mirza, A. J. Rhodes, J. Allison, M. J. Thompson: J. Physique **42**, C 4–659 (1981)

4.52　D. P. Turner, I. P. Thomas, J. H. Allison, M. J. Thompson, A. J. Rhodes, I. G. Austin, T. M. Searle: AIP Conf. Proc. **73**, 47 (1981)

4.53　T. D. Moustakas: Solar Energy Mat. **8**, 187 (1982)

4.54　S. Oguz, D. A. Anderson, W. Paul, H. J. Stein: Phys. Rev. B **22**, 880 (1980)

4.55　H. R. Shanks, F. R. Jeffrey, M. E. Lowry: J. Physique **42**, C 4–773 (1981)

4.56　G. A. N. Connell, J. P. Pawlik: Phys. Rev. B **13**, 787 (1976)

4.57　E. C. Freeman, W. Paul: Phys. Rev. B **20**, 716 (1979)

4.58　a) G. D. Cody, T. Tiedje, B. Abeles, B. Brooks, Y. Goldstein: Phys. Rev. Lett. **47**, 1480 (1981)
　　　b) T. M. Searle, T. S. Nashashibi, I. G. Austin, R. Devonshire, G. Lockwood: Phil. Mag. **39**, 389 (1979)

4.59　A. J. Rhodes, T. M. Searle, I. G. Austin, J. Allison: *Sheffield Magnetron Sputtered Data*, to be published

4.60　W. B. Jackson, N. Amer: Phys. Rev. **25**, 5559 (1982)

4.61　W. B. Jackson, M. J. Thompson: Physica B 2 C **117** & **118**, 883 (1983)

4.62　I. G. Austin, K. Richards, T. M. Searle, M. J. Thompson, M. M. Alkaisi, I. P. Thomas, J. Allison: Inst. of Phys. Conf. Series No. 43, ed. by B. L. H. Wilson (1978) p. 1155

4.63　W. Paul, D. A. Anderson: Solar Energy Mat. **5**, 229 (1981)

4.64　P. K. Bhat, G. Diprose, T. M. Searle, I. G. Austin, P. G. LeComber, W. E. Spear: Physica B 2 C **117** & **118**, 917 (1983)

4.65　P. K. Bhat, A. J. Rhodes, T. M. Searle, I. G. Austin, J. Allison: To be published

4.66　D. A. Anderson, G. Moddel, R. W. Collins, W. Paul: Solid State Commun. **31**, 677 (1979)

4.67　R. A. Street, D. K. Biegelsen, J. Stuke: Phil. Mag. B **40**, 451 (1979)

4.68　S. Depinna, B. C. Cavenett, I. G. Austin: Phil. Mag. B **46**, 473 (1982)

4.69　A. J. Snell, K. D. Mackenzie, P. G. LeComber, W. E. Spear: J. Non-Cryst. Solids **35–36**, 593 (1980)

4.70　J. Beichler, W. Fuks, H. Mell, H. M. Welsch: J. Non-Cryst. Solids **35–36**, 587 (1980)

4.71　M. Shur, W. Czubatyi, A. Madan: J. Non-Cryst. Solids **35–36**, 731 (1980)

4.72　T. Tiedje, C. R. Wronski, J. M. Cebulka: J. Non-Cryst. Solids **35–36**, 743 (1980)

4.73　P. Viktorovitch, D. Jousee: J. Non-Cryst. Solids **35–36**, 569 (1980)

4.74　H. Fernandez-Canque, J. Allison, M. J. Thompson: J. Appl. Phys. (in press)

4.75　P. Viktorovitch: J. Appl. Phys. **52**, 1392 (1981)

4.76　T. Tiedje, C. R. Wronski, B. Abeles, J. M. Cebulka: Solar Cells **2**, 301 (1980)

4.77　W. E. Spear, P. G. LeComber: J. Non-Cryst. Solids **8–10**, 727 (1972)

4.78　M. Powell: Phil. Mag. **43**, 93 (1981)

4.79　R. Weisfield, D. A. Anderson: Phil. Mag. B **44**, 83 (1981)

4.80　N. B. Goodman, H. Fritzsche, H. Azaki: J. Non-Cryst. Solids **35–36**, 500 (1980)

4.81　N. B. Goodman, H. Fritzsche: Phil. Mag. B **42**, 149 (1980)

4.82　R. L. Weisfield, P. Viktorovitch, D. A. Anderson, W. Paul: Appl. Phys. Lett. **39**, 263 (1981)

4.83　P. Bräunlich (ed.): *Thermally Stimulated Relaxation in Solids*, Topics Appl. Phys., Vol. 37 (Springer, Berlin, Heidelberg, New York 1979) Chap. 3

J. Bourgain, M. Lannao: *Point Defects in Semiconductors* II, Springer Ser. Solid-State Sci., Vol. 35 (Springer, Berlin, Heidelberg, New York 1983)

4.84 a) R. S. Crandall: J. Physique **42**, C4–413 (1981)
b) J. D. Cohen, D. V. Lang, J. P. Harbison, A. M. Sergent: J. Physique **42**, C4–371 (1981)

4.85 Yu. N. Molin, K. M. Salikhov, K. I. Zamaraev: *Spin Exchange*, Springer Ser. Chem. Phys., Vol. 8 (Springer, Berlin, Heidelberg, New York 1980)

4.86 a) A. J. Lewis Jr., G. A. N. Connell, W. Paul, J. R. Pawlik, R. J. Temkin: AIP Conf. Proc. **20**, 27 (1974)
b) J. R. Pawlik, W. Paul: Proc. 7th Intern. Conf. of Amorphous and Liquid Semiconductors, Edinburgh, ed. by W. E. Spear (1977) p. 437

4.87 W. Paul, D. A. Anderson: Solar Energy Mat. **5**, 229 (1981)

4.88 D. A. Anderson, W. Paul: Phil. Mag. B **44**, 187 (1980)

4.89 M. H. Tanielian, M. Chatani, H. Fritzsche, V. Smid, P. D. Persons: J. Non-Cryst. Solids **35–36**, 575 (1980)

4.90 T. Tiedje, T. D. Moustakas, J. M. Cebulka: J. Physique **42**, C4–155 (1981)

4.91 T. D. Moustakas, C. R. Wronski, T. Tiedje: Appl. Phys. Lett. **39**, 721 (1981)

4.92 M. Suzuki, T. Maekawa, Y. Kakimoto, T. Bardow: J. Physique **42**, C4–623 (1981)

4.93 M. J. Thompson, J. Allison, M. M. Alkaisi, I. P. Thomas: Rev. Phys. Appl. **13**, 625 (1978)

4.94 M. M. Alkaisi, M. J. Thompson: Solar Cells **1**, 91 (1979/1980)

4.95 M. J. Thompson, M. M. Alkaisi, J. Allison: IEE Proc. **127**, 213 (1980)

4.96 T. D. Moustakas, C. R. Wronski, D. L. Morel: J. Non-Cryst. Solids **35–36**, 719 (1980)

4.97 L. Vieux-Rochaz, A. Chenevas-Paule: Proc. 2nd EC Photovoltaic Solar Energy Conference, Berlin (Reidel, Holland 1979) p. 295

4.98 T. D. Moustakas, R. Friedman: Appl. Phys. Lett. **40**, 515 (1982)

4.99 W. Paul, A. J. Lewis, G. A. N. Connell, T. D. Moustakas: Solid State Commun. **20**, 969 (1976)

4.100 R. A. Street, D. K. Biegelsen, J. C. Knights: Phys. Rev. B **24**, 269 (1981)

4.101 T. D. Moustakas, R. Friedman, B. R. Weinberger: Appl. Phys. Lett. **40**, 587 (1982)

4.102 M. J. Thompson, M. M. Alkaisi, J. Allison: Proc. 2nd EC Photovoltaic Solar Engergy Conference, Berlin (Reidel, Holland 1979) p. 303

4.103 J. Baixeras, D. Moncaraglia, P. Andro: Phil. Mag. B **37**, 403 (1978)

4.104 M. Toulemonde, J. J. Grob, J. C. Bruyere, A. Deneuville, H. Hamdi, P. Siffert: J. Physique **42**, C4–799 (1981)

4.105 D. A. Anderson, W. Paul: Phil. Mag. B **45**, 1 (1982)

4.106 M. J. Thompson, J. Allison, M. Alkaisi, S. J. Barber: Proc. of Photovoltaic Solar Energy Conference, Luxembourg (Reidel, Holland 1977)

4.107 N. van Dong, T. Q. Hai: Phys. Stat. Sol. **88**, 555 (1980)

4.108 K. Tanaka, S. Yamasaki, K. Nagawa, A. Matsuda, K. Okushi, M. Matsumura, S. Iizima: J. Non-Cryst. Solids **35/36**, 183 (1980)

5. CVD Material

Daniel Kaplan

With 18 Figures

This chapter deals with amorphous silicon films prepared by thermal decomposition of a gaseous silicon compound, in most cases silane (SiH_4). This preparation method is generally termed chemical vapor deposition (CVD). In principle, the more widely used glow-discharge decomposition process is also a CVD method, plasma-assisted CVD, but most authors studying amorphous silicon refer to CVD in the restricted sense of thermally assisted CVD and we shall use this convention.

CVD deposition is widely used for preparing crystalline silicon in the electronic industry, both in the form of epitaxial layers on monocrystalline substrates and polycrystalline films, e.g., gate electrodes in MOS transistor technology. Typical process temperatures in this case are in the 900°–1100 °C range. Lower temperatures are required if the deposits are to be amorphous. The choice of gaseous source compound is then dictated by the necessity of achieving reasonable deposition rates at temperatures sufficiently low to avoid crystallisation. With silane, at a temperature of 600 °C which is close to the crystallisation limit, deposition rates are of order 1 μm/h [5.1, 2]. Other compounds, such as chlorosilanes [5.1] or fluorosilanes with the exception of SiF_2 [5.3, 4], decompose less readily and are thus not suited. On the contrary, higher silanes (Si_2H_6, Si_3H_8, etc.) are less stable chemically and can be used for CVD deposition at temperatures below 450 °C [5.5], so that crystallisation is no longer a concern.

Another path for low-temperature deposition is the HOMOCVD (homogeneous CVD) technique, recently introduced by *Scott* et al. [5.6]; a cooled substrate is inserted in a hot reactor in which silane is homogeneously predecomposed. Room temperature deposition has been demonstrated by this method [5.7].

Despite these recent developments towards low temperature CVD deposition, the main work has been on material deposited just below the crystallisation temperature. Part of the motivation for studying such a high temperature deposited material comes from the expectation that it may a priori be anneal stable, well reconstructed, and constitute a prototype of "perfect" amorphous silicon. The price to pay is a low hydrogen content and a correlated high density of dangling bonds. As was the case for vacuum deposited a-Si [5.8], it has proved to be impossible to produce pure a-Si without dangling bonds.

On the other hand, plasma post-hydrogenation techniques [5.9, 10] allow hydrogen to be introduced after deposition. One is then in a position to make independent studies of the effect of hydrogen which have implications for other methods of preparation of amorphous silicon.

Several extended studies of the CVD material have already been published, in particular from Hiroshima University [5.1], the University of Arizona [5.11] and Thomson-CSF [5.10, 12]. The present chapter will attempt to synthetize these, together with more recent data. We have tried to extract and compare the main results. The reader is referred to the original articles for details on experiments and models.

The chapter starts with discussions of deposition processes, crystallisation and hydrogen incorporation in relation to dangling bonds. It goes on to describe optical and transport properties. Special emphasis is given to the effect of doping which is treated as a separate subject. Finally, some of the considered applications are briefly described.

5.1 Deposition and Structure

5.1.1 Reactor Configurations

Figure 5.1 gives three examples of reactor configurations. Reactor A is a "cold wall" atmospheric pressure type in which substrates are placed on an rf heated susceptor, contained in a bell jar. Reactor B is a "hot wall", low pressure type (LPCVD). In this case the reaction chamber is a quartz tube inserted in a furnace. Substrates are stacked parallel to each other in the tube. Deposition occurs both on the substrates and on the tube wall. Operating at low pressures enhances gas diffusion rates, permitting uniform deposition on the stacked substrates. Configuration C is the HOMOCVD configuration. It is similar to B except that the stacked substrate arrangement is replaced by a substrate holder, efficiently cooled by a nitrogen gas flow [5.6].

5.1.2 Chemical Reactions

The main chain of reactions in the gas phase is thought to be the following [5.6]:

$$\text{SiH}_4 \rightleftarrows \text{SiH}_2 + \text{H}_2 , \tag{5.1}$$

$$\text{SiH}_2 + \text{SiH}_4 \rightleftarrows \text{Si}_2\text{H}_6 , \tag{5.2}$$

$$\text{Si}_2\text{H}_6 + \text{SiH}_2 \rightleftarrows \text{Si}_3\text{H}_8 , \quad \text{etc.} \tag{5.3}$$

The first of these reactions has been well characterized thermodynamically [5.13]. The chain can lead eventually to homogeneous nucleation of

A

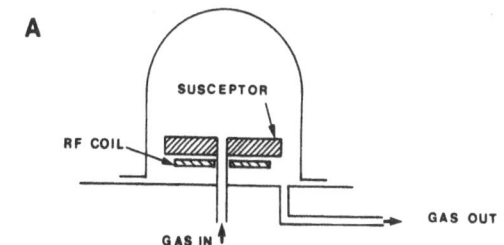

Fig. 5.1. Schematics of CVD
reactors
(A) cold wall atmospheric
pressure system,
(B) hot wall LPCVD sys-
tem,
(C) HOMOCVD   system
[5.6]

B

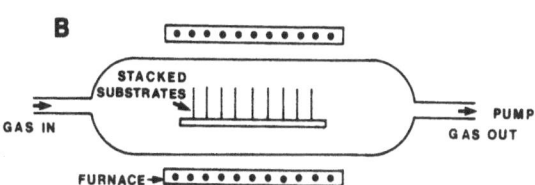

C

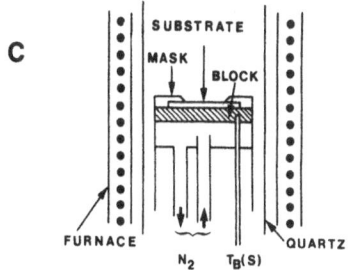

solid polymers which manifests itself by the appearance of "snow" in the
reactor. This is an important consideration, particularly in the case of "hot
wall" configurations. In order to prevent homogeneous nucleation, one has
to lower the silane partial pressure, either by working at a low total pressure
(LPCVD) or by dilution in the case of atmospheric-pressure systems. The
more frequently used dilutant gases are hydrogen, nitrogen and argon.

Since only silicon hydrides are produced in the homogeneous reactions, it
is generally considered that heterogeneous reactions are needed to form the
silicon. There have been several studies [5.14, 15] attempting to understand
the complete deposition process, but the matter is still unclear, especially in
the range of temperatures of interest here. It is not, for instance, unambiguously known whether silane is directly decomposed at the surface, or whether
it is some other precursor formed in the gas phase. Some information on the
latter type of process may be inferred from the recent HOMOCVD experiments [5.6, 16].

Figure 5.2 from *Scott* et al. [5.16] shows measurements, in the configuration of Fig. 5.1 C, of deposition rates as a function of (i) furnace temperature
at constant substrate temperature and (ii) substrate temperature at constant
furnace temperature. These results are interpreted as follows.

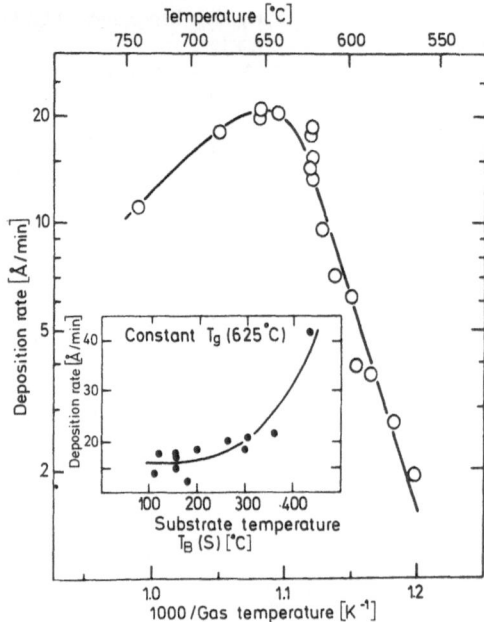

Fig. 5.2. Deposition rates in HOMOCVD experiments from *Scott* et al. [5.16]. The main curve (OOO) shows the effect of gas temperature at constant substrate temperature and the inset (●●●) the effect of substrate temperature at constant gas temperature

a) The furnace temperature dependence shows a maximum due to the competition between decomposition and homogeneous nucleation [5.17]. Below this maximum, the activation energy of 53.9 kcal/mol corresponds closely to that of the rate of reaction (5.1) [5.13], which should be the limiting step.

b) The deposition rate does not depend upon substrate temperature below 300 °C. The only precursor that could account for this is the radical SiH_2.

c) The mechanism thus involves the homogeneous formation of the radical SiH_2, which decomposes at the surface by a heterogeneous reaction that eliminates the extra hydrogen. It should be noted that the SiH_2 does not necessarily come primarily from reaction (5.1), but may be provided by the reverse reactions (5.2, 3), etc.

We turn now to speculations on the operative mechanism in ordinary silane reactors. Figure 5.3 shows data on the deposition rate as a function of temperature for a reactor of the type of Fig. 5.1 A, using silane diluted by hydrogen. The activation energy of 47 kcal/mol is close to that of the rate of (5.1). This is not, however, the general case and other results, corresponding to different reactor configurations, give lower activation energies, e.g., 34.6 kcal/mol reported by *Hirose* [5.1] for a reactor of different geometry using nitrogen as a carrier gas. A reaction path similar to HOMOCVD and involving SiH_2 as an intermediate has been proposed [5.1]. However, *Scott* et al. [5.16] interpreted the rapid rise above 300 °C of the deposition rate in HOMOCVD as a function of substrate temperature as implying the onset of

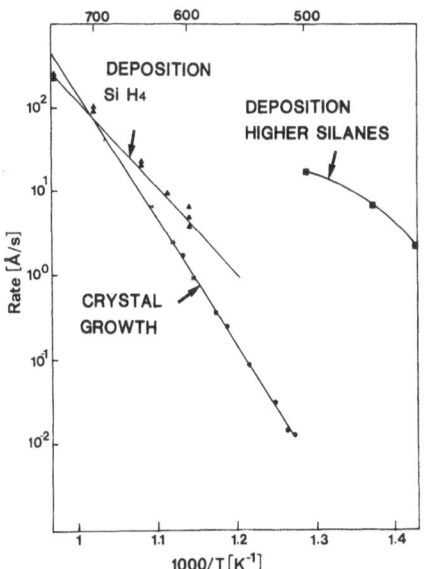

Fig. 5.3. Temperature dependence of the deposition rate for a cold wall atmospheric pressure CVD system using monosilane diluted in hydrogen and for a system using polysilanes [5.5]. The rate of crystal growth [5.20] is shown for comparison

a different *heterogeneous* decomposition mechanism. A possibility is that this heterogeneous decomposition does not concern SiH_4, but rather some of the higher silanes formed in the gas phase.

There is current interest in using higher silanes as the source gas for depositing at low temperatures. *Gau* et al. [5.5] have studied this type of deposition, using a higher silane mixture prepared by a chemical reaction of Mg_2Si with HCl. The deposition rates they report are plotted for comparison in Fig. 5.3. Note the three order of magnitude enhancement at a temperature of 500 °C. In addition to their practical interest, one may hope that studies of higher silane decomposition will help clarify the mechanism involved in ordinary silane reactors.

Janaï et al. [5.3, 4] have reported CVD deposition from SiF_2 gas produced by a SiF_4 reaction with silicon around 1100 °C. Deposition rates of order 1 μm/h are obtained for a 600 °C deposition temperature. The films contain typically 1 at.% of fluorine. Few data exist on physical properties of these films [5.2] which do not appear significantly different from those produced by silane decomposition with regard to their optical gap and electrical conductivity.

5.2 Amorphous to Crystalline Transition

Because of the broad use of CVD deposition in semiconductor technology, there are many reports on the transition from an amorphous to a crystalline

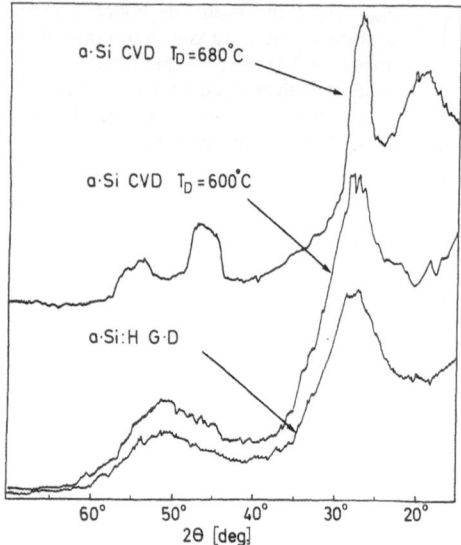

a·Si CVD T$_D$=680°C

a·Si CVD T$_D$=600°C

a·Si:H G·D

60°   50°   40°   30°   20°
2θ [deg]

Fig. 5.4. Glancing angle x-ray diffraction for films deposited by CVD at temperatures above (680 °C) and below (600 °C) the crystallization threshold. An amorphous film deposited by glow discharge (gd), in the vicinity of 250 °C, is also shown for comparison

structure as one increases the deposition temperature. *Janaï* et al. [5.11] have presented an extensive study of films prepared by atmospheric pressure CVD and their paper gives a comprehensive bibliography. In the context of this presentation of CVD amorphous silicon, we are mainly concerned with ascertaining that the films are truly amorphous.

Figure 5.4 gives a comparison of x-ray diffraction spectra for films grown at 600° and 680 °C. Diffraction from an a-Si film grown by the glow discharge process is also shown for comparison. All three films are on fused silica substrates. Copper Cu K_α radiation is used with a glancing angle of incidence. The 600 °C CVD deposited film gives a spectrum similar to the gd film and characteristic of the amorphous structure. The most significant angular domain is the region corresponding to the ⟨220⟩ and ⟨311⟩ crystalline diffraction peaks. These peaks are absent for the amorphous films and replaced by a halo roughly halfway between the peaks. From this type of diagram, it is possible to infer that less than 10% of the volume of 600 °C deposited CVD films is crystallized. A confirmation comes from Raman scattering studies [5.18] which indicate no sign of a sharp structure associated with the zone-center optical phonon in the crystal.

Signs of crystallisation are generally reported around 670 °C for atmospheric pressure CVD [5.11] and around 575 °C for LPCVD [5.19]. This to be contrasted with high-vacuum evaporation for which crystallised films have been observed at temperatures as low as 450 °C [5.8] for similar deposition rates.

To analyse these discrepancies, one should consider that there are two possible types of processes leading to crystallisation:
a) crystallisation at the time of deposition by surface rearrangements and

b) crystallisation after deposition of an amorphous material; the first layers to be deposited crystallize by subsequent annealing during the time needed to grow the complete film.

In Fig. 5.3 we have plotted recent results by *Zellama* et al. [5.20] on the rate of crystal growth in CVD and evaporated amorphous silicon films. The two types of films yield similar results in this respect. Comparison with the deposition rates indicates that, in the temperature range 600°–700 °C, subsequent crystal growth will play an important role if crystallites are nucleated. On the other hand, this process should be negligible for crystallisation at 450 °C during evaporation. There the process has to be crystallisation at the time of deposition. The reason why this mechanism is less operative in CVD deposition may be related to foreign atoms such as hydrogen tying up the surface and hindering surface reconstruction.

A particular situation is one in which there is a very fast crystallite nucleation process at the substrate's interface and negligible volume nucleation. In this case the structure of the film is expected to be a two-layer structure if the crystal growth rate is smaller than the deposition rate, i.e., below the 700 °C crossing point in Fig. 5.3. Such films have indeed been identified by several methods, of which the most direct is a comparison of Raman scattering from both sides of the film (for a transparent substrate).

These remarks are evidently bothersome for interpreting physical properties. Although a thin crystalline layer would not significantly alter an optical absorption measurement, it may considerably affect measurement of the resistivity in a coplanar configuration. For a deposition temperature of 600 °C or below, no sign of crystallisation has been reported for pure a-Si films, although the crystal growth rate is still typically one fourth of the deposition rate. This is presumably due to the fact that under these conditions, the nucleation rate at the substrate's interface is sufficiently low. We shall thus assume that films grown at or below 600 °C are amorphous, bearing in mind that the proper checks, e.g., by Raman scattering, have not been made in every case.

Because of their interest in designing high temperature absorbers for photothermal solar energy conversion, the group led by *Seraphin* at the University of Arizona has studied the possibility of raising crystallisation temperatures by doping with suitable impurities [5.21]. The most promising appears to be carbon; films containing 18 at.% carbon have their crystallisation characteristics shifted by more than 300 °C compared to pure films. Nitrogen was also found to retard crystallisation.

5.3 Hydrogen and Dangling Bonds

5.3.1 Hydrogen Content

Several methods of hydrogen analysis have been applied to CVD a-Si: secondary ion mass spectrometry (SIMS) [5.10, 22], nuclear reactions [5.7, 23], nuclear magnetic resonance (NMR) [5.7] and infrared absorption [5.7]. The information on hydrogen clustering that can be extracted from NMR spectra has not yet been reported and would be valuable. Figure 5.5 shows results on the hydrogen atomic fraction as a function of deposition temperature for CVD deposited films and for the HOMOCVD system [5.7]. Note the rapid decrease in hydrogen content with increasing deposition temperature. The HOMOCVD results appear to extrapolate at high temperatures to the CVD results, although considering the differences in experimental conditions (pressure, dilution, etc.), the agreement may be fortuitous.

The ir absorption is at 2000 cm^{-1} for the CVD films [5.24] and the HOMOCVD films [5.7] above 300 °C deposition temperature. Below 300 °C the HOMOCVD films show a shift towards 2100 cm^{-1}, indicating a departure from simple Si–H configurations; the origin of this departure is still controversial [5.25].

Information on hydrogen content for films grown from the higher silanes is not available at the time of writing. It would be interesting to see whether it fits with the above data in the intermediate range of temperatures.

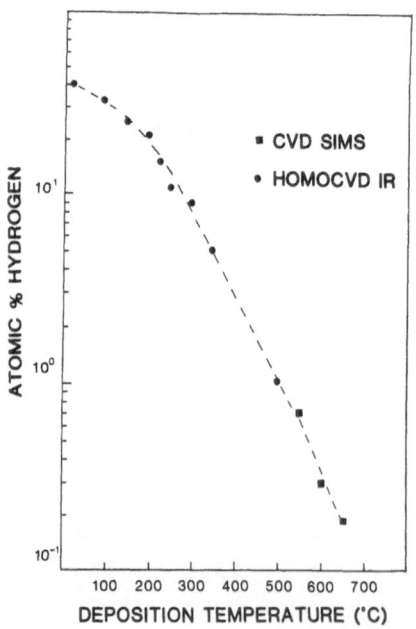

Fig. 5.5. Hydrogen content in at.% for films deposited by CVD (■■■; measurement by secondary ion mass spectrometry [5.22]) and for films deposited by HOMOCVD (●●●); measurement by ir absorption [5.7]

5.3.2 Dangling Bonds

Evidence for dangling bonds is obtained from electron spin resonance (ESR). We consider the intense resonance with a g-value of 2.0055 (observed in films prepared by evaporation in high vacuum) as due to a dangling silicon bond in its neutral charge state. Figure 5.6 shows results of spin densities obtained by integrating the ESR spectrum and expressed in at.%. The CVD films show an increase in spin density with increasing deposition temperature [5.19, 26, 27]. The HOMOCVD films grown below 300 °C have much smaller spin densities, inferior to 10^{-4} at.%, although such low values should be taken with caution since even a slight amount of residual doping can affect their magnitude, as we discuss later.

On Fig. 5.6 we have indicated for comparison results obtained with films prepared by evaporation in ultrahigh vacuum [5.8]. For these films the spin density is a slowly decreasing function of deposition temperature, probably indicating an improvement in bond reconstruction with increasing temperature. For the CVD films, the variation is opposite and is clearly to be related to the variation in hydrogen content. Note that the CVD films with the lower hydrogen content have a spin density which is of the same order of magnitude as that of pure a-Si prepared by evaporation at high temperature. Similar spin densities can also be obtained by heat treatment of glow discharge films which removes the hydrogen [5.28]. ESR is not sensitive enough to demonstrate any differences in structure between these different types of pure or weakly hydrogenated films. The effect of hydrogen tying up the dangling bonds appears to be the main consideration.

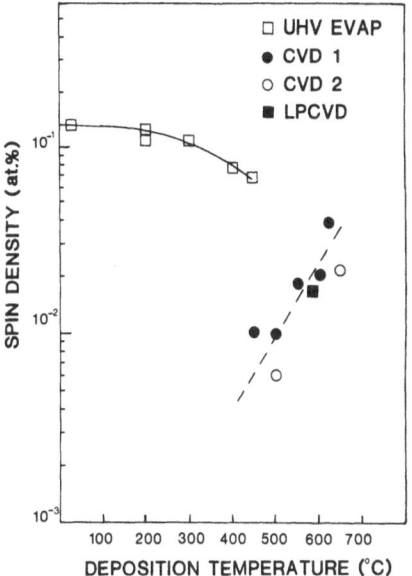

Fig. 5.6. Spin density in atomic percent for CVD and LPCVD films as a function of deposition temperature. CVD 1 is from [5.26]; CVD 2 is from [5.27]; LPCVD is from [5.19]. Results for films grown by evaporation in UHV [5.8] are shown for comparison

5.3.3 Post-Hydrogenation

By treating the films in a hydrogen plasma, it is possible to eliminate most of the dangling bonds which are present in CVD films deposited at high temperature. Different plasma systems have been used which differ by the mode of excitation of the discharge: dc, rf or microwave. Systems using a microwave cavity are particularly convenient since they allow work at low pressures and limit the region of the discharge to a well-defined zone within the cavity. One can then place the sample outside this zone in which case hydrogenation is achieved mostly by atomic hydrogen diffusing from the discharge to the sample's surface. This avoids complications associated with plasma-surface interactions.

Figure 5.7 shows a depth profile measured by SIMS after such post-hydrogenation treatment. In this case, deuterium was used instead of hydrogen to facilitate the analysis. This profile can be fitted to an expression of the form $C_0 \mathrm{erfc}[x/2(Dt)^{1/2}]$, where t is the time, x the depth, C_0 the concentration at the surface and D the diffusion constant. The quality of this fit to a complementary error function over several orders of magnitude in concentration indicates that the diffusion constant is, in a first approximation, independent of hydrogen concentration, a result which was not to be assumed a priori.

The surface concentration is dependent upon plasma conditions and treatment temperature. For example, C_0 under given plasma conditions may vary between 20 at.% for a treatment at 350 °C to 1 at.% for a treatment at 450 °C. The diffusion coefficient has an activated temperature dependence. In Table 5.1 we have compared results obtained by fitting profiles such as the

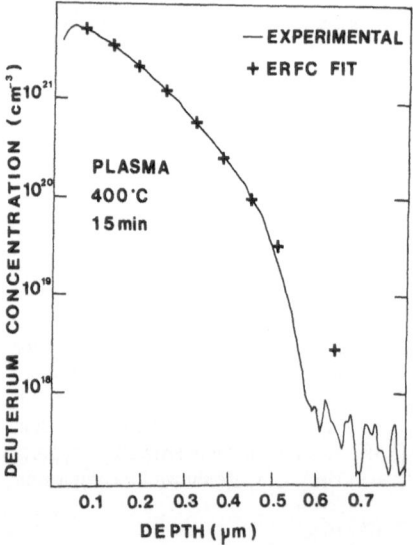

Fig. 5.7. Typical deuterium profile obtained by plasma treatment (400 °C, 15 min). A fit to a complementary error function profile (ERFC) is presented. The observed deviation at low hydrogen content is systematic in this type of experiment

Table 5.1.

	Glow discharge Exodiffusion Hydrogen	Glow discharge Interdiffusion Deuterium	CVD Indiffusion Hydrogen
D (400 °C) [cm²/s]	1×10^{-13}	0.3×10^{-13}	1.3 to 0.6×10^{-13}
Activation Energy [eV]	1.4	1.4	1.2

one of Fig. 5.7 in CVD material to exodiffusion studies in glow discharge hydrogenated a-Si by *Zellama* et al. [5.29], and interdiffusion studies in deuterated glow discharge a-Si by *Carlson* and *Magee* [5.30].

Considering the differences in methods and the errors involved, it is not possible to infer any significant differences between the materials.

Elimination of the dangling bonds by post-hydrogenation has been studied by ESR [5.10, 12]. The signal decreases with increasing plasma treatment time until sufficient hydrogen is diffused throughout the film. When these results are analysed, together with the SIMS profiles, one concludes that an additional hydrogen concentration of the order of 0.1 at.% or less is sufficient to eliminate most of the dangling bonds. The picture that emerges from these results is that the first hydrogen atoms that diffuse to a certain depth saturate the dangling bonds, but a much larger amount is later inserted into sites that are not directly related to the original dangling bonds.

Post-hydrogenation is relatively efficient in removing dangling bonds: proper treatment can reduce the spin density in undoped CVD films to the vicinity of 10^{17} cm^{-3}. There are, however, some problems inherent to the method. Very strong plasma intensities will result in etching away the a-Si layers. Also hydrogen will, in some cases, concentrate at the substrate's interface causing imperfections in the form of bubbles, cracks, etc. For these reasons it is difficult to diffuse in large and uniform concentrations of hydrogen. Many of the physical results on post-hydrogenated films, to be described later, concern films with nonuniform hydrogen concentrations. Therefore, in general, they can only be analysed qualitatively.

In situ post-hydrogenation may add some flexibility by permitting the sample to be deposited in several stages, with post-hydrogenation treatment in between. Preliminary experiments have been performed by *Bustarret* et al. [5.23] using a capacitive geometry deposition chamber.

Post-hydrogenation has also been accomplished by ion implantation of hydrogen [5.31]. The amount of existing data is not sufficient to appreciate the efficiency of this method. Although it may overcome some of the difficulties of the plasma method mentioned above, it creates additional irradiation defects. An annealing treatment is necessary to eliminate them.

5.4 Optical Properties

5.4.1 Optical Absorption and Refractive Index

Several groups [5.1, 11, 32] have measured the optical properties in the visible and near ir of CVD films deposited in the vicinity of 600 °C. The data are in good agreement, despite differences in reactor configurations. There is some scatter in the reported optical gaps E_0, obtained by extrapolating to zero absorption plots of $(\alpha h \nu)^{1/2}$ as a function of $h\nu$ (Tauc's plot). E_0 ranges from 1.6 to 1.45 eV [5.1]. This scatter is probably due to different wavelength ranges used when performing the extrapolation. If one uses a definition of the gap as the photon energy where the optical absorption is 10^4 cm^{-1}, all reported data yield a value $E_{04} = 1.7 \pm 0.02$ eV. This is to be compared with a typical value of 1.9 eV [5.33] for glow discharge films grown in the vicinity of 250 °C. Annealing glow discharge films to temperatures of 600 °C reduces the gap by typically 0.15 to 0.2 eV [5.33], leading to values close to that of CVD films.

The index of refraction at long wavelength ($\lambda = 2$ µm) is reported to be between 3.7 and 3.5 [5.1–11]. This is to be compared with values between 3.3

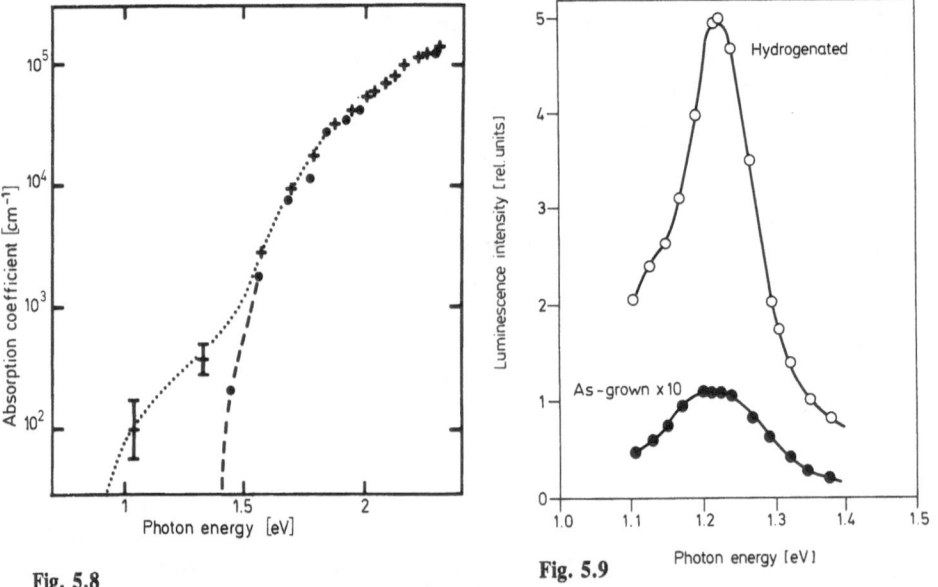

Fig. 5.8

Fig. 5.9

Fig. 5.8. Optical absorption coefficient as a function of energy for a CVD film as-deposited (×××) and after post-hydrogenation (●●●). Error bars are indicated for the two points at lower energy which are derived from interference maxima. No absorption is detected in this range after post-hydrogenation

Fig. 5.9. Luminescence spectra for as-grown and plasma hydrogenated CVD films [5.1]

and 3.6 for glow discharge films grown at 250 °C and annealed at 600 °C [5.33].

Measurements of the temperature dependence of the optical absorption [5.32] give a variation of the optical gap $dE_0/dT_m \simeq -5 \times 10^{-4}$ eV/K between $T_m = 0$ °C and $T_m = 500$ °C.

In Fig. 5.8 we show a comparison of the optical absorption as a function of photon energy between a CVD film deposited at 600 °C, and the same film after a post-hydrogenation treatment in a plasma at 400 °C. Above 1.5 eV we detect no significant differences between the two curves, indicating that this particular plasma treatment was mild and did not introduce a large concentration of hydrogen. Below 1.5 eV the two points shown correspond to interference maxima in the transmission curve, for which errors made in the determination of small absorption coefficients are minimal. After post-hydrogenation, the absorption becomes too small to be measured.

The above behaviour can be understood by assuming that the absorption below photon energies of 1.4 eV is due to dangling-bond states. These are removed by post-hydrogenation. A similar conclusion has been reached by *Jackson* and *Amer* [5.34] from their measurements on glow discharge films using a highly sensitive photothermal deflection technique. They found a linear relationship between the ESR signal and optical absorption in a series of film. Extrapolating their variation, one estimates an optical absorption of 2×10^2 cm^{-1} for the CVD film, in reasonable agreement with the data of Fig. 5.9.

5.4.2 Photoluminescence

Photoluminescence spectra reported by *Hirose* [5.1] showed a peak at 1.2 eV. As shown in Fig. 5.9, the luminescence increases upon hydrogenation, while the peak position remains the same. There is no sign of a peak at lower energy which could be assigned to dangling bonds [5.35]. In glow discharge undoped a-Si, the luminescence peak is generally found at 1.4 eV. It is tempting to assign the observed shift to the change in band gap of 0.2 eV, but the existing data is too scarce to draw a firm conclusion.

Very recently efficient luminescence in the visible has been observed in a-Si prepared by HOMOCVD near room temperature [5.36]. This confirms the low density of defects in HOMOCVD prepared films, already suggested by the low spin densities.

5.5 Transport Properties of Undoped Films

5.5.1 Conductivity

Amorphous silicon deposited by CVD around 600 °C has a room tempera-
ture conductivity in the range 10^{-7} to 10^{-8} Ω^{-1} cm^{-1}, irrespective of reactor
systems [5.1, 12, 19]. The temperature dependence, shown in Fig. 5.10
(undoped), is not linear in an Arrhenius plot. Post-hydrogenation reduces
the room temperature conductivity by typically an order of magnitude [5.12]
and the temperature dependence becomes simply activated, with activation
energies of order 0.6 eV.

A natural explanation of this behaviour is that the conductivity in as-
deposited films contains two contributions:

a) transport in states near the Fermi level, presumably associated with
dangling bonds, which dominates at low temperatures because of the relatively small activation energies involved;

b) transport in conduction band states, yielding a contribution of the form
$\sigma_0 \exp(-E_a/kT)$ which dominates at higher temperatures. The activation
energy is related, although not necessarily equal, to the energy difference
between the conduction band mobility edge and the Fermi level. By analysis
of this high temperature region, values of E_a between 0.7 and 0.8 eV have
been derived [5.1, 5.7]. Values of the pre-exponential factor σ_0 are between
10^3 and 10^4 Ω^{-1} cm^{-1}, a correct order of magnitude for mobility edge conduction.

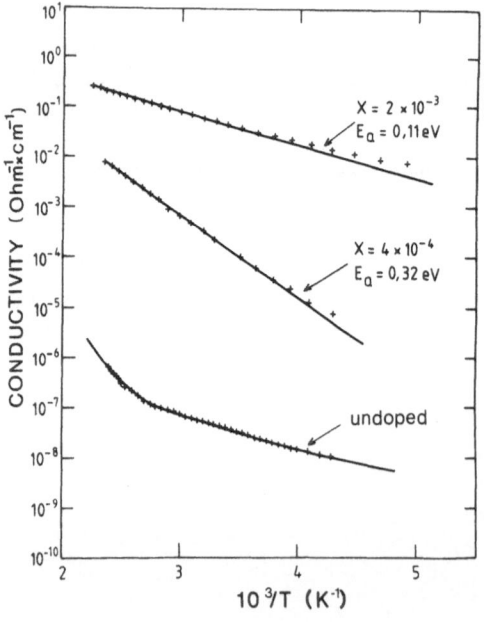

Fig. 5.10. Arrhenius plots of the conductivity as a function of temperature for undoped and phosphorus
doped CVD films. The quantity X
refers to the concentration ratio
[PH$_3$]/[SiH$_4$] in the gas mixture used
for deposition. In the case of doped
films an activation energy E_a is obtained from a least square fit to the
data

Post-hydrogenation, by removing dangling-bond states, suppresses the first contribution·to the conductivity related to these defects. There is also probably an associated Fermi level shift, but the experimental data are too limited to determine it.

5.5.2 Photoconductivity

Measurements of photoconductivity have been reported by *Taniguchi* et al. [5.37] for CVD deposited films and *Harbeke* et al. [5.19] for LPCVD. Typical films, grown around 600 °C, show a change of conductivity under AM 1 solar spectrum conditions of order 0.2%, i.e., of order $10^{-9} \, \Omega^{-1} \, cm^{-1}$. In general, the temperature dependence of photoconductivity is nonlinear on an Arrhenius plot.

Taniguchi et al. [5.37] reported that for films deposited at lower temperatures, e.g., 490 °C, photoconductivity increased by more than an order of magnitude and the temperature dependence was modified. HOMOCVD films grown around 300 °C [5.7] showed AM 1 photoconductivities between 10^{-4} and $10^{-5} \, \Omega^{-1} \, cm^{-1}$. This general trend of increasing photoconductivities with decreasing deposition temperatures should be related to the increase in hydrogen content and decrease in dangling-bond density, resulting in a lifetime increase. Even higher photoconductivities, of order $10^{-3} \, \Omega^{-1} \, cm^{-1}$, are reported for post-hydrogenated films [5.1], confirming the role played by hydrogen.

5.6 Doping

5.6.1 Generalities

Doping with donor or acceptor impurities produces several order of magnitude changes of conductivity in a-Si prepared by CVD [5.1]. This is to be contrasted with the failure to observe such changes in a-Si prepared by evaporation. This failure had raised the question that dopants might be inactive because the random network could accommodate them in their normal chemical coordination rather than forcing them to be fourfold coordinated, as is the case for substitutional impurities in the crystal. In fact, efficient pinning of the Fermi level by dangling bonds and experimental difficulties in introducing the dopant combine to make it hard to achieve variation in the transport properties in evaporated films. With CVD systems, on the other hand, doping is easily achieved by adding a suitable compound in the gas mixture, commonly phosphine (PH_3) [5.37] for *n*-type doping and diborane (B_2H_6) [5.38] for *p*-type doping. Deposition rates increase considerably for diborane doping. The change is smaller and not monotonous for phosphine.

The main problem with doping studies is to acquire information about the amount of dopant atoms actually incorporated in the material. The analysis methods are too heavy for systematic use, applicable in a limited range of concentrations and sometimes encounter difficulties in hydrogenated amorphous silicon such as the similarity between masses of $Si^{28}H_3$, $Si^{29}H_2$, $Si^{30}H$ and P^{31} in SIMS analysis. Moreover, one suspects that the chemistry of dopant incorporation will be reactor dependent so that calibration from one laboratory of the incorporated concentration with respect to the gas mixture concentration may not be valid in other cases. Care must thus be exercised when comparing results expressed in terms of gas mixture concentrations. As an example, for a reactor of the type of Fig. 5.1 A using hydrogen dilution at a temperature of 600 °C, high mass resolution SIMS has yielded the following results for concentration ratios:

$[P]/[Si]$ in the solid $= 7 \times 10^{-3}$ for $[PH_3]/[SiH_4] = 10^{-3}$,

$[B]/[Si]$ in the solid $= 7 \times 10^{-3}$ for $[B_2H_6]/[SiH_4] = 3 \times 10^{-4}$.

5.6.2 ESR and Transport Properties

Systematic modifications of the ESR signal with doping are observed. The behaviour is similar in the *n*-type [5.1, 12, 27, 39] and *p*-type [5.12, 38] cases, as shown in Fig. 5.11. At low doping levels, expressed in relative dopant concentration x in the gas mixture, the dangling-bond ESR signal ($g = 2.0058$) decreases with increasing dopant concentration. At a critical dopant concentration the ESR signal becomes very small, in some cases below the sensitivity limit ($\sim 10^{17}$ cm^{-3}). Estimations from SIMS data of the dopant concentration in the solid, at this point, yield a value of 2×10^{-3} both for phosphorus and boron doping.

Above this dopant concentration an ESR signal is again observed, but an abrupt [5.12] transition in *g*-factor occurs. It becomes $g = 2.0043$ for *n*-type doping and $g = 2.013$ for *p*-type doping. Signals with similar *g*-values have been observed in glow discharge a-Si and ascribed to carriers in conduction band and valence band-tail states [5.40]. We shall use this denomination, although little is known on the specific nature of these states.

The conductivity behaviour is closely correlated with the ESR behaviour. Figure 5.12 shows a comparison of ESR and room temperature conductivity variations for phosphorus doped samples. There are only small variations of conductivity in the range where the dangling-bond signal is observed. Above the critical concentration where the shift in *g*-factor occurs, the conductivity quickly rises to reach values exceeding 10^{-1} Ω^{-1} cm^{-1}. In this doping range, as exemplified by the two curves for $x = 4 \times 10^{-4}$ and $x = 2 \times 10^{-3}$ in Fig. 5.10, the temperature dependence is simply activated. Comparing this with similar data for glow discharge a-Si grown in the vicinity of 250 °C, we

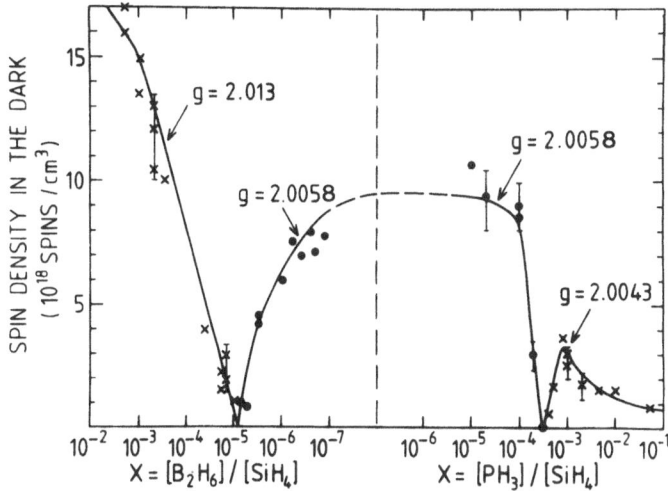

Fig. 5.11. ESR spin density for doped CVD films. The quantity X refers to dopant proportion in the gas mixture. The circles refer to the signal associated with dangling bonds ($g = 2.0058$). The crosses refer to the signals associated with conduction band ($g = 2.0043$) and valence band ($g = 2.013$) tail states

note that the maximum conductivities are higher for CVD films by an order of magnitude; the minimum activation energies are less by 0.1 eV.

When the dangling bonds are removed by post-hydrogenation, the electrical conductivity becomes sensitive to doping even for low dopant concentration, as exemplified in Fig. 5.12 for phosphorus doping. More complex effects of post-hydrogenation are observed in boron doped films and will be discussed later.

Before entering into a discussion of doping mechanisms, several additional experimental observations should be mentioned.

a) Light-induced ESR in phosphorus doped samples has been reported by *Friederich* and *Kaplan* [5.41]. They found in doped samples for which the ESR is observed at $g = 2.0043$ in the dark that a light-induced signal was produced by above-band gap radiation with a g-value of 2.0055 corresponding to the dangling bonds.

b) Photoconductivity is strongly enhanced by phosphorus doping [5.37]. The temperature dependence shows a maximum near room temperature for films doped above the threshold of the conductivity increase.

c) Field effect measurements have been performed for phosphorus and boron doped samples [5.1, 38]. The results have been analysed in terms of band conduction in a manner similar to previous studies of glow discharge a-Si. The results are summarized in Fig. 5.13, taken from *Hirose's* review [5.1]. The deduced gap state density has a sharply decreasing tail, both on the conduction band side for phosphorus doped samples and on the valence band

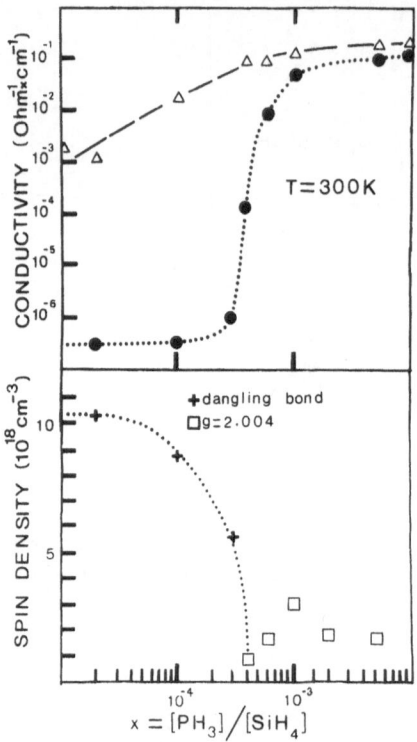

Fig. 5.12. Comparison of the conductivity (●●●) and spin-density variations (×××, □□□) with n-type doping (in as-deposited CVD films). The quantity X refers to the dopant proportion in the gas mixture. The dangling bond ESR data are shown as crosses and the tail-state ESR data as squares. Conductivity after post-hydrogenation is also shown for comparison (△△△)

side for boron doped samples. The position of this tail moves with doping, approaching the band edges as the doping is increased.

5.6.3 Discussion of the Effect of Doping

The most simple way to interpret the results of Figs. 5.11, 5.12 is to assume that a fraction of the incorporated dopants act as donors and acceptors. The corresponding electrons and holes fill states in the center of the gap that act as deep compensators. Once all these states are filled, additional doping will produce a shift of the Fermi level towards the bands and a rapid increase in the conductivity. The ESR behaviour points to the dangling bonds as the origin of the compensating states: paramagnetic neutral dangling bonds D^0 are converted into nonparamagnetic D^- and D^+ charged dangling bonds by doping. The center of the gap region can thus be described schematically by the density of states of Fig. 5.14, where the conduction band and valence band-tail states are, respectively, responsible for the $g = 2.0043$ and $g = 2.013$ ESR signals.

Within this model it is interesting to compare the critical dopant concentration for which the ESR signal is minimal, estimated to be above 2×10^{-3}

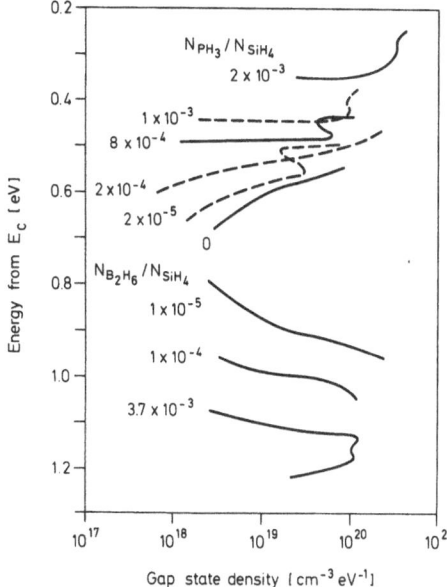

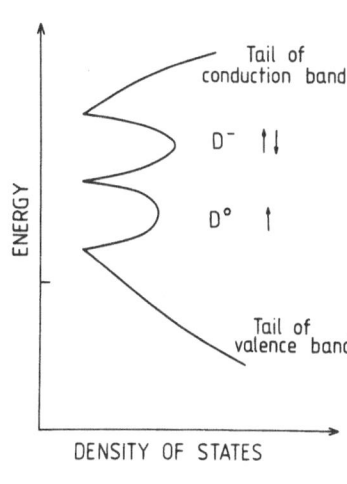

Fig. 5.13. Density of gap states, deduced from field effect measurements, in a series of *n*-type and *p*-type doped films [5.1]

Fig. 5.14. Schematic density of gap states showing two charge states D^0 and D^- of the dangling bonds with the associated spin configurations

in the solid, with the dangling-bond density in undoped films which is of order 2×10^{-4} (Fig. 5.6). This mismatch may imply some of the following:
a) only a fraction of the dopant atoms are electrically active.
b) doping induces additional dangling bonds, as has been suggested for gd a-Si [5.42];
c) additional nonparamagnetic states exist in the center of the gap.

An attempt has been made to estimate doping efficiency in post-hydrogenated films. *Szydlo* et al. [5.43] have made Schottky-diode structures on weakly phosphorus doped $(x < 5 \times 10^{-6})$ post-hydrogenated films. It is argued that for this doped material in which the Fermi level lies in the steep conduction band tail, the analysis of ordinary diodes on crystalline material applies approximately. A density of active impurities is derived. Using SIMS calibration, about 50% of the phosphorus atoms are found to be active.

If this result applies to as-deposited specimens, which is not obvious in view of the effects of hydrogen on electrical activity to be discussed later, then arguments (b) or (c) have to be used to resolve the mismatch. It should be pointed out, however, that large uncertainties of a factor of two or more exist for all the numbers involved and only order of magnitude agreements can be hoped for.

A different description of the effect of doping has been proposed by several authors involving the reduction of the density of dangling bonds with increasing dopant concentration. In addition to electronic compensation reducing the spin signal, one can consider a chemical compensation mechanism eliminating dangling bonds. The main arguments are as follows.

a) An initial decrease in conductivity with increasing dopant concentration before the minimum of the ESR signal is often observed [5.37]. This could indicate a reduction in the density of the states near the Fermi level which play a role in conductivity (Sect. 5.5.1).

b) The enhancement of the photoconductivity [5.37] with doping may be related to an increase in lifetime due to dangling bond elimination.

c) Field effect results of Fig. 5.13 show a strong reduction of the density of states in the center of the gap with increased doping.

d) A reduction in the optical absorption in the below band gap region for a doped film has been reported [5.37].

On the other hand, the optically induced ESR results [5.41] show that a fraction of the dangling bonds are still present, so that chemical compensation is at least not completely efficient.

In the opinion of the author, the arguments given in favor of chemical compensation are not thoroughly conclusive.

a) Recent data on the density of states of glow discharge a-Si indicate a minimum located somewhat above the dangling-bond states at an energy of 0.4 eV below the conduction band mobility edge. This may explain the conductivity minimum observed in phosphorus doped films.

b) Increasing photoconductivity with increasing n-type doping is observed in glow discharge films [5.44] for which phosphorus is considered to increase the density of dangling bonds rather than reduce it. Changes in photoconductivity are interpreted in terms of the variation of defect state occupancy rather than defect numbers.

c) Interpretation of the field effect data for states near the center of the gaps is somewhat suspect since the conductivity is not solely due to band conduction. The analysis corrects for contributions to the conductivity arising from states near the Fermi level, but neglects any field dependence of this conductivity. Such field dependence could account for apparent low densities of states.

d) The reduction in optical absorption is still preliminary data. If confirmed it could prove to be the strongest evidence in favor of chemical compensation, since the behaviour of glow discharge a-Si is opposite: below band gap absorption increases with doping [5.45]. This is a crucial method to test for systematic variations of gap state density with doping.

5.6.4 Effect of Post-Hydrogenation on Boron-Doped Films

Two effects of post-hydrogenation have been discussed so far, both related to the elimination of dangling bonds:

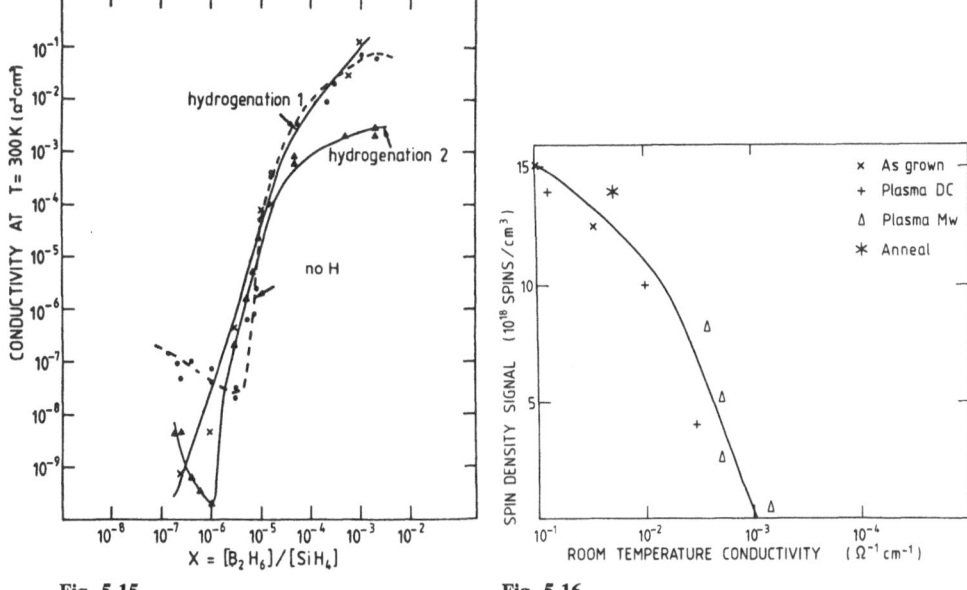

Fig. 5.15. Room temperature conductivity for as-deposited (●●●, ---) and post-hydrogenated boron-doped CVD films. X refers to the diborane proportion in the gas mixture. Hydrogenation 1 (×××) corresponds to hydrogenation by a weak plasma (dc system) and hydrogenation 2 (△△△) to a strong plasma (microwave system)

Fig. 5.16. Correlation between tail-state ESR spin density ($g = 2.013$) and room temperature conductivity in strongly boron-doped CVD films. The different points correspond to different doping levels and different post-treatments, as indicated. DC and Mw refer to hydrogenation in a dc or microwave plasma system, respectively. Anneal refers to annealing in vacuum after hydrogenation

a) reduction of the conductivity mechanism associated with states near the Fermi level;

b) an increase in Fermi level shifts at small dopant concentrations.

The behaviour is more complex in boron-doped films [5.12], as exemplified in Fig. 5.15 which compares room temperature conductivity of as-deposited films with films exposed to a weak plasma (hydrogenation 1) and to a stronger plasma (hydrogenation 2). The hydrogenation 1 curve can be interpreted in terms of dangling-bond elimination, taking into account that at small boron doping, the Fermi level shifts towards the center of the gap so that both (a) and (b) lead to a decrease in conductivity. The hydrogenation 2 curve, however, poses a problem since it appears that increasing the plasma strength, i.e., the hydrogen content, shifts the conductivity increase towards higher doping and characteristically decreases the maximum conductivities by two orders of magnitude. This decrease can be recovered by annealing for a few minutes in vacuum at $T \sim 300$ °C. This indicates that strongly doping with boron has the additional effect of increasing the hydrogen diffusion coefficient.

Related behaviour is seen for the tail-state ESR ($g = 2.013$): it decreases upon hydrogenation in the stronger plasma. The correlation between tail-state ESR and conductivity is demonstrated in Fig. 5.16 where one is plotted against the other for a number of films with varied dopings and submitted to varied hydrogenation treatments.

This simple correlation between conductivity and band-tail ESR is a strong indication that what happens with strong hydrogenation is a reduction in the number of electrically active boron atoms as opposed to a modification of the structure of the density of states keeping the boron activity constant. The origin of this effect of hydrogen on the boron activity is very much a matter of conjecture. Presumably boron-hydrogen complexes are formed which account for both the change in activity and the modified diffusion coefficient. Recent ir data indicate the existence of boron-hydrogen bonds in glow discharge a-Si. NMR results suggest the existence of two kinds of boron sites.

The question may be asked as to whether a similar effect exists with phosphorus. It has never been reported, but it should be pointed out that if the diffusion coefficient of hydrogen remains low in phosphorus doped samples, it is difficult to hydrogenate strongly the complete depth of the sample and the effect may be masked. Recent data of *Zellama* et al. [5.46] on annealing studies of glow discharge a-Si show increases in conductivity by an order of magnitude at temperatures slightly below crystallization where most of the hydrogen has diffused out.

5.7 Applications

5.7.1 Solar Absorbers

The first proposed application for CVD amorphous silicon has been as one of the components of a solar absorber stack [5.47]. A proposed stack structure is silicon nitride/a-Si/silver/substrate. The silicon layer acts as an absorber in the visible photon energy range and is transparent in the infrared range. The silver layer ensures that the stack is reflecting in the infrared, thus minimizing radiation losses. The silicon nitride layer acts as an anti-reflection coating in the visible. It is needed because of the high refractive index of silicon. Replacement of crystalline silicon by amorphous silicon which has a higher absorption constant allows the use of thinner layers. This kind of selective absorber is especially attractive for high temperature photothermal conversion. However, in this context, recrystallization of amorphous silicon presents a potential problem. The Tucson group [5.21] has shown that suitable dopants can stabilize the amorphous structure while maintaining adequate optical properties. As mentioned earlier, films containing 18% carbon will crystallize around 950 °C and should be stable for several decades at 700 °C.

Amorphous silicon is also a good optical match to metals like molybdenum [5.21] for use in tandem structures in which no further anti-reflection coating is needed.

5.7.2 Fast Photoconductor

As-deposited CVD amorphous silicon has been demonstrated to be a fast, if not sensitive, photoconductor [5.48]. The general concept is that, if the mobility in the extended states is sufficiently high, photoconduction will occur mainly during the short time an excited carrier remains above the mobility edge. After this, due to the relatively high density of defects in this material, the carrier should be rapidly trapped into deeper gap states and participate no further in the conduction process. A picosecond-pulse experiment has been reported by *Auston* et al. [5.48]. The a-Si films were arranged in a strip line configuration and the photoconduction current was measured using a sampling oscilloscope. The pulse width was of the order of 20 to 30 ps, a figure comparable to the oscilloscope response time. Later experiments using correlation techniques confirmed this order of magnitude of the photoconduction peak, a value of the mobility of the order of $1 \text{ cm}^2(\text{V s})^{-1}$ could be inferred for transport in the extended states. Because of the simplicity of their fabrication, this type of detector may play a role in future picosecond pulse experiments.

5.7.3 Diodes and Photovoltaic Applications

Schottky barrier diodes have been fabricated on post-hydrogenated CVD a-Si [5.43, 49]. Typically, on a conductive substrate a phosphorus doped n^+ layer is first deposited, then a lightly doped or intrinsic layer. The sample is then post-hydrogenated around 400 °C in a plasma. Finally, a Pt contact is deposited and usually annealed at a temperature in the vicinity of 200 °C.

I–V characteristics for two such devices are shown in Fig. 5.17. Doping usually increases the currents and the ideality factor n. Ideality factors as low as 1.15 have been reported for undoped active layers.

Weakly post-hydrogenated CVD a-Si has a smaller band gap than glow discharge a-Si and could thus be interesting in photovoltaic applications by collecting a larger fraction of the solar spectrum. Figure 5.18 shows examples of collection efficiencies as a function of wavelength obtained by *Szydlo* et al. [5.43] in Schottky diodes with semi-transparent platinum electrodes. The collection at long wavelength decreases with increasing doping. This should be related to space charge width variations. The best results are obtained for undoped active layers [5.43, 49]. They are comparable to similar data for

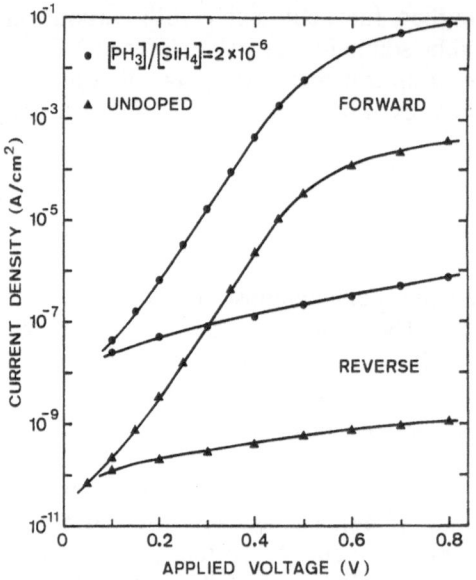

Fig. 5.17. Current versus voltage curve for Pt/a-Si Schottky diodes, with an undoped ($\triangle\triangle\triangle$) and phosphorus doped ($\bullet\bullet\bullet$, doping indicated) active layer

glow discharge a-Si so that, with the present quality of the films, advantage has not been gained from the increased optical absorption.

p-i-n structures have also been fabricated [5.50]. Rectification ratios of order 10^6 are obtained. The main feature of these diodes is their ability to accommodate large forward currents of order 50 A cm^{-2} in dc, together with inverse voltages of order 20 V. They can thus rectify relatively large powers. The large forward currents may be related to the large conductivities observed in n^+ and p^+ layers.

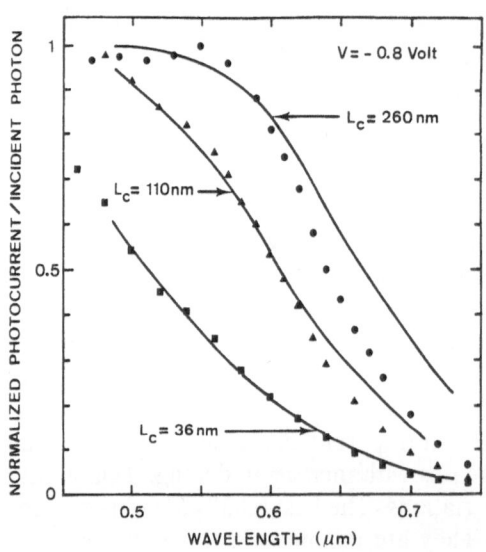

Fig. 5.18. Photocurrent versus wavelength in Schottky photodiodes of various dopings [5.43]. A fit (——) is made to obtain the collection lengths L_c indicated, which decrease with increasing doping level

5.8 Conclusion

CVD a-Si has been given less attention than glow discharge a-Si. The reason is the large density of native dangling-bond defects which make it less suitable for electronic applications. This situation may be modified in the future by use of the post-hydrogenation technique and/or extension of the low temperature deposition methods, using higher silanes or HOMOCVD. In the long term, specific advantages of CVD (or LPCVD) systems for designing large production systems may become important. In the near term, it is not likely to be a strong competitor for those applications which require the lowest density of gap states, e.g., solar cells. By other criterion, however, the material appears to be superior, for instance, in the ability to obtain high conductivity n^+ or p^+ layers. It may find its niche in specific applications.

From a fundamental point of view it can be considered as a reference material, relatively independent of preparation details, and a good vehicle to investigate basic questions concerning a-Si. In this respect, the problem of the relationship between doping and defect formation appears to be central at the time of writing and deserves further investigations.

Another question is that of the microstructure. Perhaps the most striking observation for CVD material is the fact that hydrogen diffusion coefficients appear to be independent of hydrogen content and similar to those observed in glow discharge samples. This gives a strong presumption that the hydrogen microstructures, e.g., clustered and diluted regions, do not depend on whether the hydrogen is grown into the film as in glow discharge deposition or introduced a posteriori as in plasma post-hydrogenated CVD films. This is an incentive to pursue the idea that these microstructures are intrinsic to the amorphous network and a basic property of tetrahedral materials.

References

5.1 M. Hirose: J. Physique **42**, C4–705 (1981)
5.2 J. Bloem, L. J. Giling: In *Current Topics in Materials Science* 1, (North-Holland, Amsterdam 1978) Chap. 4, p. 147
5.3 M. Janaï, R. Weil, K. H. Levin, B. Pratt, R. Kalish, G. Braunstein, M. Teicher: J. Appl. Phys. **52**, 3622 (1981)
5.4 M. Janaï, S. Aftergood, R. B. Weil, B. Pratt: J. Electrochem. Soc. **128**, 2661 (1981)
5.5 S. C. Gau, B. R. Weinberger, M. Akhtar, Z. Kiss, A. G. MacDiarmid: Appl. Phys. Lett. **39**, 436 (1981)
5.6 B. A. Scott, R. M. Plecenik, E. E. Simonyi: Appl. Phys. Lett. **39**, 73 (1981)
5.7 B. A. Scott, J. A. Reimer, R. M. Plecenik, E. E. Simonyi, W. Reuter: Appl. Phys. Lett. **40**, 973 (1982)
5.8 P. A. Thomas, M. H. Brodsky, D. Kaplan, D. Lépine: Phys. Rev. B **18**, 3059 (1978)
5.9 D. Kaplan, N. Sol, G. Velasco, P. A. Thomas: Appl. Phys. Lett. **35**, 440 (1978)
5.10 N. Sol, D. Kaplan, D. Dieumegard, D. Dubreuil: J. Non-Cryst. Solids **35–36**, 291 (1980)
5.11 M. Janaï, D. D. Allred, D. C. Booth, B. O. Seraphin: Solar Energy Mat. **1**, 11 (1979)

5.12 J. Magariño, D. Kaplan, A. Friederich, A. Deneuville: Phil. Mag. B **45**, 285 (1982)
5.13 J. H. Purnell, R. Walsh: Proc. R. Soc. A **293**, 543 (1966)
5.14 M. L. Hitchman, J. Kane, A. E. Widmer: Thin Solid Films **59**, 231 (1979)
5.15 W. A. Bryant: Thin Solid Films **60**, 19 (1979)
5.16 B. A. Scott, R. M. Plecenik, E. E. Simonyi: J. Physique **42**, C4–635 (1981)
5.17 K. J. Sladek: J. Electrochem. Soc. **118**, 655 (1971)
5.18 J. F. Morhange: Private communication
5.19 G. Harbeke, A. E. Widmer, J. Stuke: J. Phys. Soc. Jpn. **49**, Suppl. A, 1229 (1980)
5.20 K. Zellama, P. Germain, S. Squelard, J. C. Bourgoin, P. A. Thomas: J. Appl. Phys. **50**, 6995 (1979) and private communication
5.21 D. C. Booth, D. D. Allred, B. O. Seraphin: Solar Energy Mat. **2**, 107 (1979)
5.22 C. W. Magee: cited in [5.19]
5.23 E. Bustarret, J. C. Bruyère, A. Deneuville, J. F. Currie, P. Depelsenaire, R. Groleau: Proc. of the CVD "81" Conf., ed. by J. M. Blocher, G. E. Vuillard, Electrochem. Soc. (1981) p. 347
5.24 M. Olivier, A. Chevenas-Paule: Private communication
5.25 W. Paul: Solid State Commun. **34**, 283 (1980)
5.26 D. Kaplan: Physica Scripta **24**, 396 (1981)
5.27 S. Hasegawa, T. Kasajima, T. Shimizu: Solid State Commun. **29**, 13 (1979)
5.28 H. Fritzsche: Solar Energy Mat. **3**, 447 (1980)
5.29 K. Zellama, P. Germain, S. Squelard, B. Bourdon, J. Fontenille, R. Daneliou: Phys. Rev. **B 22**, 6648 (1981) and private communication
5.30 D. E. Carlson, C. W. Magee: Appl. Phys. Lett. **33**, 81 (1978)
5.31 T. Suzuki, M. Hirose, Y. Osaka: Japan J. Appl. Phys. **19**, Suppl. 19-2, 91 (1979)
5.32 M. Janaï, B. Karlsson: Solar Energy Mat. **1**, 387 (1979);
A. Divrechy, B. Yous, J. M. Berger, J. P. Ferraton, J. Robin, A. Donnadieu: Thin Solid Films **78**, 235 (1981)
5.33 C. C. Tsai, H. Fritzsche: Solar Energy Mat. **1**, 29 (1979)
5.34 W. B. Jackson, N. M. Amer: J. Physique **42**, C4–293 (1981)
5.35 R. A. Street, B. K. Biegelsen: Solid State Commun. **33**, 1159 (1980)
5.36 D. J. Wolford, B. A. Scott, J. A. Reimer, R. M. Plecenik, J. A. Bradley: Bull. Am. Phys. Soc. **27**, 145 (1982)
5.37 M. Taniguchi, M. Hirose, Y. Osaka: J. Cryst. Growth **45**, 126 (1978)
5.38 T. Nakashita, M. Hirose, Y. Osaka: Jpn. J. Appl. Phys. **20**, 471 (1981)
5.39 M. Hirose, M. Taniguchi, T. Nakashita, Y. Osaka, T. Suzuki, S. Hasegawa, T. Shimizu: J. Non-Cryst. Solids **35–36**, 297 (1980)
5.40 H. Dersch, J. Stuke, J. Beichler: Phys. Stat. Sol. (b) **105**, 265 (1981)
5.41 A. Friederich, D. Kaplan: J. Phys. Soc. Jpn. **49**, Suppl. A, 1233 (1980)
5.42 R. A. Street: Phys. Rev. Lett. **49**, 1187 (1982)
5.43 N. Szydlo, J. Magariño, D. Kaplan: J. Appl. Phys. **53** (7), 5044 (1982)
5.44 D. A. Anderson, W. E. Spear: Phil. Mag. **36**, 695 (1977)
5.45 B. Jackson, N. M. Amer: Phys. Rev. B **25**, 5559 (1982)
5.46 K. Zellama, P. Germain, S. Squelard, J. Magariño, B. Bourdon: Submitted for publication
5.47 B. O. Seraphin: J. Vac. Sci. Tech. **16** (2), 193 (1979)
5.48 D. H. Auston, P. Lavallard, N. Sol, D. Kaplan: Appl. Phys. Lett. **36** (1), 66 (1980)
5.49 Y. Mishima, M. Hirose, Y. Osaka: Japan J. Appl. Phys. **20**, 593 (1981)
5.50 N. Szydlo, E. Chartier, N. Proust, J. Magariño, D. Kaplan: Appl. Phys. Lett. **40** (11), 988 (1982)

6. Solar Energy Conversion

David E. Carlson

With 29 Figures

Hydrogenated amorphous silicon (a-Si : H) is a glassy semiconducting material that is not likely to occur in nature. The material is a disordered alloy of silicon and hydrogen whose composition may vary from about 1 at.% H to about 50 at.% H, depending on the method of formation. Scientific interest in this material has increased enormously in the past few years so that today there are several hundred scientists worldwide investigating a-Si : H. The reason for this large research effort is that a-Si : H promises low-cost, nonpolluting electrical power via the photovoltaic effect.

This chapter will focus on those properties of a-Si : H that are relevant to photovoltaic energy conversion and on the various procedures and techniques used to fabricate a-Si : H solar cells.

6.1 Background

6.1.1 Photovoltaic Energy Conversion Efficiency

A solar cell converts sunlight directly into electricity via the photovoltaic effect. The conversion efficiency (η) is defined as that percentage of the total power in sunlight that is converted into electrical power and may be expressed as

$$\eta = \frac{J_m V_m}{P_i} = \frac{(FF) J_{sc} V_{oc}}{P_i} , \tag{6.1}$$

where J_m and V_m are the output current density and voltage of a solar cell operating under maximum power conditions and P_i is the incident power density of the sunlight ($P_i \simeq 1$ kW m^{-2} for the sun directly overhead on a clear day; AM 1 illumination). J_{sc} is the short-circuit current density and V_{oc} is the open-circuit voltage. The fill factor (FF) is defined by (6.1) and is determined by finding the maximum area under a plot of photocurrent versus photovoltage (Fig. 6.1). The J-V characteristic shown in Fig. 6.1 is obtained by either varying the value of a load resistor in series with the cell or by varying the potential of a low impedance power supply in series with the cell.

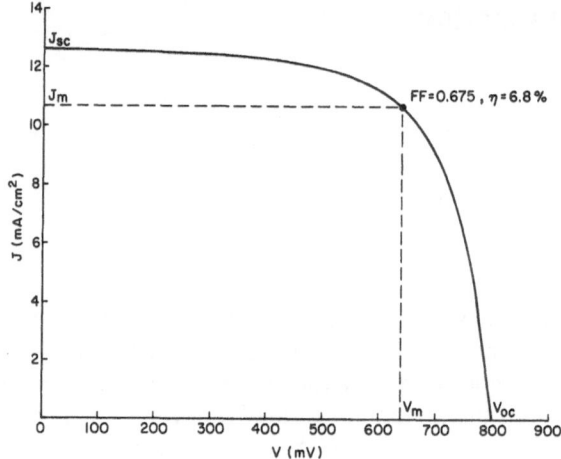

Fig. 6.1. Current density as a function of voltage for an a-Si:H solar cell exposed to 100 mW cm⁻² of simulated sunlight

There are several conditions that a thin-film solar cell must satisfy in order to exhibit efficient photovoltaic energy conversion:
a) the optical absorption coefficient (α) of the active film material must be large enough so that a significant fraction of the energy in sunlight is absorbed within the thickness of the film;
b) the photogenerated electrons and holes must be collected efficiently by the contacting electrodes on both sides of the active film material;
c) the solar cell must have a large built-in potential provided, for example, by a semiconductor junction;
d) the total resistance in series with the cell (excluding the load) must be small to minimize power losses (Joule heating) during operation;
e) the thin film structure must be uniform over the entire active area of the cell so that shorts and shunts do not degrade performance.

We will discuss in the sections to follow each one of these conditions as they apply to a-Si:H solar cells.

6.1.2 A Brief Review of a-Si:H History

Hydrogenated amorphous silicon was first made and investigated by *Chittick* et al. [6.1] in 1969, but the role of hydrogen was not appreciated until about 1975 [6.2, 3]. Pure or nonhydrogenated amorphous silicon (a-Si) contains many dangling or broken bonds that create a large density of localized states throughout the energy gap of the semiconductor. Hydrogen can terminate these dangling bonds and remove the localized states from the energy gap. Consequently, a-Si:H exhibits many properties not evident in pure a-Si.

Chittick et al. [6.1] observed a large photoconductive effect in their films whereas the photoconductivity of sputtered or evaporated a-Si films is negligible. In 1972, *Spear* and *LeComber* [6.4] demonstrated that the Fermi level in a-Si:H could be moved by the electric field generated in a MOS transistor-like structure. In pure a-Si, the density of gap states is so large that the Fermi level is pinned near midgap. In 1974, *Engemann* and *Fisher* [6.5] observed relatively efficient photoluminescence in a-Si:H at a temperature of 77 K; photoluminescence is not measurable in pure a-Si because the nonradiative recombination lifetime is very short.

The first electronic devices utilizing a-Si:H were made in 1974 at RCA Laboratories [6.6]. In 1975, *Spear* and *LeComber* [6.7] published a detailed study of the substitutional doping of a-Si:H. Prior to that time it was thought that one could not dope amorphous semiconductors because the amorphous material would locally arrange itself to satisfy the normal valence state of the dopant atom [6.8]. The first published work on a-Si:H devices appeared in 1976 [6.9, 10], and since then there has been an explosive growth in the number of publications dealing with a-Si:H.

6.2 Properties of a-Si:H Relevant to Photovoltaics

A semiconducting material must exhibit certain properties before it can be considered for application in a thin-film solar cell. In this section we consider those properties of a-Si:H that are relevant to photovoltaic energy conversion.

6.2.1 Optical Absorption

Semiconductors absorb radiation mainly by exciting electrons from the valence band to the conduction band, thus creating mobile electron-hole pairs. These mobile carriers constitute the photocurrent in a solar cell. The bandgap of a semiconductor determines the maximum possible photocurrent that a solar cell can deliver in the short-circuit mode. For example, the short-circuit current density (J_{sc}) of a semiconductor with a bandgap of 1.7 eV cannot exceed ~ 23 mA cm^{-2} in AM 1 illumination (~ 100 mW cm^{-2}) [6.11]. This current density limit is determined by the photon flux in the solar spectrum for photon energies greater than 1.7 eV. Since the photons with energy less than 1.7 eV cannot be absorbed in this case (or at least do not generate mobile electron-hole pairs), they do not contribute to J_{sc}.

In a thin-film solar cell, the short-circuit current density may be significantly less than that estimated by the above procedure depending on the thickness of the active layer and the variation of the absorption coefficient (α) with photon energy ($h\nu$). Figure 6.2 shows the absorption coefficient as a function of photon energy for device quality a-Si:H, single crystal silicon and gallium arsenide (GaAs). Single crystal silicon is an indirect bandgap semiconductor and consequently, the optical absorption is relatively weak for an extended range above the indirect bandgap of 1.1 eV. Gallium arsenide is a

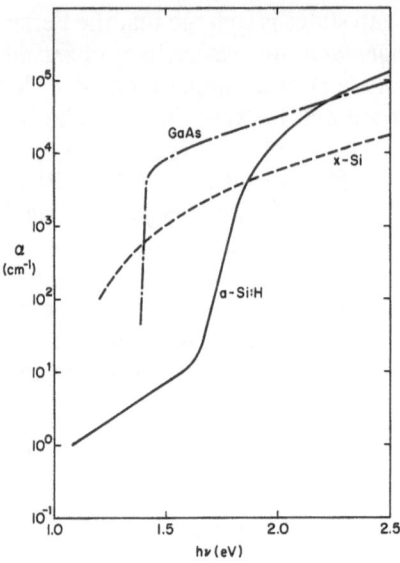

Fig. 6.2. Optical absorption coefficient as a function of photon energy for GaAs, crystalline Si and undoped a-Si:H deposited at ∼ 300 °C in a dc proximity discharge

direct bandgap semiconductor and $\alpha(h\nu)$ is relatively large above the bandgap (E_g) of 1.43 eV. Clearly, the optical absorption coefficient of a-Si:H exhibits a behavior more like that of GaAs than that of crystalline Si.

An optical gap (ε_{opt}) can be defined for a-Si:H by plotting $(\alpha h\nu)^{1/2}$ as a function of $h\nu$, as shown in Fig. 6.3. This dependence is predicted for amorphous semiconductors if the band edges are parabolic and the matrix elements for the optical transitions are energy independent [6.12]. Optical

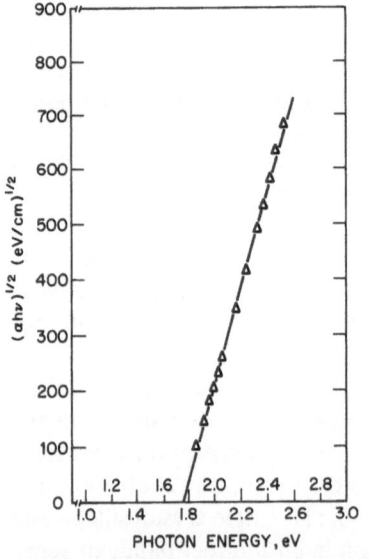

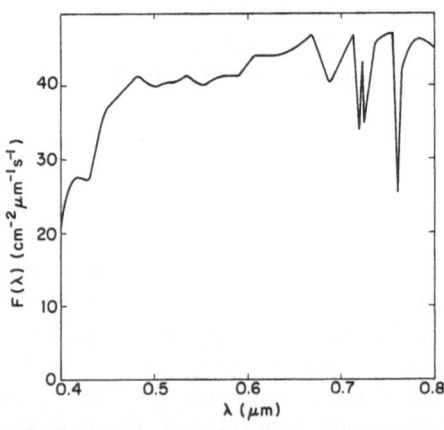

Fig. 6.3. A plot of $(\alpha h\nu)^{1/2}$ versus $h\nu$ for an undoped a-Si:H film

Fig. 6.4. Photon flux as a function of photon wavelength for AM1 sunlight (∼ 100 mW cm^{-2})

gaps for device quality a-Si : H are usually in the range of 1.65–1.80 eV. The optical gap of a-Si : H varies with deposition conditions primarily because of changes in the hydrogen content [6.13]. The hydrogen content always decreases with increasing deposition temperature, but the variation with other deposition conditions is more complicated [6.14]. Device quality a-Si : H typically contains ~ 5 to 15 at.% hydrogen.

Generally, a semiconductor should have a bandgap somewhere in the range of 1.0–2.0 eV to qualify as an active layer material for an efficient solar cell [6.15]. However, in a thin-film cell, the material must also satisfy the condition $\alpha d \gtrsim 1$ over most of the visible light range where d is the thickness of the active layer. Referring to Fig. 6.2 we see that a-Si : H and GaAs satisfy the condition if $d = 1$ μm ($\alpha \gtrsim 10^4$ cm^{-1}), but crystalline Si does not. This condition follows from the observation that the maximum J_{sc} in a thin film of thickness d is given by

$$J_{sc} = q \int_{E_g}^{\infty} F(\varepsilon)\{1 - \exp[-\alpha(\varepsilon)d]\}d\varepsilon , \qquad (6.2)$$

where $F(\varepsilon)$ is the photon flux spectrum for sunlight and where we assume no reflection losses and that light transmitted through the film is absorbed at the back contact. A similar expression can be written as an integral over wavelength using $F(\lambda)$ and $\alpha(\lambda)$. $F(\lambda)$ is shown in Fig. 6.4.

In Fig. 6.5 we show how the maximum J_{sc} varies with thickness for a-Si : H films with $\varepsilon_{opt} \simeq 1.58$ and 1.80 eV (deposited at 420 and 215 °C, respectively). It is apparent from Fig. 6.5 that significant current densities (~ 10–16 mA cm^{-2}) are predicted for films only a few hundred nanometers thick.

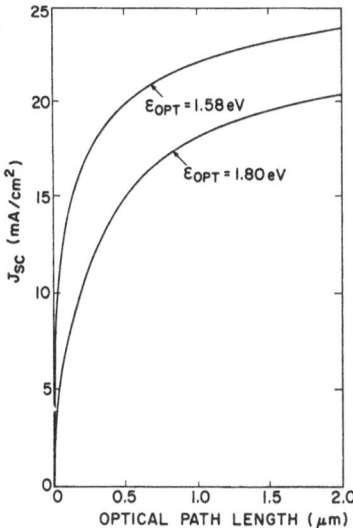

Fig. 6.5. The maximum possible short-circuit current density as a function of optical path length for a-Si : H films with $\varepsilon_{opt} \simeq 1.58$ and 1.80 eV

There are two other aspects of optical absorption data that are relevant to the operation of an a-Si : H solar cell. First, as shown in Fig. 6.2, the absorption coefficient of a-Si : H exhibits an exponential tail that has a characteristic energy of ~ 0.05–0.07 eV [6.16]. This absorption tail is apparently caused by intrinsic states associated with the disorder in the material. As discussed in Sect. 6.7.3, these tail states can influence the built-in potential of a solar cell.

The other point, also shown in Fig. 6.2, is that some optical absorption is always detectable even at relatively low photon energies ($hv \lesssim 1.5$ eV). Such small values of α ($\sim 10^{-1}$–10 cm^{-1}) can be measured using techniques such as photothermal deflection spectroscopy [6.17] or photoacoustic spectroscopy [6.18]. Both techniques have been used to show that the low energy absorption correlates with the electron spin density; in other words, α for $hv \lesssim 1.2$ eV is related to defects (with unpaired electrons) that give rise to localized states within the energy gap. These deep gap states act as recombination centers that limit the lifetimes of photogenerated carriers.

6.2.2 Mobilities and Lifetimes of Photogenerated Carriers

Early drift mobility measurements indicated an electron drift mobility of $\sim 10^{-1}$ cm^2/V s at room temperature with an activation energy of ~ 0.19 eV [6.19]. The data were interpreted in terms of a trap-controlled transport with a mobility of electrons in the extended states of ~ 10 cm^2/V s. The drift mobility of holes in boron-doped a-Si : H was estimated to be $\sim 6 \times 10^{-4}$ cm^2/V s at room temperature with an activation energy of ~ 0.35 eV [6.20]. More recent drift mobility measurements have shown the electron and hole drift mobilities to be ~ 1 and $\sim 10^{-3}$ cm^2/V s at room temperature with extended state values of 13 and 0.67 cm^2/V s, respectively [6.21].

The Hall mobility for electrons has been determined to be $\sim 10^{-1}$ cm^2/V s in undoped a-Si : H (slightly n-type) at room temperature [6.22]. As shown in Fig. 6.6, this mobility is roughly temperature independent up to ~ 100 °C and then exhibits an activation energy of ~ 0.13 eV for higher temperatures. These results indicate that the conduction of electrons at room temperature takes place by some tunneling or temperature-independent jumping process while thermally activated hopping occurs above ~ 100 °C. The sign of the Hall effect was anomalous, as found in earlier work on phosphorus-doped a-Si : H films [6.23]. Recent Hall mobility measurements on a-Si : H films lightly doped with boron indicate a mobility of $\sim 10^{-1}$ cm^2/V s for holes [6.24].

The lifetimes of excess carriers in a-Si : H have been determined by the junction recover method and are generally in the range of 10–30 μs for injection current densities of ~ 10 mA cm^{-2} [6.25]. This study showed that the lifetimes were almost inversely proportional to the injection current level over more than 2 orders of magnitude.

In some low mobility materials such as amorphous selenium, the lifetime of photogenerated carriers is limited by geminate recombination [6.26]. This

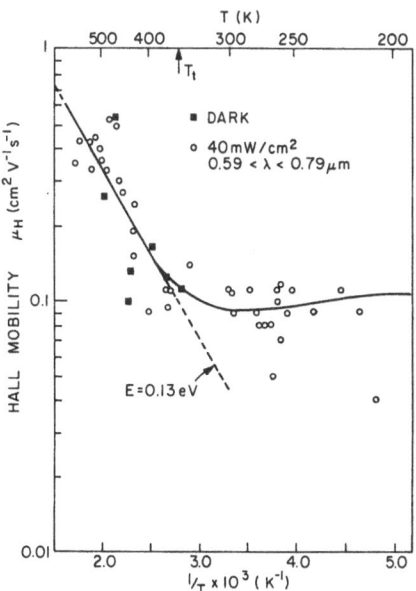

Fig. 6.6. Hall mobility of electrons in undoped a-Si:H under illumination and in the dark

is a process where an electron and a hole are created by the absorption of a photon and then recombine before they can escape from their mutual coulombic well. While some results for a-Si:H have been interpreted in terms of geminate recombination [6.27], it is clear that such effects must be small in good solar cells since the collection efficiency is close to 90% (for $\lambda \simeq 0.55$ mm) without correcting for reflection losses or absorption losses in the top doped layer [6.28].

For solar cells, the most important parameter is the product of the mobility and lifetime for the excess carriers at one sun illumination (~ 100 mW/cm^2) or at current densities of ~ 10–15 mA/cm^2. The $\mu\tau$ product enters into the determination of cell performance (see Sect. 6.6) either through the drift length $\mu\tau E$ (where E is the electric field), or through the diffusion length $(kT\mu\tau/q)^{1/2}$.

Moore [6.29] used the photoelectromagnetic effect to measure the diffusion length in undoped a-Si:H and has determined a value of ~ 0.1 μm. Recently, *Dresner* et al. [6.30] used a surface photovoltage technique to measure diffusion lengths as large as ~ 0.5 μm. In this technique, the surface photovoltage is held constant by adjusting the photon flux while the wavelength of the light is varied. The data can be plotted as shown in Fig. 6.7 to determine the diffusion length.

As shown in Fig. 6.8, the hole diffusion length exhibits a thermal activation energy of ~ 0.28 eV [6.31]. This temperature dependence is apparently related to the variation of the hole mobility with temperature since the temperature dependence of the lifetime should be weak as it is in crystalline silicon. More recently, measurements on numerous undoped films have

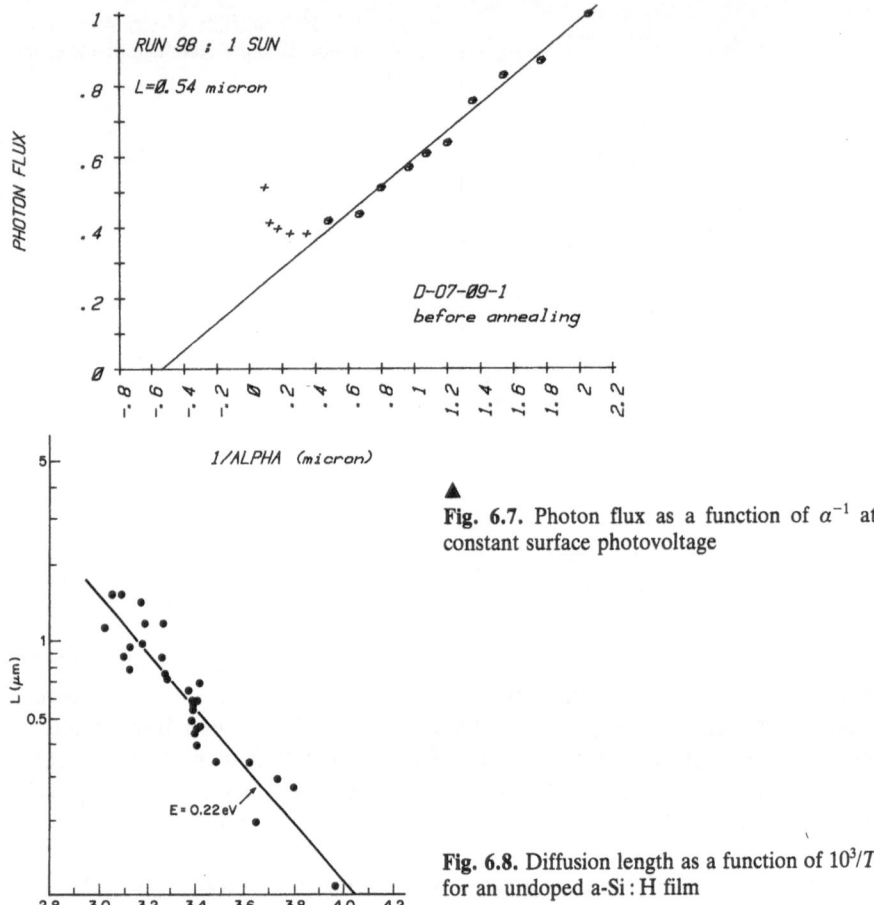

Fig. 6.7. Photon flux as a function of α^{-1} at constant surface photovoltage

Fig. 6.8. Diffusion length as a function of $10^3/T$ for an undoped a-Si:H film

yielded values of 0.5–0.8 μm while illuminated with constant white light of ~100 mW cm^{-2} [6.32].

6.2.3 Properties of Doped Films

We will now address those properties of doped a-Si:H films that are relevant to photovoltaic energy conversion.

In a *p-i-n* solar cell (Sect. 6.4), the light must traverse either a *p* or *n* layer before reaching the active *i* (intrinsic) layer. For heavily doped layers containing a few at.% of either boron or phosphorus, the recombination lifetime appears to be so short that these layers are essentially "dead"; i.e., photons absorbed in the doped layers do not contribute to the collected photocurrent [6.33].

Optical absorption data are shown in Fig. 6.9 for boron and phosphorus-doped a-Si:H films deposited in a dc proximity discharge at a substrate temperature of ~ 300 °C. Both doped films exhibit enhanced optical absorption (especially at low photon energies), as compared to that for an undoped film deposited under similar conditions. For a *p-i-n* cell with the light incident through a "dead" *p* layer (~ 10 nm thick), the short-circuit current can be reduced by ~ 20%, as compared to the maximum possible value predicted by (6.2) [6.34].

The doping of a-Si:H appears to create many defect states in addition to the desired acceptor or donor levels [6.14, 35]. Some of these defect states are dangling bonds [6.35] while others seem to be associated with dopant-

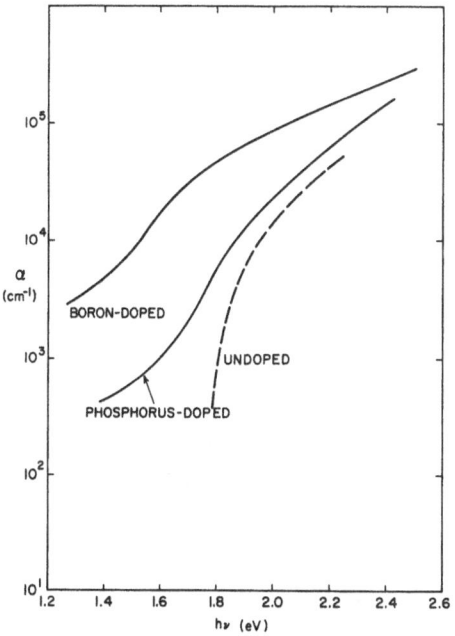

Fig. 6.9. Optical absorption coefficient as a function of photon energy for a boron-doped a-Si:H film (~ 6 at.% boron), a phosphorus-doped a-Si:H film (~ 7 at.% phosphorus) and an undoped a-Si:H film

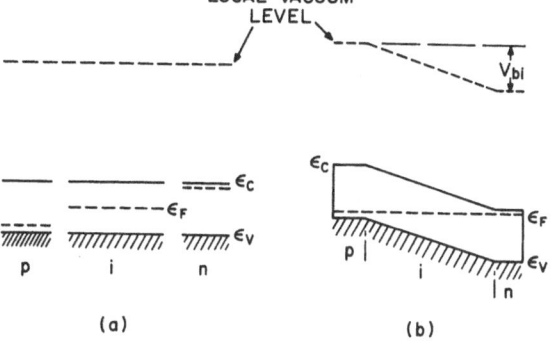

Fig. 6.10. (a) The energy band edges and Fermi levels of isolated *p*, *i* and *n* layers relative to the local vacuum level. (b) The energy band edges and Fermi level of a *p-i-n* structure in equilibrium in the dark

hydrogen complexes [6.14]. The relatively large infrared absorption tails shown in Fig. 6.9 are attributed to the large density of gap states in the doped films.

The doped layers in a p-i-n solar cell are mainly responsible for determining the built-in potential V_{bi} (and consequently, the open-circuit voltage V_{oc}). In Fig. 6.10 a we show the energy band edges and Fermi levels of isolated p, i and n layers relative to the local vacuum level. When these layers are in contact with each other (Fig. 6.10 b), the Fermi levels are in equilibrium (for no applied bias) and an internal electric field is developed. The built-in potential depends on the relative position of the Fermi levels in the p and n layers with respect to the local vacuum level.

From conductivity and thermopower measurements, the Fermi level in phosphorus-doped films is found to be ~ 0.2 eV below the conduction band [6.36] while in boron-doped films it is ~ 0.5 eV above the valence band [6.37]. Since the optical gap of undoped a-Si : H is ~ 1.65–1.80 eV, we might expect the built-in potential of p-i-n devices to be in the range of ~ 1.0–1.1 eV, and indeed this is the case [6.38, 39].

The conductivity of the doped layers is important for solar cells since it is associated with the position of the Fermi level and is probably related to the contact resistance at doped layer/electrode interfaces. Thus, improving the conductivity of a doped layer (without changing the bandgap) should lead to a larger built-in potential and a smaller contact resistance.

Recently, two new approaches for modifying the properties of doped layers have been reported. In one case, the conductivity of the doped layers has been increased from $\sim 10^{-2}(\Omega \text{ cm})^{-1}$ for conventional doped a-Si : H to $> 10(\Omega \text{ cm})^{-1}$ by inducing the formation of a microcrystalline phase [6.40]. Microcrystalline Si : H films can be formed by diluting the SiH_4 in H_2 and increasing the discharge power. The other approach involves increasing the optical gap by alloying the boron-doped a-Si : H layer with carbon [6.41]. This approach has increased the optical gap from ~ 1.4 eV to ~ 2.0 eV and has resulted in the development of relatively efficient a-Si : C : H/a-Si : H heterojunction solar cells [6.42] (Sects. 6.4.3, 6.5.1).

6.3 Deposition Conditions for Fabricating Solar Cells

Hydrogenated amorphous silicon films have been made by a variety of techniques such as glow-discharge deposition in SiH_4 [6.1], sputtering in an Ar–H_2 atmosphere [6.2], chemical vapor deposition from SiH_4 followed by plasma hydrogenation [6.43] and chemical vapor deposition from higher silanes [6.44]. Since relatively efficient solar cells have been produced only by glow-discharge and sputter deposition, we will confine our discussion of deposition conditions to these two processes.

6.3.1 Substrate Preparation

Most a-Si : H solar cells have been fabricated either on stainless steel sheet or on glass substrates coated with a conductive film. In the latter case, the conductive film is often a transparent conductive oxide such as indium-tin-oxide or tin oxide so that the cell can be illuminated through the glass. Many other conductive materials such as Cr, W, Ta, Ni, Mo, V, Ti, etc., can be used as substrates for a-Si : H solar cells either in sheet form or by utilizing thin films on glass [6.45]. However, some metals such as Au and Al are not suitable substrates since they interdiffuse with amorphous silicon at low temperatures and induce crystallization [6.46, 47]. Metals such as Cu and Ag are also not suitable substrates since the a-Si : H films do not adhere well to these materials.

In general, the substrate must be smooth and relatively free of dust in order to avoid shunts and shorts in large area devices. Contaminants such as fingerprints and organic deposits can be removed by conventional cleaning techniques such as rinsing in deionized water and degreasing in methanol vapor.

Sputter etching can also be used to clean substrates such as stainless steel. Since sputter cleaning can often be done in the same system used to deposit the a-Si : H, this technique can remove even oxide layers from the surface of the substrate.

6.3.2 dc Glow Discharges

The first a-Si : H solar cells were made in a SiH_4 glow discharge operating in a dc cathodic mode [6.6], i.e., the substrate was acting as the cathode (Fig. 6.11). The anode is usually a metal plate or screen located a few centimeters above the substrate. For discharges in pure SiH_4, the pressures are typically 0.5–1.0 torr with flow rates of ~ 30 sccm for cathode areas of ~ 100 cm^2. The substrate temperature is usually in the range of 200–400 °C. The deposition rate can vary from ~ 50 to ~ 1000 nm/min as the power density is increased from 0.1 to 2.0 W cm^{-2} at the cathode substrate. Thus, a-Si : H films can be deposited rapidly in this discharge mode.

However, positive ion bombardment of the growing a-Si : H film can create defects that act as recombination centers [6.48]. Bombardment damage can be minimized by locating a cathode screen just above the substrate (dc proximity method) or by operating in an anodic mode. Unlike the cathodic or anodic modes, the dc proximity method can be used to coat insulating substrates. One disadvantage is that the film growing on the cathode screen may flake off during long runs and debris may accumulate on the substrate.

For dc glow discharges, the most common discharge atmosphere has been pure SiH_4 [6.38, 45], but atmospheres such as SiF_4 and H_2 [6.49], 10% SiH_4 in

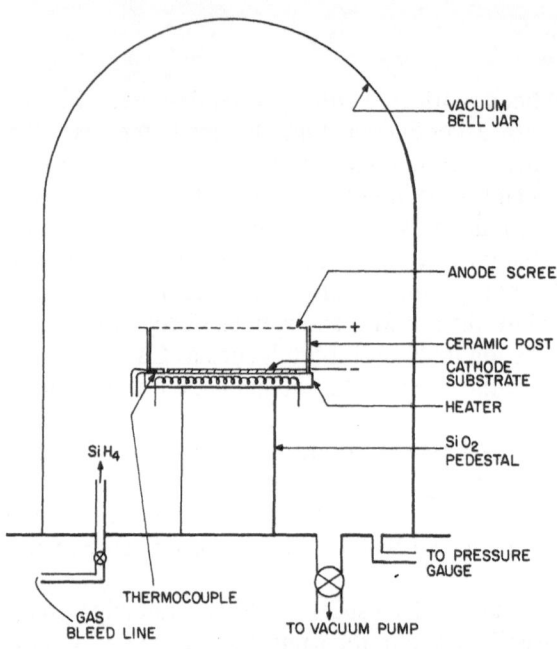

Fig. 6.11. dc cathodic glow-discharge deposition system

Ar [6.50] and 10% SiH$_4$ in H$_2$ [6.51] have also been used. Doped films for solar cells are usually produced from SiH$_4$ discharges containing ~ 1 vol.% of B$_2$H$_6$ in SiH$_4$ for *p*-type films or ~ 1 vol.% of PH$_3$ in SiH$_4$ for *n*-type films [6.10]. Recent work at RCA Laboratories has shown that BCl$_3$ and PF$_5$ can also be used as doping gases in a dc silane discharge.

6.3.3 rf Glow Discharges

The majority of researchers have used rf glow discharges to make a-Si : H. *Chittick* et al. [6.1] used an rf electrodeless system where the SiH$_4$ gas flowed

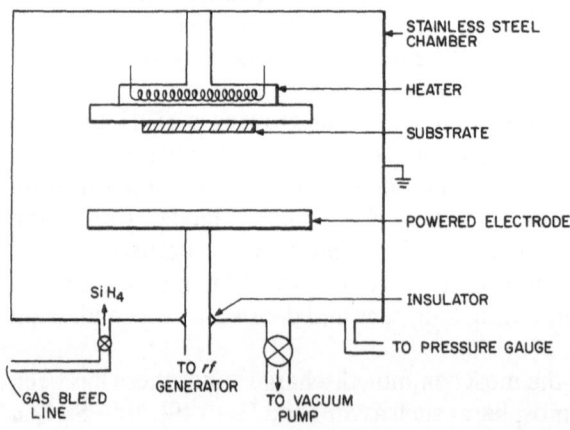

Fig. 6.12. rf capacitive glow-discharge deposition system

through a glass tube and an external coil powered the discharge. Generally, this type of system is relatively small and the uniformity of the a-Si:H films is adversely affected by the proximity of the glass walls.

The most common rf discharge geometry utilizes internal parallel plate electrodes as shown in Fig. 6.12 [6.52]. These rf capacitive discharge systems are similar to those used in rf diode sputtering. In most cases only one electrode is powered and the substrate is located on the grounded electrode. These systems usually operate at 13.56 MHz and at pressures in the range of 5 to 250 mtorr for pure SiH_4; pressures as high as a few torr may be used if the SiH_4 is diluted in H_2 or an inert gas. Power densities are typically 0.1 to 2.0 W cm^{-2} with the best films usually obtained at low power densities [6.53]. Deposition rates are generally in the range of 5 to 50 nm/min. Substrate temperatures and flow rates are comparable to those used in dc glow-discharge systems.

Some rf discharge systems employ two powered electrodes so that the dc bias potential on the substrate can be varied over a wide range [6.53]. Increasing the bias on the substrate appears to reduce the defect density of a-Si:H films deposited at high rf powers. The rf capacitive discharge systems are capable of depositing uniform a-Si:H films over large areas and commercial equipment developed for sputtering applications is readily available.

Magnetic fields have also been used in conjunction with rf glow discharges [6.54, 55]. The principle advantage of magnetic confinement is that impurities can be reduced by keeping the discharge away from the chamber walls. For an rf capacitive discharge system, one can use a geometry similar to that used in magnetron sputtering [6.55]. This deposition mode increases the deposition rate, but either the substrate or the electrode must be moved to obtain good uniformity.

All the discharge atmospheres used for dc glow discharges have also been used for rf discharges (see Sect. 6.3.2). In addition, *Knights* et al. [6.56] have deposited a-Si:H films in rf discharges using SiH_4 diluted in either He, Ne, Ar or Xe. Their films exhibited a columnar microstructure and a defect density that increased as the size of the inert gas atom increased. The films with the lowest density of unpaired spins and the highest photoluminescence intensity were made in pure SiH_4 and did not exhibit the columnar microstructure.

6.3.4 rf Sputter-Deposition

The first sputtered a-Si:H films were made by scientists at Harvard University in 1974 [6.2]. The rf diode sputtering system is similar to that shown in Fig. 6.11. However, a polycrystalline Si target is located on the powered electrode and the sputtering atmosphere is a mixture of Ar and H_2 [6.57]. The hydrogen content is typically 3 to 20 vol.% and the total pressure is usually in the range of 5 to 100 mtorr. Power densities at the target are ~ 0.5

to 4 W cm^{-2} and the deposition rates usually fall in the range of 3 to 20 nm/ min. In order to obtain a-Si:H films with relatively low defect densities, the substrate temperature should be in the range of ~ 200 to 300 °C.

Sputtered a-Si:H films have been doped by either adding gases such as B_2H_6 or PH_3 to the sputtering atmosphere [6.58] or by using doped polycrystalline Si targets [6.59].

6.4 Device Structures

A photovoltaic device must be designed so that it will efficiently absorb light in an active semiconducting layer and deliver electrical power to an external load with a minimum of internal losses. As mentioned in Sect. 6.1.1, the device must possess a semiconductor junction with a large built-in potential. In addition, the top contacting electrode must be relatively transparent to sunlight and the contact resistance of both electrodes must be small. We now consider the various device structures that have been used to make a-Si:H solar cells.

6.4.1 Schottky-Barrier and MIS Cells

One of the simplest photovoltaic devices is the Schottky-barrier cell. Since undoped a-Si:H is usually slightly n-type, a Schottky barrier is formed by depositing a high work function metal such as Pt, Rh, Pd, etc., on top of the a-Si:H film [6.60]. The high work function metal creates a positive space charge region or depletion layer within the a-Si:H film. If the undoped a-Si:H is deposited on a relatively low work function metal such as Mo, Ti, Nb, etc., then the barrier at that interface will be relatively small and the electrical properties of the device will be determined by the high work function contact on top of the a-Si:H.

A relatively simple Schottky-barrier structure to fabricate is Pd (~ 5–10 nm)/undoped a-Si:H (~ 0.3–1.0 μm)/Mo sheet, since the Pd film can be thermally evaporated. The performance of the device can be improved by first depositing a thin (10–30 nm) phosphorus-doped layer (PH$_3$/ SiH$_4 \approx 10^{-2}$) on the Mo substrate before depositing the undoped layer (Fig. 6.13a). The built-in potential (and hence V_{oc}) of a Schottky-barrier cell increases with the work function of the top metal film, but even with a Pt contact, V_{oc} is usually limited to ~ 600 mV.

Schottky-barrier cells have also been fabricated by evaporating relatively low work function metals such as Al and Cr onto p-type a-Si:H (B_2H_6/SiH$_4$ $\lesssim 10^{-4}$). In these cells, a thin p^+ layer (B_2H_6/SiH $\simeq 10^{-2}$) was first deposited on Cr-coated glass substrates. The maximum value of V_{oc} for these structures was ~ 550 mV.

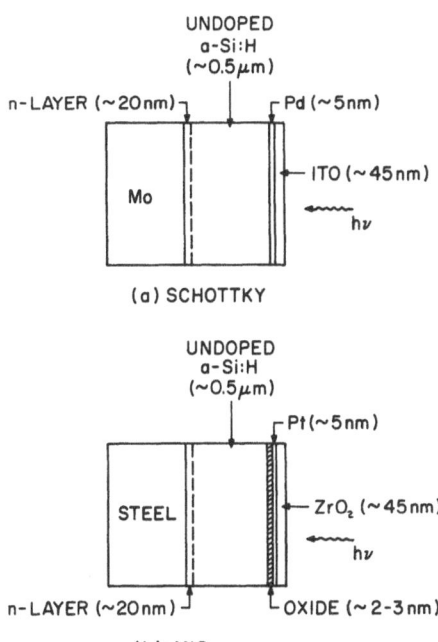

UNDOPED
a-Si:H
(~0.5 μm)

n-LAYER (~20nm)

Mo

(a) SCHOTTKY

Pd (~5nm)

ITO (~45nm)

hν

UNDOPED
a-Si:H
(~0.5 μm)

STEEL

n-LAYER (~20nm)

(b) MIS

Pt (~5nm)

ZrO₂ (~45nm)

hν

OXIDE (~2-3nm)

Fig. 6.13. (a) A Schottky-barrier solar cell structure. (b) An MIS solar cell structure

One way to increase V_{oc} is to modify the Schottky-barrier structure by placing a thin (~ 2–3 nm) insulating layer between the a-Si:H and the top metal contact (Fig. 6.13 b). This modified structure is called an MIS (metal-insulator-semiconductor) device. There are several mechanisms [6.61] by which a thin insulating layer can increase V_{oc}: interface states may influence the effective barrier height by trapping photogenerated carriers; the insulating layer may act as a barrier to the majority carriers but not the minority carriers; the insulating layer may contain fixed charges that increase the effective barrier height.

Several a-Si:H MIS structures have been fabricated by various investigators. In all cases, the a-Si:H was slightly n-type. One structure employed a thin nasient oxide grown by heating the a-Si:H for 15 min at 350 °C in air [6.62]. With a top contact of Pt(~ 5 nm), open-circuit voltage as large as 875 mV have been observed. Amorphous silicon MIS cells have also been fabricated with the insulating layers of TiO₂ [6.63], Si₃N₄ [6.64] and Nb₂O₅ [6.65].

Both Schottky-barrier and MIS cells must be coated with an antireflection layer to increase the transmission of light into the active region of the cell. An antireflection layer can increase the short-circuit current of a Schottky-barrier or MIS cell by ~ 70–100%. This layer typically consists of ~ 35–45 nm of a material such as ZrO₂, ZnS, TiO₂, Si₃N₄ or ITO (indium-tin-oxide). For large area cells, a current collection grid must be deposited on the top metal film before depositing the antireflection layer. The grid may be deposited on top of the antireflection layer only if the layer is conductive (e.g., ITO).

6.4.2 p-i-n Cells

The *p-i-n* structure is inherently more rugged than the Schottky-barrier or MIS structure since the junction of a *p-i-n* cell is buried. However, since the front doped layer of a *p-i-n* cell is usually very thin (~ 10 nm), the device characteristics may be influenced by the front contacting electrode and by states at that interface. The front doped layer must be thin in order to minimize light absorption in that layer (Sect. 6.2.3).

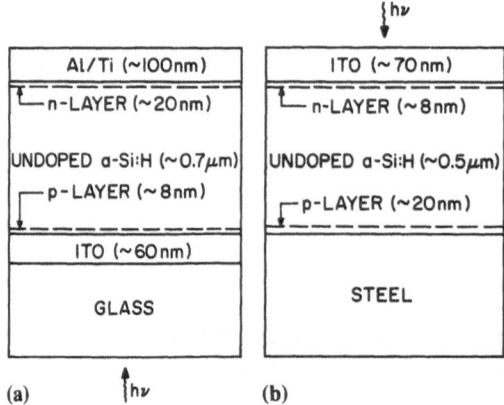

Fig. 6.14. (a) A *p-i-n* solar cell structure on a glass substrate. (b) A *p-i-n* solar cell structure on a steel substrate

The first *p-i-n* were fabricated on ITO-coated glass (Fig. 6.14 a) and the back contact was formed by evaporating Al [6.10]. The *p* layer in this type of structure is deposited from a discharge containing ~ 1 vol.% B_2H_6 in SiH_4, while the discharge atmosphere for the *n* layer contains ~ 1 vol.% PH_3 in SiH_4. Generally, better performance is obtained when the *p* layer (~ 10 nm) is deposited first, possibly due to the effect of interface states. However, another possible explanation is that when the *n* layer is deposited first, trace amounts of phosphorus may contaminate the "undoped" layer and degrade performance; the presence of small amounts of boron ($< 10^{18}$ cm^{-3}) in the *i* or "intrinsic" layer does not appear to seriously degrade performance. The thickness of the *i* layer is typically 0.5 μm, while the back doped layer is usually 10–50 nm thick.

Since the resistivity of doped a-Si:H is ~ 10^2 Ω cm, the sheet resistance of a layer 10 nm thick is ~ 10^8 Ω/□. Thus, unlike crystalline Si solar cells, an a-Si:H *p-i-n* cell must utilize a transparent conductive layer such as ITO or SnO_2 on the illuminated side of the cell in order to collect the current.

Another type of *p-i-n* structure is shown in Fig. 6.14 b where the substrate is stainless steel and the front contacting layer is formed by the electron-beam deposition of ITO. In this structure, the n layer is the front doped layer. The short-circuit current density is generally larger for this structure than for the one shown in Fig. 6.14 a since the optical absorption of phosphorus-doped a-Si : H is less than that of boron-doped a-Si : H (Fig. 6.9). The front ITO layer is ~ 70 nm thick and acts as both a contacting electrode and an antireflection layer.

A modification of the structure shown in Fig. 6.14 a involves the deposition of a thin (~ 10 nm) co-sputtered Pt–SiO$_2$ cermet onto the ITO-coated glass substrate before depositing the p layer [6.66]. Values of V_{oc} as high as 920 mV have been obtained with this structure. The cermet appears to provide a better contact to the p layer than ITO. However, the cermet layer may absorb ~ 3–10% of the incident light while absorption in the ITO is ~ 1–3% (the thin metal films used in Schottky-barrier or MIS cells will absorb $\sim 10\%$ of the incident light).

6.4.3 Heterojunction Cells

As mentioned in Sect. 6.2.3, one limitation of p-i-n cells is that some of the incident light is absorbed in the front doped layer and does not contribute to the photocurrent. This limitation can be removed if the front doped layer is replaced by a wide bandgap semiconductor since such a layer will be relatively transparent. The photovoltaic performance of a heterojunction device depends on the band structure of both semiconductors, the positions of their Fermi levels and the nature of any interface states [6.67].

The first a-Si : H heterojunctions involved the use of ITO or SnO$_2$ as wide bandgap semiconductors [6.68]. However, since these materials are degenerate n-type semiconductors and the undoped a-Si : H was slightly n-type, these heterojunction cells produced relatively low values of V_{oc} (< 500 mV). The situation is actually more complicated since it is likely that a thin silicon oxide layer is formed in the interfacial region and thus these cells may be more accurately described as SIS (semiconductor-insulator-semiconductor) cells [6.69].

A promising new heterojunction structure involves the use of an amorphous silicon-carbon-hydrogen (a-Si : C : H) alloy as a wide bandgap, p-type material in conjunction with a-Si : H [6.41, 42, 70a]. The device structure is similar to that shown in Fig. 6.14 a where the p layer is now boron-doped a-Si : C : H. The highest conversion efficiency reported to date, $\eta \simeq 10.1\%$, was recently obtained with this type of structure [6.70b].

6.4.4 Stacked Junction Cells

The conversion efficiency of an ideal solar cell is limited by the discrete nature of the band gap, i.e., photons with energy less than E_g are not absorbed and for energetic photons that are absorbed, the energy in excess of the bandgap is converted into heat. One method of improving the overall conversion efficiency is to construct a stacked or multiple-junction structure where the bandgaps of the junctions are increased systematically, from the back of the structure to the front [6.71]. The bandgaps and thicknesses of the junctions are tailored so that the photocurrents produced in each junction are the same. Since the junctions are in series, the photovoltages are additive.

The first a-Si:H stacked junction cells employed the same material for each p-i-n junction and the thicknesses of the undoped layers were selected so as to match the photocurrents [6.72, 73]. These cells exhibit open-circuit voltages that are larger than those of a single junction cell by a factor or almost N, where N is the number of junctions in the stack. However, since the photocurrents are reduced by roughly the same factor, the conversion efficiency is not enhanced.

As mentioned above, in order to improve the performance, the bandgaps of the various junctions must be tailored so that light is first incident on the widest bandgap material and the bandgap decreases for each subsequent junction. *Nakamura* et al. [6.74] constructed a stacked junction cell where the front p-i-n junction consisted of a-Si:H with an optical gap of ~ 1.85 eV and the back p-i-n junction utilized an i-layer of a-Si$_{0.75}$Ge$_{0.25}$:H with an optical gap of ~ 1.65 eV. While the conversion efficiency of this type of cell was only 2.1% in 1980 [6.74], efficiencies as high as 7.7% have been reported recently [6.75].

6.4.5 Monolithic Series-Connected Panels

Conventional crystalline Si solar cells use a current collecting grid as a front electrode and this grid must become thicker and/or wider as the cell size increases since the photocurrent is proportional to the active area. Thus, as the cell size increases, the cost of the grid increases and the area masked by the grid (typically 7–12% of the total area) also increases. Crystalline Si solar cells are usually made from Si wafers with diameters in the range of 5–10 cm. Conventional solar-cell panels are fabricated by wiring together as many as 100 individual cells.

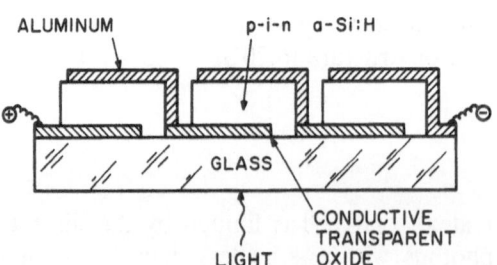

Fig. 6.15. A diagram of a monolithic series-connected panel

Similar technology can be used to fabricate a-Si : H solar-cell panels but a more cost-effective scheme is to fabricate an array of series-connected cells on a single large glass substrate [6.76]. A schematic diagram of such a monolithic series-connected panel is shown in Fig. 6.15. The panel is fabricated by first patterning the transparent conductive oxide (TCO) layer, then depositing the a-Si : H junction and performing another patterning step, and then depositing the back metal contact and performing a final patterning step. The patterning is performed in such a way that the back metal electrode of one cell contacts the edge of the front TCO electrode of the adjacent cell and thus all the cells are connected in series.

The monolithic series-connceted panel has several advantages over the conventional solar-cell panel. First, if the patterning is performed using a high resolution technology such as photolithography, then the inactive area due to the interconnects can be kept below 1% of the total area. Also, the amount of material used in a monolithic panel is small since the contacting layers are of the order of 0.1 μm thick, the a-Si : H is ~ 0.5–1.0 μm thick, and no current-collecting grid is needed. Since the interconnects are made along the entire length of the panel, the probability of interconnect failure is very small. Moreover, the panel will deliver power at a high output voltage (~ 70–140 V/m of panel width) which is desirable for many applications. Recently, Sanyo Corp. has reported a conversion efficiency of 7.31% for a 100 cm^2 panel consisting of nine cells in series [6.77].

6.5 Characteristics of a-Si : H Solar Cells

Of all the parameters used to characterize solar cells, the conversion efficiency is clearly the most important since it determines the size of a solar-cell array needed for a specific application (and hence strongly influences the cost). However, many other characteristics must be considered and evaluated in order to improve the performance of a solar cell during the research and development phase. In this section we consider the more important characteristics of a-Si : H solar cells.

6.5.1 Current-Voltage Characteristics

As mentioned in Sect. 6.1.1, the conversion efficiency (η) can be determined from the current-voltage characteristic measured under illumination. One can also readily determine the open-circuit voltage (V_{oc}), the short-circuit density (J_{sc}) and the fill factor (FF). These photovoltaic parameters are related by (6.1) and together they form the most common basis for comparing the performance of various types of solar cells.

In Table 6.1 we list the photovoltaic parameters of a variety of a-Si:H solar-cell structures fabricated by several different research organizations. It is worth noting that conversion efficiencies of ~7–10% have been achieved for several different device structures and as pointed out in the second half of this chapter, similar performance has been attained in devices utilizing a-Si:F:H alloys. Moreover, these results have been obtained using different glow discharge atmospheres and different discharge modes.

Table 6.1. Performance of a-Si:H solar cells

Type	Configuration	Area [cm^2]	V_{oc} [mV]	J_{sc} [mA cm^{-2}]	FF	η [%]	Ref.
p-i-n	ITO/*n* μc-Si:H/*i-p* a-Si:H/steel	1.2	860	13.9	0.655	7.8	6.78
p-i-n	ITO/*n-i-p* a-Si:H/steel	0.04	857	13.0	0.62	6.9	6.79
p-i-n	Al/Ti/*n-i-p* a-Si:H/SnO$_2$/glass	0.1	801	12.55	0.675	6.8	6.80
HJ	Al/*n-i* a-Si:H/*p* a-Si:C:H/ITO/glass	0.033	880	15.2	0.601	8.0	6.70
HJ	Al/*n-i* a-Si:H/*p* a-Si:C:H/SnO$_2$/glass	1.0	880	14.1	0.624	7.7	6.70
SJ	ITO/*n-i-p-n* a-Si:H/*i* a-Si:Ge:H/*p* a-Si:H/steel	0.25	1410	9.6	0.57	7.7	6.75
HJ	ITO/*p* a-Si:H/*i* a-Si:Ge:H/*n* a-Si:H/steel	1.0	635	7.2	0.57	2.6	6.74
HJ	ITO/*p* a-B:Si:H/*i-n* a-Si:H/steel	0.05	670	6.9	0.35	1.6	6.81
HJ	Pd/a-B:H/*i-n* a-Si:H/steel	100	800	6.0	0.55	2.6	6.82
HJ	Ag/*m-i* a-Si:H/*p* a-Si:C:H/SnO$_2$/glass	1.09	840	17.8	0.676	10.1	6.70 b

HJ: heterojunction
SJ: stacked junctions

Most of the structures listed in Table 6.1 were described in Sect. 6.4. However, three of the structures listed in the table were not discussed earlier, and they represent some of the relatively recent attempts to make new heterojunction structures. It is evident that the present performance of these heterojunctions is relatively poor.

As shown in Table 6.1, the highest reported conversion efficiency (10.1%) and short-circuit density (~17.8 mA cm^{-2}) have been obtained with a heterojunction structure utilizing a-Si:C:H [6.70b]. The highest value of V_{oc} to date (933 mV) was obtained with a similar device structure [6.83]. However, the best value of the fill factor (0.728) was attained with a *p-i-n* cell [6.80], although values as high as 0.72 have been obtained with heterojunction structures utilizing a-Si:C:H [6.84]. It is interesting to note that a conversion efficiency of ~12.1% would result if the best values of V_{oc}, J_{sc} and FF were obtained in the same cell.

The current density of a photovoltaic device in the dark obeys the relation

$$J = J_0 \left[\exp\left(\frac{qV}{nkT}\right) - 1 \right] , \tag{6.3}$$

where J_0 is the saturation current density and n is the diode quality factor; $n = 1.0$ for an ideal diode while $n = 2.0$ if recombination and generation effects dominate. Some representative current-voltage characteristics are shown in Fig. 6.16 for different types of a-Si:H solar cells. The Schottky-barrier cell exhibits a diode quality factor of ~ 1.07 while the MIS cell exhibits one of ~ 1.57. The diode quality factor of the *p-i-n* cell changes from ~ 1.8 at low voltages to ~ 1.5 for voltages greater than ~ 0.6 V. This current-voltage characteristic is representative of *p-i-n* cells with conversion efficiencies of about 6.0%. Low performance *p-i-n* cells ($\eta < 3.0\%$) may exhibit diode quality factors greater than 2.0 over the entire voltage range [6.38].

The saturation current density J_0 is a measure of the built-in potential or barrier height of a semiconductor junction. Generally, an efficient photovoltaic device will exhibit a small value of J_0. For example, in a-Si:H Schottky-barrier devices,

$$J_0 \propto \exp(-\phi_B/kT) , \tag{6.4}$$

where ϕ_B is the barrier height [6.85]. A large barrier height insures not only a small value of J_0, but also a large value of V_{oc}. The open-circuit voltage is

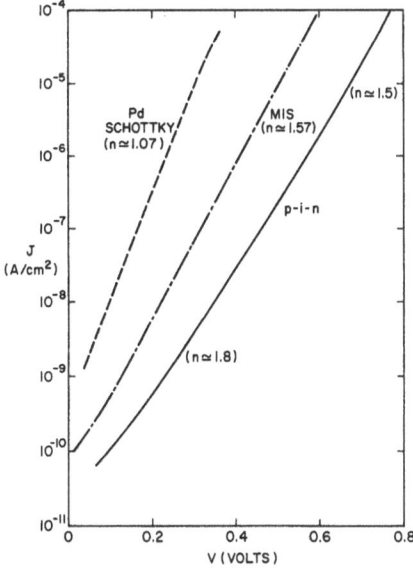

Fig. 6.16. Current-voltage characteristics in the dark for a Pd Schottky-barrier (SB) cell, a Pt MIS cell and a *p-i-n* cell deposited on steel (Fig. 6.14 b)

related to the saturation current density and J_{sc} by

$$V_{oc} = \frac{n'kT}{q} \ln\left(\frac{J_{sc}}{J_0} + 1\right),$$ (6.5)

where n' is the diode quality factor under illumination. Thus, decreasing J_0 increases V_{oc} provided n' does not change. From the data in Fig. 6.16 we can predict that the *p-i-n* device should exhibit a large value of V_{oc} while this would not be true of the Schottky-barrier device.

Some a-Si:H cells do not exhibit good diode behavior in the dark, i.e., their current-voltage characteristic is either not described by (6.3) or $n > 2$. However, these cells may still perform reasonably well under illumination. This situation may occur if the undoped a-Si:H layer is so resistive that the dielectric relaxation time is longer than the carrier lifetime. The device is then operating in the relaxation semiconduction regime and the current-voltage characteristics will be nonideal [6.86]. Illumination increases the conductivity so that then the device operates in the normal lifetime regime. In such cases, (6.5) may still be valid but J_0 cannot be determined from the characteristics in the dark.

The current-voltage characteristics can also be affected by poor contacts such as the presence of a small Schottky barrier at a metal/doped layer contact. The presence of interface states associated with thin interfacial layers can also influence the device characteristics [6.61].

For forward biases $\gtrsim 1$ V in the dark, the current-voltage characteristics are not described by (6.3) and the behavior varies depending on the deposition conditions and the resulting density of state distribution in the gap. In some a-Si:H cells, the current becomes series resistance-limited due to the existence of a quasi-neutral region [6.62]. Other a-Si:H cells exhibit high-field characteristics that appear to be described by space-charge-limited conduction [6.87], while in other cases, the Poole-Frenkel mechanism (field-assisted ionization of carriers from coulombic traps) describes the current-voltage behavior reasonably well [6.88].

As shown by (6.5), a plot of $\ln J_{sc}$ as a function of V_{oc} allows one to determine the diode quality factor under illumination. Figure 6.17 shows some representative data for an MIS cell, a low performance *p-i-n* cell ($\eta \simeq 2.5\%$) and a high performance *p-i-n* cell ($\eta \simeq 6.0\%$). Generally, the high performance cells exhibit values of n' close to unity, as expected if recombination and interface effects are small.

One can also analyze the variation of the forward-bias I–V characteristics with light intensity to determine the series resistance of the device [6.89]. Amorphous silicon solar cells with conversion efficiencies in the 6–7% range typically exhibit series resistances of ~ 3–5 Ω cm^2 [6.90].

Another useful procedure involves an analysis of the wavelength dependence of the slope of the I–V curve near J_{sc} [6.91]. This analysis can provide

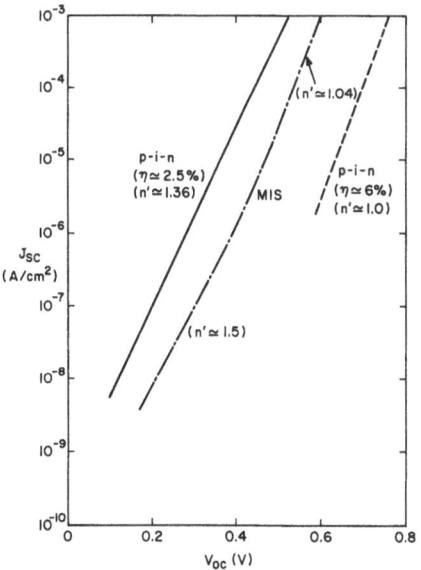

Fig. 6.17. Short-circuit current density as a function of open-circuit voltage for a Pt MIS cell, a low performance *p-i-n* cell ($\eta \simeq 2.5\%$) and a high performance *p-i-n* cell ($\eta \simeq 6.0\%$)

information about the field distribution in the cell including the width of the space-charge region and the magnitude of the built-in potential. High performance a-Si:H cells ($\gtrsim 6.0\%$) are usually fully depleted under short-circuit conditions.

6.5.2 Spectral Response

The spectral response curve of a solar cell is a plot of the collection efficiency as a function of the photon wavelength; the collection efficiency is defined as the percentage of electron-hole pairs collected by the external circuit per incident photon. Typical spectral response data are shown in Fig. 6.18 for a *p-i-n* cell fabricated on steel (Fig. 6.14b). The spectral response curve changes only slightly as the illumination level increases. However, the spectral response of some a-Si:H solar cells can change significantly with light intensity [6.92]. The presence of constant AM1 illumination (in conjunction with chopped monochromatic light) assures that the cells are characterized under normal operating conditions.

The decrease in collection efficiency at long wavelengths is primarily caused by the decreasing absorption coefficient of undoped a-Si:H (Fig. 6.2). For a fully depleted *p-i-n* cell, the long wavelength response is described by the expression

$$\eta_{\text{coll}}(\lambda) \simeq [1 - R(\lambda)][\exp(-\alpha_p d_p)][1 - \exp(-\alpha_i d_i)]$$
$$[1 + R'(\lambda)\exp(-2\,\alpha_n d_n - \alpha_i d_i)]\,, \tag{6.6}$$

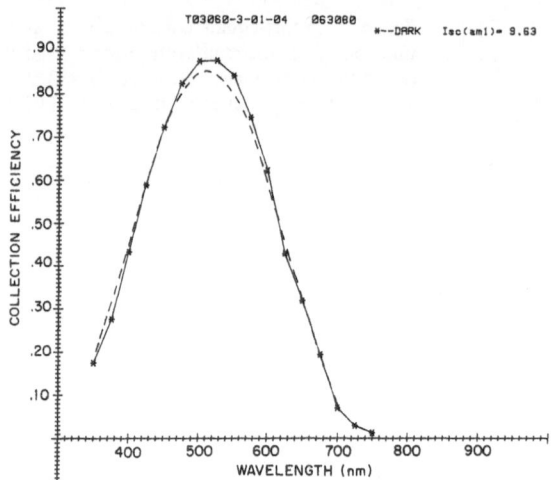

Fig. 6.18. The collection efficiency as a function of wavelength for monochromatic light only (——) and with constant AM1 illumination present (– – –)

where $R(\lambda)$ and $R'(\lambda)$ are the reflectances at the incident surface and the back contact, respectively, and α_p, α_i, and α_n are the absorption coefficients of the p, i and n layers, respectively; the doped layers are assumed to be "dead" and absorption losses in the top contact layer are assumed to be negligible. Moreover, $R(\lambda)$ is assumed to be small in the near infrared so that multiple internal reflections can be ignored. This expression was derived for a cell where the light is first incident on the p layer. If the light is first incident on the n layer, $\alpha_p d_p$ and $\alpha_n d_n$ are interchanged in (6.6).

There are two factors in (6.6) that can contribute to the fall-off in collection efficiency at short wavelengths. One is the absorption of light in the top doped layer and the other factor is that in most cells the reflection from the top surface increases at short wavelengths [6.34]. The short wavelength response may also be decreased by absorption in the top contact layer; for an ITO contact, the response at $\lambda \simeq 0.4$ µm can be reduced by several percent or more. The response at short wavelengths may also be reduced by losses associated with the back diffusion of local minority carriers [6.93].

The collection efficiency at all wavelengths may be reduced by recombination in the bulk. As discussed in Sect. 6.2.2, geminate recombination does not appear to be significant in good quality a-Si:H. Also, recombination associated with gap states is small provided the diffusion length is larger than the width of the undoped layer. For high performance cells ($\eta \gtrsim 6.0\%$), the spectral response at zero bias can be modeled without assuming significant bulk recombination [6.94].

Figure 6.19 shows spectral response data as a function of an applied bias for a p-i-n cell (same structure as shown in Fig. 6.14b). The small variation between the 0 and -2 V bias indicates that bulk recombination is not significant in this bias range. However, the spectral response does decrease as the cell is forward biased, reflecting the decrease in photocurrent (Fig. 6.1). The

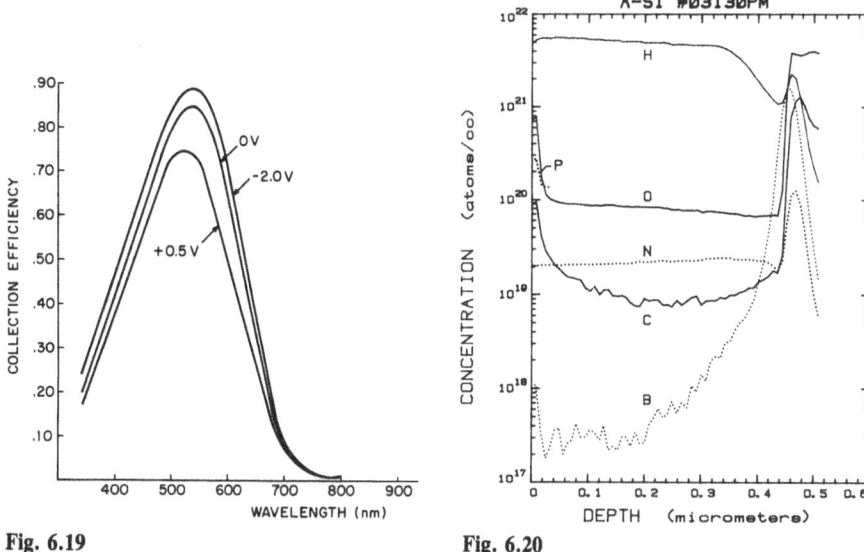

Fig. 6.19

Fig. 6.20

Fig. 6.19. Collection efficiency as a function of wavelength for applied biases varying from − 2 to + 0.5 V

Fig. 6.20. A compositional profile of a *p-i-n* cell deposited on stainless steel (Fig. 6.14 b)

spectral response is reduced at all wavelengths, but the decrease is somewhat larger for long wavelengths indicating that the electron transport might be adversely affected by residual boron (Fig. 6.20). The losses at short wavelengths may be associated with the back diffusion of holes to the top *n* layer. However, another possibility is that holes moving toward the back contact recombine with electrons in a quasi-neutral region somewhere in the center of the cell.

6.5.3 Effect of Impurities

Common impurities in a-Si:H are oxygen, carbon and nitrogen with concentrations typically in the range of 10^{18}–10^{20} cm^{-3} [6.28, 95]. These impurities are readily detected by secondary ion mass spectroscopy (SIMS) [6.95] and a compositional profile of a *p-i-n* cell ($\eta \simeq 5.6\%$) is shown in Fig. 6.20. As shown, oxygen is generally the most common contaminant and the source is either an air leak or adsorbed water vapor on the walls of the vacuum system. The source of nitrogen is also either an air leak or adsorbed N_2 on the walls of the system. The carbon may come from adsorbed CO_2 on the walls, but hydrocarbons from pump oil are a more common source.

Increasing the oxygen content to $\gtrsim 10^{21}$ cm^{-3} by adding 2000 ppm H_2O or 1000 ppm O_2 to a SiH_4 discharge does not drastically degrade the conversion efficiency [6.81, 96]. (This statement only applies to cells measured shortly after the deposition of the a-Si : H or after annealing the cell at $\sim$ 150–200 °C since the addition of impurities can cause the efficiency to decrease under prolonged illumination; see Sect. 6.5.4.) Similarly, the addition of $\lesssim 1$ vol.% N_2 to a SiH_4 discharge does not cause significant changes in efficiency [6.81, 96], but the addition of 10 vol.% N_2 or 1 vol.% NH_3 can reduce the efficiency by more than a factor of four [6.81].

Delahoy and *Griffith* [6.96] have found that the presence of both oxygen and nitrogen in the SiH_4 discharge gives rise to a synergistic effect which causes a significant reduction in the photovoltaic parameters of an a-Si : H device as compared to that caused by the presence of either gas by itself. This observation indicates that the recombination centers might be nitrogen-oxygen complexes, and indeed species such as NO_2 have been observed by ESR in x-irradiated a-Si : H films [6.97].

Table 6.2 shows how the photovoltaic parameters of *p-i-n* cells (deposited on steel) are affected when various impurity gases are added to the SiH_4 discharge only during the deposition of the undoped layer. Adding 10 to 20 vol.% CH_4 to the SiH_4 discharge does not significantly affect V_{oc}, but both J_{sc} and the fill factor are reduced. Part of the decrease in J_{sc} is due to a reduction in the optical absorption as carbon opens the optical gap [6.98]. Adding SiF_4 to the SiH_4 discharge reduces the efficiency somewhat, but this reduction may actually be due to other contaminants (such as oxygen and carbon) whose concentration increases significantly when SiF_4 is added. This effect is apparently due to a "scrubbing" of the walls of the vacuum system by reactive fluorine species [6.99].

Other impurities that have been shown to adversely affect a-Si : H solar cell performance are Ge and S (if present in relatively large quantities; $\gtrsim 1$ vol.% GeH_4 or H_2S) [6.100] and Cl ($\sim$ 500 ppm in the film) [6.96]. Monochlorosilane (SiH_3Cl) has been detected as an impurity gas in some SiH_4 gas cylinders [6.96], and another contaminant that has been detected in SiH_4 is tetrahydrofurane (C_4H_8O) [6.101]. Other gases that are commonly found in SiH_4 gas cylinders are H_2 ($\lesssim 750$ ppm), N_2 (< 150 ppm), Si_2H_6 (< 500 ppm) and He (< 400 ppm) [6.102].

Table 6.2. Effects of impurities on device performance

Impurity gas	V_{oc} [mV]	J_{sc} [mA/cm^2]	FF	η [%]
None, control cell	814	10.8	0.575	5.0
17% SiF_4	759	9.9	0.533	4.2
1% N_2	767	10.9	0.519	4.3
10% N_2	630	3.6	0.458	1.1
1% NH_3	198	4.3	0.328	0.3
5% NH_3	184	2.4	0.326	0.1
0.2% H_2O	755	11.7	0.474	4.2
10% CH_4	816	9.0	0.458	3.4
20% CH_4	788	6.9	0.285	1.6

Vanier et al. [6.103] found Al ($\lesssim 2.2$ ppm) and Cr ($\lesssim 40$ ppm) in their a-Si : H films, probably from metal parts in their discharge system. They also looked for impurities such as Fe, Ni, Zn and Cu, but none were detected. Dopants such as phosphorus and boron may degrade device performance if present in the *i* layer of an a-Si : H cell. *Delahoy* and *Griffith* [6.96] found that the short-circuit current density of MIS cells was decreased by ~40% when 10 ppm PH_3 was added to the SiH_4 discharge. An earlier study [6.100] showed that both J_{sc} and V_{oc} fell by a factor of four or more when 600 ppm PH was added to the SiH discharge during the fabrication of a Pt Schottky-barrier cell. Recently, *Kuwano* et al. [6.79] reported a significant improvement in the performance of *p-i-n* cells when they deposited each layer in a separate deposition chamber, thus preventing contamination of the *i* layer by boron or phosphorus.

There is considerable evidence that dopants such as phosphorus and boron create deep states in a-Si : H in addition to the expected shallow donor and acceptor levels [6.14, 104, 105]. Some of these deep centers are associated with dangling bond-type defects, and vacancy-impurity complexes may be quite common in a-Si : H [6.104]. However, many of the impurity atoms may be completely bonded into the a-Si : H structure (e.g., Si–O–Si) and not give rise to deep gap states.

Before concluding this section, we should mention that there are defect states which have been identified in a-Si : H that are not related to impurities. A common defect in a-Si : H is a dangling bond which can result from hydrogen out-diffusion during deposition. Other dangling bonds are associated with short polymer chains or $(SiH_2)_n$ groups, and a strong correlation has been found between the concentration of $(SiH_2)_n$ groups and the presence of a columnar morphology in some a-Si : H films [6.106].

6.5.4 Stability

There are several mechanisms that may limit the life of an a-Si : H solar cell. First, unless the cell is encapsulated, normal weathering processes may corrode the metal contacts. Similarly, humidity may adversely affect the interfacial regions of unencapsulated a-Si : H solar cells [6.48]. High operating temperatures ($\gtrsim 150$ °C) may cause degradation when metals such as Al are used as contacts [6.107]. Finally, some a-Si : H films contain a high density of metastable defects which can cause degradation during illumination [6.108].

Encapsulation is most likely a necessity for any type of solar cell in order to obtain long life. A variety of encapsulants have been developed for crystalline Si solar cells [6.109] and some of these should be applicable to a-Si : H solar cells. Thus, degradation effects due to weathering or humidity should be negligible in a well-designed a-Si : H solar panel.

At normal operating temperatures ($\lesssim 60$ °C), degradation associated with diffusion effects should also be negligible. Even at 100 °C, hydrogen diffusion should be so small that degradation should not be noticeable until after more than 10^4 years [6.110]. If Al is used as a contact electrode, then the device performance can be adversely affected by induced crystallization that may occur at temperatures close to 100 °C [6.47].

For ITO contacts on crystalline Si solar cells, there is some evidence that a resistive silicon oxide layer develops at the interface at elevated temperatures [6.111]. However, the ITO/a-Si:H interface should be more stable since a-Si:H is more resistant to oxidation than crystalline Si [6.112]. Many other electrode materials such as Mo, Cr, Ta, etc., provide stable contacts to a-Si:H, and even at elevated temperatures, the diffusion coefficients are relatively small [6.81] (see Table 6.3). The diffusion coefficients of the dopants are also small at elevated temperatures so that there should not be any degradation associated with the diffusion of dopants at normal operating temperatures.

Light-induced effects in a-Si:H were first studied by *Staebler* and *Wronski* in 1977 [6.113]. They showed that the dark conductivity and photoconductivity of a-Si:H could both be reduced by prolonged illumination and that these effects could be reversed by annealing at ~ 200 °C. In 1980, *Morigaki* et al. [6.114] reported on a fatigue effect in the photoluminescence of a-Si:H, and subsequent studies have shown that prolonged illumination creates new dangling bonds [6.115]. More recently, *Dresner* et al. [6.30] have shown that prolonged illumination could reduce the diffusion length in some a-Si:H films and that annealing at ~ 200 °C would reverse the degradation.

Light-induced effects are also evident in some a-Si:H solar cells. Figure 6.21 shows how the current-voltage characteristic of an unstable *p-i-n* cell (Fig. 6.14b) changed after 48 h of illumination (~ 100 mW cm^{-2}) [6.108]. The largest change in the spectral response curve occurred at short wavelengths. This degradation could also be reversed by annealing the cell at ~ 200 °C. Moreover, operating the cells in far reverse bias inhibited the light-induced degradation [6.108].

Table 6.3. Diffusion coefficients in a-Si:H

Diffusing species	T[°C]	D[cm^2/s]
Mo	450	$< 10^{-18}$
Cr	350	$< 10^{-18}$
Nb	350	$< 10^{-18}$
Ta	350	$< 10^{-18}$
Fe	400	$\sim 2 \times 10^{-15}$
Fe	350	$\sim 10^{-17}$
Fe	300	$< 10^{-17}$
O	450	$\sim 6 \times 10^{-18}$
B	400	$< 3 \times 10^{-17}$
P	450	$< 3 \times 10^{-17}$

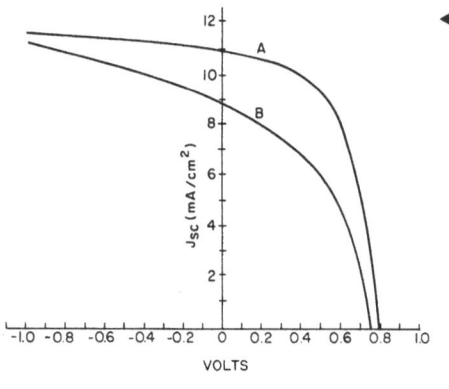

◀ **Fig. 6.21.** Current-voltage curves of an unstable *p-i-n* cell in simulated AM1 light. A is the initial curve and B is after 48 h of AM1 light

Fig. 6.22. Life tests of relatively stable *p-i-n* cells

▼

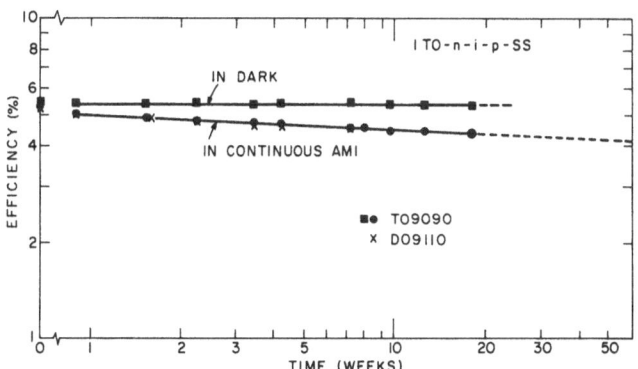

Life test data are shown in Fig. 6.22 for some relatively stable *p-i-n* cells deposited on steel. No degradation is noticed in the dark, but the cells exposed to AM1 illumination exhibit a gradual degradation that appears linear on a log-log plot. An extrapolation of these data predicts a decrease of only ~20% after 20 years in sunlight [6.108]. *Kuwano* and *Ohnishi* [6.116] tested some a-Si:H *p-i-n* cells for ~1 year under illumination and found that the conversion efficiency only fell from ~3.7% to ~3.5%.

Crandall [6.117] has used deep-level trap spectroscopy (DLTS) to study metastable states in *p-i-n* cells deposited on steel. He detected a metastable state that captured electrons with a thermal activation energy of about 1.0 eV. The activation energy for electron emission from the trap was 0.93 eV, surprisingly close to the activation energy for capture.

In one series of experiments, the metastable state has been associated with impurities [6.81]. A series of *p-i-n* cells were made with varying amounts of air bled into the discharge atmosphere and the performance was measured just after fabrication and then after 48 h of exposure to white light (~ 100 mW cm^{-2}). The conversion efficiency was ~4% initially but after light soaking, the efficiency decreased with increasing air bleed rate

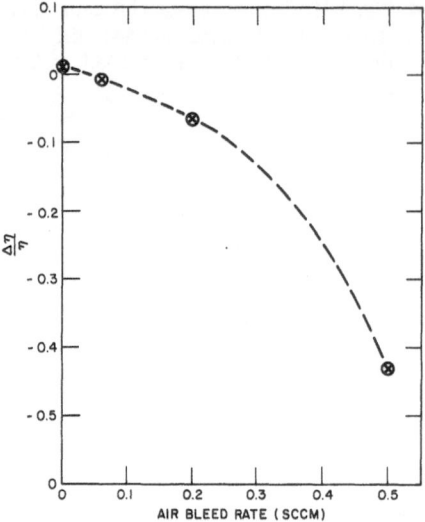

Fig. 6.23. Fractional change in conversion efficiency after light soaking (48 h in AM 1 light) as a function of the air bleed present during the deposition of the a-Si:H

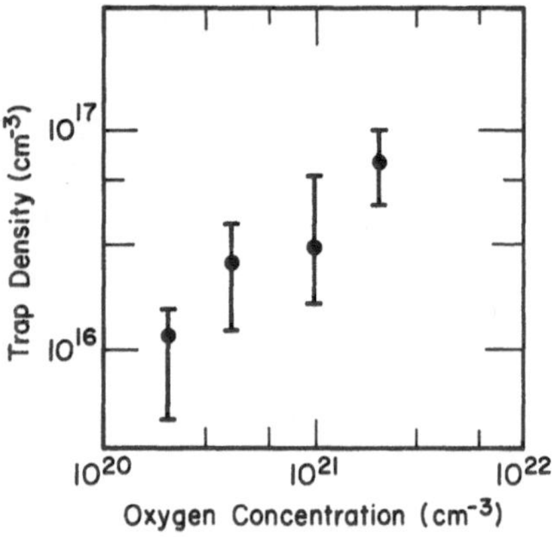

Fig. 6.24. Electron trap density as a function of oxygen concentration in the *i* layer of *p-i-n* solar cells

(Fig. 6.23). Annealing for several minutes at ~ 200 °C reversed the degradation.

DLTS measurements [6.117] showed that the density of metastable centers increased almost linearly with the concentration of oxygen and nitrogen in the a-Si:H (Fig. 6.24). The nitrogen concentration was $\sim 10\%$ of the oxygen concentration in all samples. In some a-Si:H material the density of metastable centers is as low as 2×10^{14} cm^{-3} [6.117]. Light-induced effects in a-Si:H solar cells become significant when the density of these electron traps is $\gtrsim 10^{16}$ cm^{-3}. The identity of the metastable center is not presently known, but it may be a nitrogen-oxygen complex (Sect. 6.5.3).

6.6 Theoretical Modeling

In most cases the modeling of solar-cell characteristics starts with Poisson's equation and the continuity equations for electrons and holes. Poisson's equation is

$$\varepsilon \frac{\partial E}{\partial x} = q\,(p - n + N_D - N_A)\,, \tag{6.7}$$

where E is the electric field, p is the hole density, n is the electron density, N_D is the density of ionized donors and N_A is the density of ionized acceptors.

The continuity equations for electrons and holes are

$$\frac{\partial n}{\partial t} = G - R_n + \frac{1}{q}\frac{\partial J_n}{\partial x}, \tag{6.8}$$

$$\frac{\partial p}{\partial t} = G - R_p - \frac{1}{q}\frac{\partial J_p}{\partial x}, \tag{6.9}$$

where

$$J_n = nq\mu_n E + kT\mu_n\frac{\partial n}{\partial x}, \tag{6.10}$$

$$J_p = pq\mu_p E - kT\mu_p\frac{\partial p}{\partial x}. \tag{6.11}$$

In these expressions, G is the photogeneration rate, R_n and R_p are the electron and hole recombination rates and μ_n and μ_p are the electron and hole mobilities. In general, these coupled sets of equations do not have an analytical solution so that one either makes simplifying assumptions or uses a computer to obtain numerical solutions. At present, there is no single model that encompasses all the physics of a-Si:H solar cells and that successfully predicts all the observed characteristics. However, significant progress has been made in the last few years in explaining many of the characteristics of a-Si:H solar cells.

6.6.1 Models with Exact Solutions

In early work on a-Si:H Schottky-barrier cells, the device characteristics were interpreted in terms of the diffusion theory of metal-semiconductor rectification [6.85] and the photovoltaic expression (6.5). As mentioned in Sect. 6.5.1, this latter expression is not easy to interpret if the superposition principle does not hold, i.e., the dark current and photocurrent are not additive [6.118]. The spectral response of Schottky-barrier cells was first modeled using an expression similar to (6.2), and collection widths of 0.3 μm were determined [6.62]. The spectral response of early p-i-n cells was modeled by *Debney* [6.119] assuming a "dead" top p layer and current collection only from the field region in the i layer. Debney also showed that the fill factor of a Schottky-barrier cell would be limited to <0.64 if carriers were only collected from the depletion region (i.e., no diffusion). *Konagai* et al. [6.120] analyzed the photovoltaic behavior of a-Si:H Schottky-barrier cells and concluded that the diffusion length was <0.05 μm for cells with fill factors ≲0.58.

Recently, *Gutkowicz-Krusin* et al. showed that the short-wavelength response of Schottky-barrier or MIS cells could be explained by back-diffusion losses where some electrons generated near the surface are able to diffuse against the electric field and are lost to the metal contact [6.93]. The same authors stated that Schottky-barrier lowering due to the image force was not a large effect even at short wavelength. However, *Han* et al. [6.121] claimed that the short-wavelength response can be explained entirely by Schottky-barrier lowering.

More recently, *Gutkowicz-Krusin* [6.122] has derived exact solutions for the spectral response of Schottky-barrier cells assuming that the space-charge field is not altered by illumination and that the hole lifetime is constant throughout the cell thickness. For front contacts which absorb back-diffusing electrons, the model predicts that the short-wavelength response increases significantly as the depletion width decreases. The long-wavelength response increases dramatically as the diffusion length increases.

Chen and *Mort* [6.27] modeled the photovoltaic response of an a-Si:H Schottky-barrier cell assuming that geminate recombination was the dominant loss mechanism. However, other investigators [6.123] have concluded that geminate recombination is not a significant factor in a-Si:H.

In a more recent analysis, *Chen* and *Lee* [6.124] have modeled the current-voltage characteristics of a-Si:H diodes in the dark, assuming Shockley-Read-Hall recombination. They also assumed that the density of states could be approximated by exponential tails extending from each band edge. They showed that the diffusion length could be estimated from the change in the slope of the current-voltage curve (at high voltages, the drift/diffusion current dominates while the recombination current can dominate at low voltages depending on the built-in potential and the ratio of the hole to electron mobilities).

Crandall [6.125] has solved the coupled set of equations (6.7–9) for transport in a *p-i-n* cell assuming that the free-carrier space charge is negligible. Since the trapped charge in good *p-i-n* cells is also negligible [6.126], the electric field is uniform. He also separated the *i* layer into three regions in which the recombination rates are expressed as linear functions of the free carrier density (a constant lifetime regional approximation). The following expression for the photocurrent was derived:

$$J = qG(l_p + l_n)\{1 - \exp[-d_i/(l_p + l_n)]\} , \tag{6.12}$$

where G is the generation rate of electron-hole pairs per unit volume per second, l_p and l_n are the hole and electron drift lengths ($l_n = \mu_n\tau_n E$), and d_i is the thickness of the *i* layer. This expression shows that it is the carrier with the longest drift length that determines the current-voltage characteristic under illumination for a *p-i-n* device with a uniform field.

While the density of free carriers does not distort the electric field in a thin *p-i-n* cell ($d_i \lesssim 1$ μm), these carriers can contribute to the photocapacitance as given by

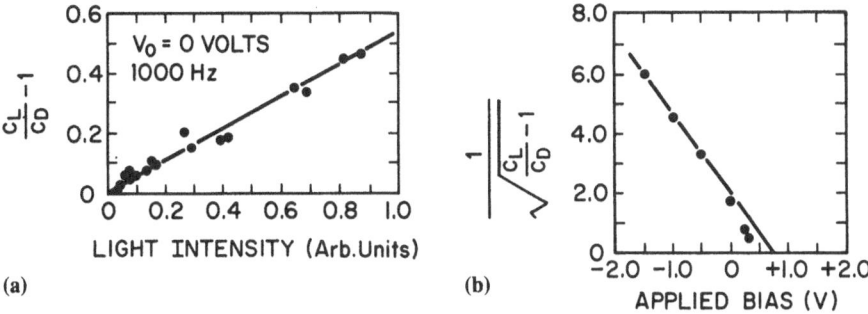

Fig. 6.25. (a) Fractional change in capacitance as a function of light intensity.
(b) $[(C_L - C_D)/C_D]^{1/2}$ as a function of applied voltage

$$C_p = \frac{qGd_i^3}{2\,V_0^2}\left(\frac{1}{\mu_n + \mu_p}\right), \tag{6.13}$$

where V_0 is the voltage across the i layer and C_p is the difference between the capacitance in the light (C_L) and the capacitance in the dark (C_D) [6.127]. Figure 6.25 a shows that C_p does vary linearly with light intensity while Fig. 6.25 b shows that experimental data for *p-i-n* cells also exhibit the correct voltage dependence [6.127].

6.6.2 Computer Modeling

Another approach to modeling a-Si : H solar cells is to solve (6.7–9) using a computer. This approach has been very successful for single crystal silicon solar cells [6.128]. *Swartz* [6.129] has modified the program to simulate a-Si : H *p-i-n* solar cells where he has used experimentally measured values of

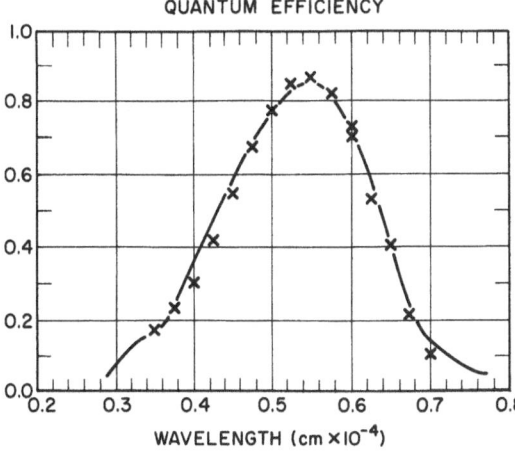

Fig. 6.26. The spectral response of an a-Si : H *p-i-n* cell; the solid line was calculated by the computer

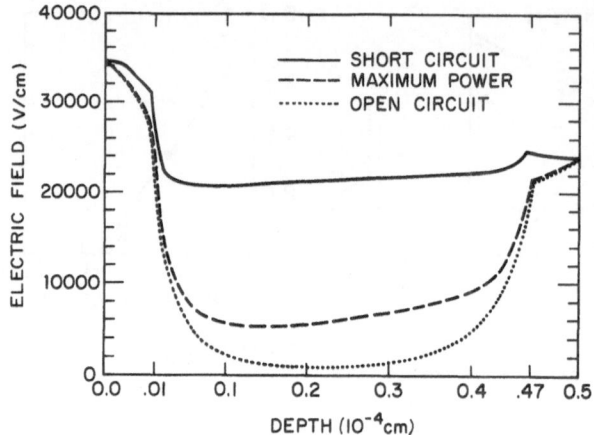

Fig. 6.27. The electric field as a function of position for a *p-i-n* cell in AM 1 illumination under various bias conditions; note the depth scale is expanded for the doped layers (*n* layer on top)

parameters such as layer thickness, mobilities, doping levels, optical absorption, etc. He has obtained excellent agreement between the observed and calculated values for the current-voltage characteristics under illumination and for the spectral response. Figure 6.26 shows both experimental data and the calculated curve for the spectral response of a *p-i-n* cell deposited on steel (Fig. 6.14 b). Geminate recombination is assumed to be negligible in this calculation, but the other recombination mechanisms mentioned in Sect. 6.5.2 are included. The model is also capable of generating spatial plots of the electric field, the carrier concentrations, the carrier currents and recombination rates. The electric field distribution is shown in Fig. 6.27 for a *p-i-n* cell in AM 1 illumination under various bias conditions [6.129].

At present, the computer program does not explicitly take into account the density of states distribution in a-Si : H, and some parameters such as the hole carrier density in the *p* layer and the intrinsic carrier density in the *i* layer are adjusted.

6.7 Approaches to Improve Performance

Since the first a-Si : H solar cell was made in 1974 [6.6], the performance has improved steadily, as shown in Fig. 6.28. This plot is an update of one published earlier [6.33] and the recent data points refer to a-Si : H cells described in Table 6.1 (except for the most recent result of 10.1% efficiency [6.70 b]). The relatively rapid improvement in performance since the end of 1979 can be attributed to a much larger effort in device processing, especially in Japan. In this section, we consider various methods that might be used to further improve the performance of a-Si : H solar cells.

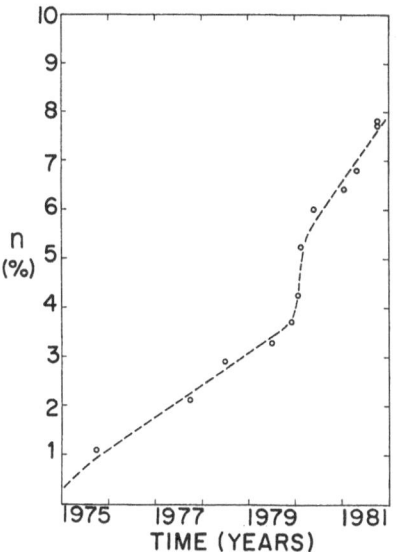

Fig. 6.28. The conversion efficiency as a function of time for a-Si : H *p-i-n* and heterojunction solar cells with active areas greater than 1.0 cm²

TIME (YEARS)

6.7.1 Increasing Carrier Lifetimes

Increasing the recombination lifetime of the carriers usually increases all the photovoltaic parameters (V_{oc}, J_{sc} and FF) to some degree. For present a-Si : H solar cells with efficiencies of ∼7 to 10%, the largest improvement should occur in the fill factor, especially if the surface recombination velocity at the contacts is small [6.125, 129]. However, for carrier absorbing contacts, computer modeling indicates that the fill factor may not be improved significantly by increasing the lifetime since back-diffusion losses can increase [6.129]. For present a-Si : H cells with $J_{sc} \simeq$ 13–17 mA cm^{-2} and $V_{oc} \simeq$ 850 mV, the same modeling indicates an improvement of only a few percent in J_{sc} and in V_{oc} upon tripling the lifetime [6.129].

The recombination lifetime of carriers can be improved by reducing the concentration of defects that create deep centers in a-Si : H. As mentioned in Sect. 6.5.3, some recombination centers appear to be associated with impurities. These types of defects have been reduced by means of separate, in-line deposition chambers [6.79]. The careful implementation of good vacuum system procedures can help to keep impurity-associated defects at a minimum. As also mentioned in Sect. 6.5.3, some recombination centers appear to be associated with microstructural defects, such as $(SiH_2)_n$ groups. These defects can be reduced by empirical optimization of the discharge conditions; for example, in rf discharge systems, the concentration of $(SiH_2)_n$ groups can be reduced by eliminating dilution in inert gases and operating at low power levels [6.56, 106].

The concentration of deep centers in present device-quality a-Si : H appears to be in the range of 10^{14}–10^{15} eV^{-1} cm^{-3} from DLTS measurements

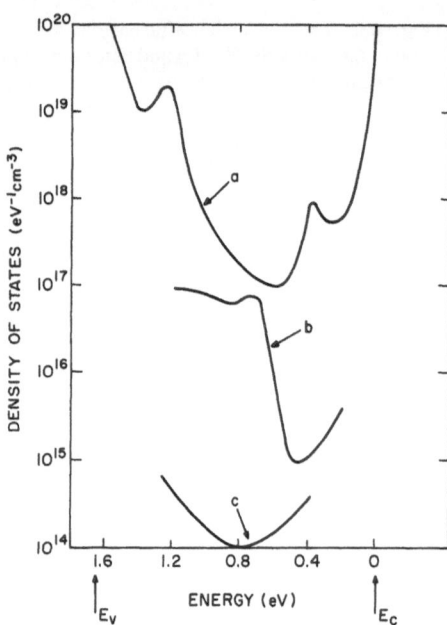

Fig. 6.29. The density of states in a-Si : H as measured by the field-technique and by DLTS

[6.126]. As shown in Fig. 6.29, field-effect measurements (Curve a) indicate a much larger density of states [6.130], but this technique probes a layer only ~ 10 nm thick near the surface and is relatively insensitive to structure in the density of states [6.131]. The field-effect values may be enhanced by the presence of surface states. Also shown in Fig. 6.29 are data for phosphorus-doped films (Curve b) that indicate the presence of deep centers [6.132] possibly due to phosphorus-hydrogen complexes [6.14].

6.7.2 Increasing the Built-In Potential

The open-circuit voltage of an a-Si : H solar cell is increased when the built-in potential is increased; both J_{sc} and the fill factor will also increase slightly [6.34, 129]. Present p-i-n cells exhibit a built-in potential of ~ 1.1 V [6.38, 39] although the optical gap is typically ~ 1.7 eV. As mentioned in Sect. 6.2.3, the built-in potential appears to be limited by the conductivity of the doped layers, but interface states may also be important, especially for thin doped layers. Moreover, the band tails in a-Si : H may ultimately limit the built-in potential.

The conductivity of the doped layers can be improved significantly [from ~ 10^{-2} to > 10 $(\Omega$ cm$)^{-1}$] by forming doped microcrystalline Si : H films [6.40]. If these layers were homogeneous, one might expect a large built-in potential since the optical gaps are relatively large (~ 1.7–1.8 eV) and the activation energies for conduction are small (~ 0.02–0.07 eV) so the Fermi levels appear to lie close to the band edges for both p and n-type microcrystalline Si : H films. However, these films are composite materials consisting of both amorphous and crystalline phases [6.133, 134] and no enhancement in V_{oc} is apparent in devices utilizing these layers [6.78] since the built-in potential is mainly determined by the phase with the lowest internal barrier.

The open-circuit voltage may also be enhanced by heterojunction contacts that reduce the losses associated with the back-diffusion of local minority carriers, i.e., a n-type, wide bandgap layer would present a barrier only to holes in the a-Si : H provided the electron affinities were similar [6.67].

6.7.3 Increasing Optical Absorption

For crystalline materials, the conversion efficiency reaches a maximum when the bandgap is $\simeq 1.5$ eV [6.135]. Thus, one might expect the efficiency of a-Si : H solar cells to improve if the bandgap could be reduced somewhat below the present range of ~ 1.65–1.80 eV for device-quality material. However, a-Si : H possesses exponential band tails with characteristic energies of ~ 0.03 eV for the tail near the conduction band and ~ 0.05 eV for the tail near the valence band, i.e., $\eta(\varepsilon) \propto \exp(-\varepsilon/\varepsilon_0)$, where ε_0 is the characteristic energy [6.21]. Therefore, the optimum gap for a-Si : H might be ~ 0.1 eV larger than in the crystalline case since the tail states will tend to limit the built-in potential. *Tiedje* [6.136] has calculated the effect of these band tails on the built-in potential and has concluded that V_{oc} would be limited to $\sim 1.0 \pm 0.1$ V for $\varepsilon_{opt} \simeq 1.7$ eV. More detailed modeling is necessary to determine the optimum gap for a-Si : H.

If the optimum gap is somewhat less than the present values of ~ 1.65–1.80 eV, then the gap can be reduced by decreasing the hydrogen content [6.13] or by alloying with Ge, but in both cases, the lifetime is usually degraded [6.137].

The stacked junction cell (Sect. 6.4.4) is perhaps the best way to increase the absorption of sunlight in a solar-cell structure and recently, a conversion of efficiency of 7.7% has been obtained in such a structure where the back undoped layer was an amorphous alloy of Si, Ge and H [6.75] (Table 6.1). Stacked junction cells have the potential of much higher conversion efficiencies than a single junction solar cell and, in fact, efficiencies greater than 50% have been calculated for ideal multi-junction cells [6.138].

6.8 Concluding Remarks

One key issue that requires more detailed study is a determination of the theoretical limit for the conversion efficiency of an a-Si : H solar cell. While a detailed theory is presently lacking, we can attempt to make a reasonable estimate based on some of the properties discussed in this artice. First, let us assume that the optical path length in the a-Si : H solar cell is 1.0 μm, i.e., the cell could be 1.0 μm thick with a nonreflecting back contact, or it could be 0.5 μm thick with total reflection of all unabsorbed light at the back contact (and no higher-order reflections). For a-Si : H with an optical gap of 1.7 eV and AM1 illumination, the theoretical limit for J_{sc} is ~ 20 mA cm^{-2} assuming no reflection losses from the front surface of the cell and no recombination losses. (The best value of J_{sc} reported to date is 17.8 mA cm^{-2} [6.70 b].)

As mentioned in Sect. 6.7.3, *Tiedje* [6.136] has estimated that recombination in the band tails of a-Si : H will limit V_{oc} to ~ 1.0 ± 0.1 V for $\varepsilon_{opt} \simeq 1.7$ eV. We believe that a limit of 1.1 V is a reasonable estimate since open-circuit voltages of this magnitude have been observed for electrolytic contacts [6.139] and for *p-i-n* cells at either high light intensities or low temperatures [6.140]. (The best value of V_{oc} obtained under AM1 illumination at room temperature is presently 933 mV [6.83].)

For an ideal diode, $n' = 1$ in (6.5), and for $V_{oc} = 1.1$ V, the fill factor is ~ 0.89 [6.141]. Using (6.1), we estimate that the theoretical limit for the conversion efficiency of a single junction a-Si : H solar cell is ~ 19.6%. In view of this estimate and the progress shown in Fig. 6.28, and in Table 6.1, it seems likely that a conversion efficiency of 12% will be achieved in the next few years. Eventually, the conversion efficiency may exceed 15% using stacked junction cells.

However, high conversion efficiencies do not guarantee commercial success of a solar-cell technology. The most important consideration is the cost of the photovoltaic energy compared to that available from other sources. There is fairly wide agreement that once solar-cell arrays are available for about $ 1 per peak watt, photovoltaics will be used in several large markets such as water pumping, remote village electrification and new residential homes. Successful penetration of these markets could lead to a billion dollars per year industry by 1990. Present crystalline Si solar cells cost about $ 10–$ 20 per peak watt and in 1980 about 4 MW of these cells were sold worldwide [6.142].

The tremendous interest in a-Si : H solar cells stems from the potential for low-cost production of these cells. Clearly, the material costs are low since the entire thin film structure is only ~ 1 μm thick and the glass substrate is the single most expensive item. Also, production costs are likely to be low since the films can be deposited over relatively large areas (perhaps several ft^2) in several minutes. Moreover, the energy required to process the films is relatively low since the substrate temperatures are < 300 °C and the glow-discharge power densities are typically ≲ 0.1 W cm^{-2}. Indeed, Sanyo projects a cost of about $ 0.25 per peak watt for the production of 8% cells at a level of 100 MW per year [6.116] (Sanyo started commercial production of a-Si : H solar cells for calculators in September 1980).

The future for a-Si : H solar cells looks very promising. It seems likely that large a-Si : H panels will be commercially available within the next few years and that low-cost power from the sun will soon be a reality.

References

6.1   R. C. Chittick, J. H. Alexander, H. F. Sterling: J. Electrochem. Soc. **116**, 77 (1969)
6.2   A. J. Lewis, G. A. N. Connell, W. Paul, J. R. Pawlik, R. J. Temkin: Proc. Intern. Conf.
      Tetrahedrally Bonded Amorphous Semiconductors, ed. by M. H. Brodsky, S. Kirkpat-
      rick, D. Weaire (AIP, New York 1974) p. 27
6.3   A. Triska, D. Dennison, H. Fritzsche: Bull. Am. Phys. Soc. **20**, 392 (1975)
6.4   W. E. Spear, P. G. LeComber: J. Non-Cryst. Solids **8–10**, 727 (1972)
6.5   D. Engemann, R. Fisher: Proc. 12th Intern. Conf. Physics of Semiconductors, ed. by M.
      H. Pilkuhn (Teubner, Stuttgart 1974) p. 1042
6.6   D. E. Carlson: U.S. Patent No. 4,064,521 (1977)
6.7   W. E. Spear, P. G. LeComber: Solid State Commun. **17**, 1193 (1975)
6.8   N. F. Mott, E. A. Davis: *Electron Processes in Noncrystalline Materials* (Clarendon Press,
      Oxford 1971)
6.9   W. E. Spear, P. G. LeComber, S. Kinmond, M. H. Brodsky: Appl. Phys. Lett. **28**, 105
      (1976)
6.10 D. E. Carlson, C. R. Wronski: Appl. Phys. Lett. **28**, 671 (1976)
6.11 H. J. Hovel: *Semiconductors and Semimetals, Vol. II, Solar Cells* (Academic Press,
      New York 1975) p. 38
6.12 E. A. Davis: *Amorphous Semiconductors*, ed. by P. G. LeComber, J. Mort (Academic
      Press, New York 1973) p. 450
6.13 P. J. Zanzucchi, C. R. Wronski, D. E. Carlson: J. Appl. Phys. **48**, 5227 (1977)
6.14 D. E. Carlson, R. W. Smith, C. W. Magee, P. J. Zanzucchi: Phil. Mag. B **35**, 51 (1982)
6.15 E. A. Perez-Albuerne, Y. S. Tyan: Science **208**, 902 (1980)
6.16 B. Abeles, C. R. Wronski, T. Tiedje, G. D. Cody: Solid State Commun. **36**, 537 (1980)
6.17 W. B. Jackson, N. M. Amer: *Tetrahedrally Bonded Amorphous Semiconductors*, ed. by
      R. A. Street, D. K. Biegelsen, J. C. Knights (AIP, New York 1981) p. 263
6.18 S. Yamasaki, T. Hata, T. Yoshida, H. Oheda, A. Matsuda, H. Okushi, K. Tanaka: J.
      Physique **42**, C4–297 (1981)
6.19 P. G. LeComber, W. E. Spear: Phys. Rev. Lett. **25**, 509 (1970)
6.20 A. R. Moore: Appl. Phys. Lett. **31**, 762 (1977)
6.21 T. Tiedje, J. M. Cebulka, D. L. Morel, B. Abeles: Phys. Rev. Lett. **46**, 1425 (1981)
6.22 J. Dresner: Appl. Phys. Lett. **37**, 742 (1980)
6.23 P. G. LeComber, D. I. Jones, W. E. Spear: Phil. Mag. **35**, 1173 (1977)
6.24 J. Dresner: J. Non-Cryst. Solids
6.25 A. J. Snell, W. E. Spear, P. G. LeComber: Phil. Mag. **43**, 407 (1981)
6.26 D. M. Pai, S. W. Ing: Phys. Rev. **173**, 899 (1968)
6.27 I. Chen, J. Mort: Appl. Phys. Lett. **37**, 952 (1980)
6.28 D. E. Carlson: 3rd E. C. Photovoltaic Solar Energy Conf., Cannes, France (1980)
      (Reidel, Dordrecht 1981) p. 294
6.29 A. R. Moore: Appl. Phys. Lett. **37**, 327 (1980)
6.30 J. Dresner, B. Goldstein, D. Szostak: Appl. Phys. Lett. **38**, 998 (1980)
6.31 B. Goldstein, J. Dresner, A. R. Moore, D. Szostak: RCA Engineer **28**, 45 (1983)
6.32 A. R. Moore: Appl. Phys. Lett. **40**, 403 (1982)
6.33 D. E. Carlson: Solar Energy Mat. **3**, 503 (1980)
6.34 D. E. Carlson: Conf. Record of 14th IEEE Photovoltaic Specialists Conf. (IEEE, New
      York 1980) p. 291
6.35 R. A. Street, D. K. Biegelsen, J. C. Knights: Phys. Rev. B **24**, 969 (1981)
6.36 W. E. Spear: Adv. Phys. **26**, 312 (1977)
6.37 Z. I. Jan, R. H. Bube, J. C. Knights: J. Appl. Phys. **51**, 3278 (1980)
6.38 D. E. Carlson, C. R. Wronski: J. Electron. Mat. **6**, 95 (1977)
6.39 R. H. Williams, R. R. Varma, W. E. Spear, P. G. LeComber: J. Phys. C **12**, L 209 (1979)

6.40   A. Matsuda, S. Yamasaki, K. Nakagawa, H. Okushi, K. Tanaka, S. Iizima, M. Matsu-
       mura, H. Yamamoto: Japan J. Appl. Phys. **19**, L 305 (1980)
6.41   J. I. Pankove: U.S. Patent No. 4,109,271 (1978)
6.42   Y. Tawada, M. Kondo, H. Okamoto, Y. Hamakawa: Conf. Record of 15th IEEE Photo-
       voltaic Specialists Conf. (IEEE, New York 1981) p. 245
6.43   D. Kaplan, N. Sol, G. Velasco, P. A. Thomas: Appl. Phys. Lett. **33**, 440 (1978)
6.44   S. C. Gau, B. R. Weinberger, M. Akhtar, Z. Kiss, A. G. MacDiarmid: Appl. Phys. Lett.
       39, 436 (1981)
6.45   D. E. Carlson, C. R. Wronski, J. I. Pankove, P. J. Zanzucchi, D. L. Staebler: RCA Rev.
       38, 211 (1977)
6.46   S. R. Herd, P. Chaudhari, M. H. Brodsky: J. Non-Cryst. Solids **7**, 309 (1972)
6.47   J. S. Maa, S. J. Lin: Thin Solid Films **64**, 63 (1979)
6.48   D. E. Carlson, C. W. Magee: 2nd E. C. Photovoltaic Solar Energy Conf., Berlin (1979)
       (Reidel, Dordrecht 1979) p. 312
6.49   M. Konagai, K. Takahashi: Appl. Phys. Lett. **36**, 599 (1980)
6.50   B. von Roedern, G. Moddel: Solid State Commun. **35**, 467 (1980)
6.51   Y. Uchida, H. Sakai, M. Nishiura, M. Miyagi, K. Maruyama: 8th Intern. Vacuum Con-
       gress, Cannes, France (1980)
6.52   J. C. Knights: Phil. Mag. **34**, 663 (1976)
6.53   J. C. Knights: Japan J. Appl. Phys. **18**, 101 (1979)
6.54   M. Taniguchi, M. Hirose, Y. Osaka: J. Non-Cryst. Solids **35 & 36**, 189 (1980)
6.55   H. Weakliem: Proc. of AIAA 13th Fluid and Plasma Dynamics Conf., Snowmass, Col-
       orado (1980)
6.56   J. C. Knights, R. A. Lujan, M. P. Rosenblum, R. A. Street, D. K. Biegelsen, J. A.
       Reimer: Appl. Phys. Lett. **38**, 331 (1981)
6.57   G. A. N. Connell, J. R. Pawlik: Phys. Rev. B **13**, 787 (1976)
6.58   W. Paul, A. J. Lewis, G. A. N. Connell, T. D. Moustakas: Solid State Commun. **20**, 969
       (1976)
6.59   M. J. Thompson, J. Allison, M. M. Alkaisi: Solid-State and Electron Devices **2**, S 11
       (1978)
6.60   C. R. Wronski, D. E. Carlson: Solid State Commun. **23**, 421 (1977)
6.61   S. J. Fonash: Thin Solid Films **36**, 387 (1976)
6.62   C. R. Wronski: Japan J. Appl. Phys. Suppl. **17-1**, 299 (1978)
6.63   J. I. B. Wilson, J. McGill, S. Kimmond: Nature **272**, 153 (1978)
6.64   W. A. Anderson, J. K. Kim, S. L. Hyland, J. Coleman: Conf. Record of 13th IEEE
       Photovoltaic Spec. Conf. (IEEE, New York 1978) p. 755
6.65   A. Madan, J. McGill, W. Czubatyi, J. Yang, S. R. Ovshinsky: Appl. Phys. Lett. **37**, 826
       (1980)
6.66   J. J. Hanak, V. Korsun: Conf. Record of 13th IEEE Photovoltaic Spec. Conf. (IEEE,
       New York 1978) p. 780
6.67   A. G. Milnes, D. L. Feucht: *Heterojunctions and Metal-Semiconductor Junctions*
       (Academic, New York 1972)
6.68   D. E. Carlson: IEEE Trans. ED-**24**, 449 (1977)
6.69   J. Shewchun, R. Singh, D. Burk, M. Spitzer, J. Loferski, J. Dubow: Conf. Record of 13th
       IEEE Photovoltaic Spec. Conf. (IEEE, New York 1978) p. 528
6.70   a) Y. Tawada, K. Tsuge, M. Kondo, H. Okamoto, Y. Hamakawa: J. Appl. Phys. **53**,
       5273 (1982)
       b) A. Catalano, R. V. D'Aiello, J. Dresner, B. Faughnan, A. Firester, J. Kane, H.
       Schade, Z. E. Smith, G. Swartz, A. Triano: Conf. Record 16th IEEE Photovoltaic
       Specialists Conf. (IEEE, New York 1982) p. 1421
6.71   J. J. Loferski: Conf. Record of 12th IEEE Photovoltaic Spec. Conf. (IEEE, New York
       1976) p. 957
6.72   Y. Hamakawa, H. Okamoto, Y. Nitta: Appl. Phys. Lett. **35**, 1871 (1979)
6.73   J. J. Hanak: J. Non-Cryst. Solids **35 & 36**, 755 (1980)

6.74 G. Nakamura, K. Sato, Y. Yukimoto, K. Shirahata: 3rd E. C. Photovoltaic Solar Energy Conf., Cannes, France (1980) (Reidel, Dordrecht 1981) p. 835
6.75 G. Nakamura, K. Sato, H. Kondo, Y. Yukimoto, K. Shirahata: *4th E. C. Photovoltaic Solar Energy Conf.*, Stresa, Italy (1982) (Reidel, Dordrecht 1982) p. 616
6.76 J. J. Hanak: Solar Energy **23**, 145 (1979)
6.77 Y. Kuwano, S. Nakano, T. Fukatsu, M. Ohnishi, H. Nishiwaki, S. Tsuda: Photovoltaics for Solar Energy **407**, to be published
6.78 Fuji Electric: Annual Meeting of J.A.A.P., Japan (1981)
6.79 Y. Kuwano, M. Ohnishi, H. Nishiwaki, S. Tsuda, H. Shibuya, S. Nakano: Conf. Record of 15th IEEE Photovoltaic Spec. Conf. (IEEE, New York 1981) p. 698
6.80 J. Dresner, J. Moles: Personal communication
6.81 D. E. Carlson: J. Vac. Sci. Technol. **20**, 290 (1982)
6.82 J. H. Coleman, J. P. Hammes, H. J. Wiesmann: Unpublished results
6.83 J. J. Hanak, J. P. Pellicane, V. Korsun: Personal communication
6.84 A. Catalano: Unpublished results
6.85 C. R. Wronski, D. E. Carlson, R. E. Daniel: Appl. Phys. Lett. **29**, 602 (1976)
6.86 G. H. Dohler, H. Heyszenan: Phys. Rev. B **12**, 641 (1975)
6.87 R. A. Gibson, W. E. Spear, P. G. LeComber, A. J. Snell: J. Non-Cryst. Solids **35 & 36**, 725 (1980)
6.88 J. I. Pankove, D. E. Carlson: Am. Rev. Mat. Sci. **10**, 43 (1980)
6.89 R. J. Hardy: Solid State Electronics **10**, 765 (1967)
6.90 G. A. Swartz: Unpublished results
6.91 G. A. Swartz, R. Williams: Conf. Record of the 14th IEEE Photovoltaic Spec. Conf. (IEEE, New York 1980) p. 1224
6.92 C. R. Wronski, B. Abeles, G. D. Cody, D. L. Morel, T. Tiedje: ibid., p. 1057
6.93 D. Gutkowicz-Krusin, C. R. Wronski, T. Tiedje: Appl. Phys. Lett. **38**, 87 (1981)
6.94 H. Schade, Z. E. Smith: To be published
6.95 C. Magee, D. E. Carlson: Solar Cells **2**, 365 (1980)
6.96 A. E. Delahoy, R. W. Griffith: Conf. Record of the 15th IEEE Photovoltaic Spec. Conf. (IEEE, New York 1981) p. 704
6.97 W. M. Pontuschka, W. E. Carlos, P. C. Taylor, R. W. Griffith: Phys. Rev. B **25**, 4362 (1982)
6.98 D. A. Anderson, W. E. Spear: Phil. Mag. **35**, 1 (1977)
6.99 D. E. Carlson, R. W. Smith: Conf. Record of the 15th IEEE Photovoltaic Spec. Conf. (IEEE, New York 1981) p. 694
6.100 D. E. Carlson: Tech. Digest of 1977 IEEE Intern. Electron Dev. Mtg., Washington, DC (IEEE, New York 1977) p. 214
6.101 A. Gallagher, J. Scott: Unpublished results
6.102 RCA Laboratories: Unpublished results
6.103 P. E. Vanier, A. E. Delahoy, R. W. Griffith: J. Appl. Phys. **52**, 5235 (1981)
6.104 R. A. Street, D. K. Biegelsen, J. C. Knights: Phys. Rev. B **24**, 969 (1981)
6.105 R. S. Crandall: Phys. Rev. Lett. **44**, 749 (1980)
6.106 J. C. Knights, G. Lucovsky, R. J. Nemanich: J. Non-Cryst. Solids **32**, 393 (1979)
6.107 J. J. Hanak: Personal communication
6.108 D. L. Staebler, R. S. Crandall, R. Williams: Conf. Record of the 15th IEEE Photovoltaic Spec. Conf. (IEEE, New York 1981) p. 249
6.109 W. E. Dennis: Conf. Record of 14th IEEE Photovoltaic Spec. Conf. (IEEE, New York 1980) p. 958
6.110 D. E. Carlson, C. W. Magee: Appl. Phys. Lett. **33**, 81 (1978)
6.111 S. M. Goodnick, J. F. Wager, C. W. Wilmsen: J. Appl. Phys. **51**, 527 (1980)
6.112 H. Fritzsche: Solar Energy Mat. **3**, 447 (1980)
6.113 D. L. Staebler, C. R. Wronski: Appl. Phys. Lett. **31**, 292 (1977)
6.114 K. Morigaki, I. Hirabayashi, N. Nakayama, S. Nitta, K. Shimakawa: Solid State Commun. **33**, 851 (1980)
6.115 I. Hirabayashi, K. Morigaki, S. Nitta: Japan J. Appl. Phys. **19**, L357 (1980)

6.116 Y. Kuwano, M. Ohnishi: Intern. J. Physique **42**, C4–1155 (1981)

6.117 R. S. Crandall: Phys. Rev. B **24**, 7457 (1981)

6.118 A. Rothwarf: Conf. Record of 13th IEEE Photovoltaic Spec. Conf. (IEEE, New York 1978) p. 1312

6.119 B. T. Debney: Solid State and Electron Devices **2**, S 15 (1978)

6.120 M. Konagai, H. Miyamoto, K. Takahashi: Japan J. Appl. Phys. **19**, 1923 (1980)

6.121 M. K. Han, W. A. Anderson, R. Lahri, J. Coleman: Appl. Phys. Lett. **39**, 325 (1981)

6.122 D. Gutkowicz-Krusin: J. Appl. Phys. **52**, 5370 (1981)

6.123 B. Abeles, C. R. Wronski, T. Tiedje, G. D. Cody: Solid State Commun. **36**, 537 (1980)

6.124 I. Chen, S. Lee: Intern. J. Physique **42**, C4–449 (1981)

6.125 R. S. Crandall: RCA Rev. **42**, 449 (1981)

6.126 R. S. Crandall: Intern. J. Physique **42**, C4–413 (1981)

6.127 R. S. Crandall: Appl. Phys. Lett. **42**, 451 (1983)

6.128 T. I. Chappell: IEEE Trans. ED-**27**, 760 (1980)

6.129 G. A. Swartz: J. Appl. Phys. **53**, 712 (1982)

6.130 A. Madan, P. G. LeComber, W. E. Spear: J. Non-Cryst. Solids **20**, 239 (1976)

6.131 N. B. Goodman, H. Fritzsche: Phil. Mag. B **42**, 149 (1980)

6.132 D. V. Lang, J. D. Cohen, J. P. Harbison: Phys. Rev. B **25**, 5285 (1982)

6.133 Y. Mishima, T. Hamasaki, H. Kurata, M. Hirose, Y. Osaka: Japan J. Appl. Phys. **20**, L 121 (1981)

6.134 Y. Uchida, T. Ichimura, M. Ueno, M. Oksawa: Intern. J. Physique **42**, C4–265 (1981)

6.135 J. J. Loferski: J. Appl. Phys. **27**, 777 (1956)

6.136 T. Tiedje: Appl. Phys. Lett. **40**, 627 (1982)

6.137 J. J. Hanak, B. Faughnan, V. Korsun, J. P. Pellicane: Conf. Record of 14th IEEE Photovoltaic Spec. Conf. (IEEE, New York 1980) p. 1209

6.138 A. Bennett, L. C. Olsen: Conf. Record of 13th IEEE Photovoltaic Spec. Conf. (IEEE, New York 1978) p. 868

6.139 R. Williams: J. Appl. Phys. **50**, 2848 (1979)

6.140 R. S. Crandall: Personal communication

6.141 H. J. Hovel: *Semiconductors and Semimetals, Vol. II, Solar Cells* (Academic Press 1975) p. 61

6.142 P. D. Maycock, E. N. Stirewalt: *Photovoltairs, Sunlight to Electricity in One Step* (Brick House Publ., Andover, MA 1981) p. 117

7. Devices Using Fluorinated Material

Arun Madan

With 28 Figures

In this chapter we shall emphasize a-Si : F : H alloys but will include some data on a-Si : H alloys for completeness. First, we shall consider some basic properties of the a-Si : F : H alloy and then concentrate on some device aspects of the material as related to photovoltaics.

7.1 Background

The present interest in amorphous silicon (a-Si) stems in part from the possibility of producing inexpensive solar cells. The interest in a-Si was initiated by *Chittick* et al. [7.1] who showed that thin films of a-Si : H could be produced by using radio-frequency (rf) glow discharge in SiH_4 gas and this was subsequently developed by *Spear* and coworkers [7.2]. They demonstrated that high quality films could be fabricated which possessed a low density of states and exhibited high photoconductivity. Further, n and p-type doping could be achieved which is vital from a device point of view. Reports of relatively high efficiency ($\sim 5.5\%$) devices appeared shortly afterwards [7.3]. Using gas mixtures of SiF_4 and H_2, we produced an amorphous Si : F : H alloy [7.4] with desirable properties for photovoltaic applications and have reported on conversion efficiencies of 6.6% over an active area of 0.73 cm^2 [7.5] using a metal-insulator-semiconductor (MIS) structure. Recent results in our laboratory have produced conversion efficiencies exceeding 7%, using a *p-i-n* type structure [7.6].

At present, substantial efforts are being made in various laboratories [7.7, 8] with the best reported efficiencies (using a-Si : H alloy) of 7.55% [7.9], but over a very small area of 0.033 cm^2 and 6.9% by the Sanyo group [7.10].

The material originally produced by the Dundee group was apparently fundamentally different in comparison with pure elemental a-Si such as that obtained by the evaporation or sputtering of elemental Si. The basic difference was pointed out by *Fritzsche* and *Tsai* [7.11] who showed that the H content of the a-Si films prepared by the rf glow discharge technique was about 10%. These materials are now appropriately referred to as a-Si : H alloys. It thus became apparent that the high density of localized gap states possessed by elemental a-Si, primarily determined by the number of dangling bonds, was reduced by orders of magnitude by the incorporation of H_2. This fact was further confirmed by *Paul's* group [7.12] who showed that high quality films could also be fabricated by reactive sputtering of Si in a H_2 environment. Films thus produced contained H of several atomic percent and the electronic and optical properties resembled those prepared by the rf glow-discharge method. Further demonstrations of the advantageous effect of H_2 have also been demonstrated by experiments with the post-hydrogenation of CVD in SiH_4 gas [7.13]. In this process, SiH_4 gas is decomposed thermally at a high temperature ($\sim 550\ °C$) and the H concentration within the film, at this stage, is about 0.2%. The quality of the film is only improved when the samples are subjected to a subsequent plasma in a reactive H environment.

Since strong evidence existed that atomic hydrogen was responsible for the reduction in dangling-bond density, it occurred to us that the density of localized states in a-Si could possibly also be reduced by the introduction of F, which has a larger electronegativity than H. We produced a-Si : F : H alloys which, according to the field effect data, indicated that the density of states decreased by about an order of magnitude in comparison with a-Si : H alloys [7.4]. Using capacitance-voltage type measurements on Schottky-barrier type cells, we have measured the depletion width to be larger (using a-Si : F : H [7.14]) than the corresponding results reported for the a-Si : H [7.15]. Further, in a relatively short period of time, we have reported on an overall conversion efficiency of 6.3% using an MIS-type structure [7.16]. Recent results in our laboratory have given a 6.6% conversion efficiency over an active area of 0.73 cm^2 [7.5]. Also, *Konagai* et al. [7.17] have obtained a 4.78% conversion efficiency with an MIS type structure using a-Si : F : H as the active material. The enhancement in V_{oc}, the open circuit voltage, with the addition of the insulator was only about 75 mV. Generally, 250 mV enhancement can be obtained, as we shall show, thus giving a potential efficiency exceeding 6% for their cells, similar to our best results [7.5].

7.2 Properties of Fluorinated Amorphous Silicon

The announcement that a-Si : F : H [7.4, 18, 19] could be fabricated with desirable properties from a photovoltaic point of view has aroused considerable interest. Various mixtures of SiF_4/H_2, $SiF_4/SiH_4/Ar$ and $SiF_4/H_2/Ar$ [7.4, 20–25] have been attempted. Sputtering of Si in SiF_4 [7.26] and CVD of SiF_2 [7.27] has also been attempted. In the following, we summarize the different results obtained at various laboratories.

7.2.1 Glow Discharge in SiF$_4$ and H$_2$

The radio-frequency glow-discharge apparatus we used to fabricate a-Si:F:H films consisted of capacitance plates [7.4]. All components of the reaction chamber and gas handling system were constructed of stainless steel and the system was generally pumped to a pressure of less than 10^{-6} torr prior to deposition. The premixed SiF$_4$ (Matheson, purity 99.50%) and H$_2$ gases were fed into the system at a constant rate and the pressure was maintained at ~ 0.5–1 torr during deposition. The plasma was generated between the plates using either a dc voltage or a radio-frequency discharge operating at 13.56 MHz. A plasma could be obtained for gas mixtures of SiF$_4$, SiF$_4$ + Ar and SiF$_4$ + H$_2$ and SiH$_4$ + SiF$_4$. However, amorphous films with properties suitable for photovoltaic applications such as a low density of gap states and high photoconductivity were produced using preferably the latter two mixtures.

The samples were deposited on heated substrates (glass, stainless steel, metallized glass, etc.) to a nominal substrate temperature T_s although the actual substrate temperature was somewhat lower. This is because the observation of T_s was made with a thermocouple somewhat removed from contact with the substrate. The actual substrate temperature was estimated to be lower by about 120 °C for T_s = 250 °C.

7.2.2 Dark Conductivity [7.4]

In this sub-section we discuss the dark conductivity σ_D of the amorphous films as a function of the premix gas ratio SiF$_4$/H$_2$ = r.

Figure 7.1 shows log σ_D versus $10^3/T$ for several samples using gas ratios r in the range 10 to 99. As the ratio r is increased the conduction mechanism changes from a well-defined activated process to one which is not activated. Samples C and D can be described by

$$\sigma_D = \sigma_0 \exp[(-\Delta E)_T/kT] \tag{7.1}$$

and the pre-exponent $\sigma_0 = e\mu_0 NkT$, where μ_0 is the mobility of the carriers, $N(E)$ the density of states in which they move, and the conductivity activation energy $(\Delta E)_T = \Delta E_0 - \beta T$, where β is assumed to be $\simeq 2 \times 10^{-4}$ eV K^{-1} [7.28]. As will be discussed later, these samples were n-type. For C, the pre-exponent $\sigma_0 \sim 10^3$ Ω^{-1} cm^{-1} which, since $\mu_0 \geq 10$ cm^2 s^{-1} V^{-1}, yields $N(E_c)$ $\geq 10^{21}$ cm^{-3} eV^{-1}. It follows that the dominant conduction mechanism is extended-state conduction at the band edge E_c.

As shown in the inset of Fig. 7.1, samples A and B can be fitted to the formula [7.29]

$$\sigma = \sigma_0' \exp[-(T_0/T)^{1/4}] , \tag{7.2}$$

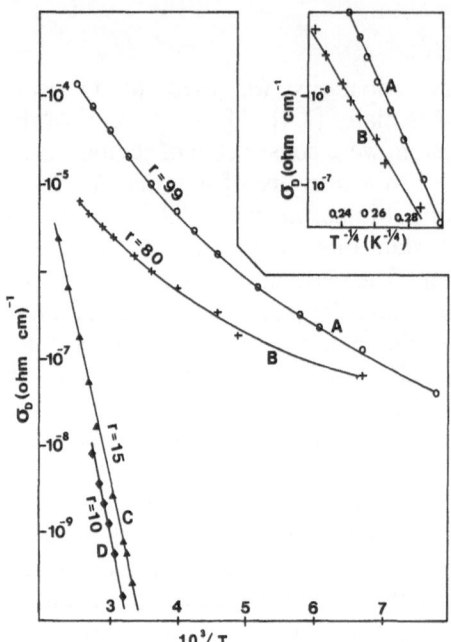

Fig. 7.1. The dark conductivity σ_D versus $10^3/T$ plotted for samples fabricated with different gas ratios r. The inset shows σ_D plotted as function of $T^{-1/4}$ for large values of r [7.4]

where $T_0 = 2[\alpha^3/N(E)k]^{1/4}$, $N(E)$ being the density of localized states, k Boltzmann's constant and α the decay of the electronic wave function. Assuming $\alpha^{-1} \simeq 10$ Å leads to an estimate of the density at the Fermi level greater than 10^{19} cm^{-3} eV^{-1}. This value is in close agreement with the density of states derived from the field effect data, as will be discussed below.

7.2.3 Localized State Density

In order to study the trapping or recombination centers, the field effect technique has so far proven [7.30–33] to be an important tool. A conventional MOSFET type structure is employed in which the application of a gate voltage (V_F) leads to an introduction of a charge Q into the material under study. This creates either an accumulation or a depletion layer, depending on the polarity of the gate voltage, which affects the source-to-drain current. For instance, if the material under study is n-type, then with an introduction of $-Q$ charge an accumulation layer of electrons in the a-Si film will form at the insulator/a-Si interface. Hence, the current i, from source to drain, will increase and the differential change in the current will depend upon the extent of the band bending. If the density of states $N(E)$ is low, then an introduction of a fixed charge $-Q$ will result in a large surface band bending which would cause a large change in the current i. Conversely, if $N(E)$ is large, then the resultant surface band bending would be small, leading to a small change in the current i. From the differential change in the current, one can, therefore, work backwards to determine the change in the surface potential and hence to the localized state density.

The analysis used to derive the density of states $N(E)$ from $i(V_F)$ curves neglects the effects of surface states and consequently the values obtained for the density of states represent an upper limit. Also, the choice of the flat-band position can somewhat alter $N(E)$. We should emphasize that the same procedure has been used for the a-Si:F:H as was used for the a-Si:H alloy. As a further check, $N(E)$ was measured for a sample deposited using SiH_4 gas at $T_s = 380\ °C$ and a power of 10 W. The electrical properties were $\sigma_D \sim 10^{-8}\ \Omega^{-1}\ cm^{-1}$ and $\Delta E = 0.60$ eV, and its density-of-states spectrum was in substantial agreement with the curve obtained for a-Si:H for a sample deposited at a nominal temperature of 570 K [7.4].

A typical source-to-drain current i as a function of gate voltage V_F for an a-Si:F:H alloy is shown in the inset of Fig. 7.2. The arrow indicates the assumed flat-band position. For samples fabricated with $5 < r < 99$, an n-type response was observed. The experimental $N(E)$ have been crudely extrapolated to the conduction-band edge E_c whose $N(E) \sim 10^{21}\ cm^{-3}\ eV^{-1}$, as estimated earlier.

The change in the conduction mechanism with r, as shown in Fig. 7.1, becomes understandable in view of the field effect data. We noted previously (Fig. 7.1), that, as r decreases from 80 to 10, the predominant transport mechanism changes from a variable range hopping type conduction to a well-defined activated process. As shown above, both samples were n-type and

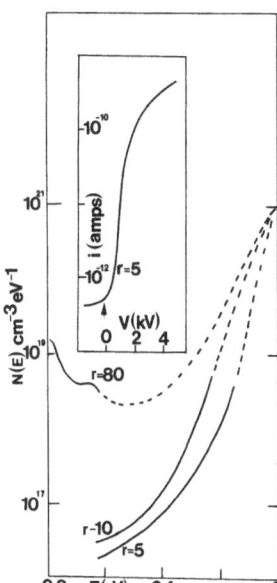

Fig. 7.2. The localized density of states $N(E)$ [$cm^{-3}\ eV^{-1}$] plotted for different gas ratios r. The inset shows a typical $i(V_F)$ curve and the arrow indicates the assumed flat-band position [7.4]

for $r = 10$, $N(E)$ is several orders of magnitude lower than that for the case $r = 80$. Since for large values of r the density of states is high, the dominant conduction is expected to be via localized states in the vicinity of the Fermi level. As r is decreased, as shown in Fig. 7.2, the density of states decreased sufficiently for the conduction to change to an activated process. The density-of-states curve indicates that (i) $N(E)$ decreased rapidly with decreasing gas ratio r, (ii) the peak in $N(E)$ observed in the Si : H alloy is absent in these films [7.31] and (iii) the density of states for this material in the upper half of the band gap is lower than in the Si : H alloy [7.31].

7.2.4 n and p-Type Doping Characteristics

The low density of states for a-Si : F : H enables this material to be readily doped. Generally, n and p-type doping is achieved by introducing PH_3 (or AsH_3) and B_2H_6 in the gas phase, respectively. In Fig. 7.3 we plot the dark conductively and the conductivity activation energy as a function of PH_3 and B_2H_6 for a-Si : F : H [7.19]. We note that the room temperature conductivity can be changed by orders of magnitude with doping. The most heavily n-type doping results in the conductivity exceeding $1(\Omega \text{ cm})^{-1}$, which suffices as a n^+ layer and provides an ohmic back contact to the solar cell structure which we shall discuss shortly. It should be mentioned that the apparent difference in the two n-type dopant curves shown in Fig. 7.3 for the Si : F : H alloy can be explained by recalling that the As–H bond is weaker than the P–H bond and hence, for a particular power employed, As would therefore be deposited in preference to P.

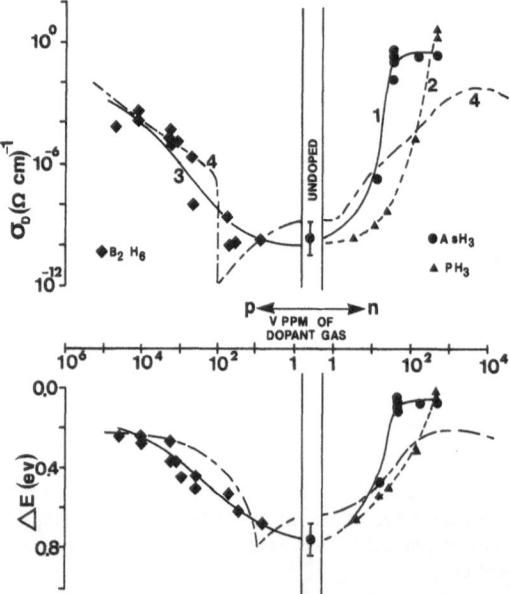

Fig. 7.3. The room temperature dark conductivity σ_D $(\Omega \text{ cm})^{-1}$ and the conductivity activation energy ΔE (eV) plotted as a function of vppm of AsH_3 (Curve 1), PH_3 (Curve 2), and B_2H_6 (Curve 3) introduced into the premix gas ratios of $r = 10/1$ and $T_s = 380$ °C. The p to n transition indicated refers to a-Si : F : H alloys. For comparison purposes (Curve 4), reported doping results for a-Si : H are also included [7.32]

For comparison purposes we have also included the n and p-type doping curve as reported by *Spear* and *LeComber* [7.34] and we note that the most heavily doped n-type layer is only $10^{-2}(\Omega\ \text{cm})^{-1}$ which is to be compared with $\sigma_{\text{D}} \sim 10(\Omega\ \text{cm})^{-1}$ for the fluorinated a-Si.

Since the integrated density of localized states for a-Si : F : H and a-Si : H is virtually the same over the mobility gap, then the reason for the improvement in the dark conductivity for the fluorinated alloy has to be sought elsewhere. It is now apparent that the highly P-doped fluorinated Si samples are microcrystalline [7.35, 36] and hence the high conductivity becomes understandable. It should be emphasized that the undoped photovoltaic material is amorphous as evidenced by the work of *Tsu* et al. [7.35]. Similar high conductivities in P-doped a-Si : H alloys have also been reproduced where high power was used to decompose SiH_4 gas [7.37, 38].

7.2.5 Photoconductivity

Amorphous films fabricated with $r > 30$ exhibited insignificant photoconductivity. This is not surprising since $N(E)$ is quite large for these types of films, as shown in Fig. 7.2. With decreasing r the photoconductivity of the samples increases markedly, as shown in Fig. 7.4. This figure shows the photoconductivity parameter $\sigma_{\text{p}} = \sigma_{\text{L}} - \sigma_{\text{D}}$, where σ_{L} is the photoconductivity under illumination for (i) incident illumination at $\lambda = 600$ nm with a photon flux of $N_0 = 10^{15}\ \text{s}^{-1}\ \text{cm}^{-1}$, and (ii) a simulated solar energy spectrum

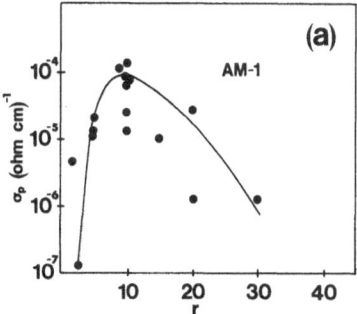

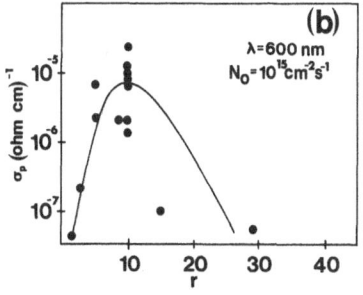

Fig. 7.4. (a) σ_{p} plotted as a function of r under AM 1 excitation of power 90 mW cm^{-2} [7.4]. (b) The photoconductivity $\sigma_{\text{p}}\ (\Omega\ \text{cm})^{-1}$ plotted as a function of r under an incident photon flux of $10^{15}\ \text{cm}^{-2}\ \text{s}^{-1}$ at $\lambda = 600$ nm

corresponding to AM1 radiation using a commercially available solar simulator.

Assuming that the photocurrent i_p is linearly dependent on illumination intensity, we can obtain an estimate of $\gamma\mu\tau$ using $i_p = (eN_0\gamma\tau/t_t)(1 - R)$ $\exp(-\alpha d)$, where N_0 is the photon flux, R the reflectivity, α the absorption coefficient, γ the quantum efficiency, τ the recombination time and t_t the transit time. Independent measurements of i_p, R and α enable us to determine $\gamma\mu\tau$. For $r = 10$, values of $\gamma\mu\tau$ of the order of $10^{-5}\mathrm{cm}^{-2}\,\mathrm{V}^{-1}$ have been obtained using $N_0 = 10^{15}\,\mathrm{cm}^{-2}\,\mathrm{s}^{-1}$ at $\lambda = 600$ nm.

From Fig. 7.4 we note that the photoconductivity of the films can exceed $10^{-4}(\Omega\,\mathrm{cm})^{-1}$ under AM1 conditions, which is sufficient for a solar cell.

7.2.6 Infrared

Figure 7.5 shows the infrared spectra for a thick (7.5 µm) sample deposited on a crystalline silicon substrate using a gas ratio of $r = 10$. From the integrated area under the peaks at 2100 and 2000 cm^{-1}, we have estimated the hydrogen content to be of the order of 0.5%. For these samples, the F content is approximately 4% as determined by Auger analysis, which is reflected in a mode of vibration at 830 cm^{-1} which we associate with a stretching of the Si–F bond.

The above estimation of H$_2$ is perhaps erroneous since from ion microprobe analysis, we find the H$_2$ concentration to be approximately 5% as shown in Sect. 7.6. The inaccuracy may be in the value assumed for the Si–H oscillator strength which is probably incorrect when F is included. *Lucovsky* [7.39] has calculated that the infrared peak corresponding to the Si–H stretch mode in a H–Si–F complex should be at 2122 cm^{-1}, which we have observed; we do not think that this peak could be attributed to the Si–H stretch mode in a dihydride (SiH$_2$) configuration [7.40].

There is a difficulty in assigning the infrared vibrational peak at 1015 cm^{-1} to the Si–O band since, as shown in Sect. 7.6, the oxygen contamination in the film is less than ~0.50%. The recent results of sputtered Si in

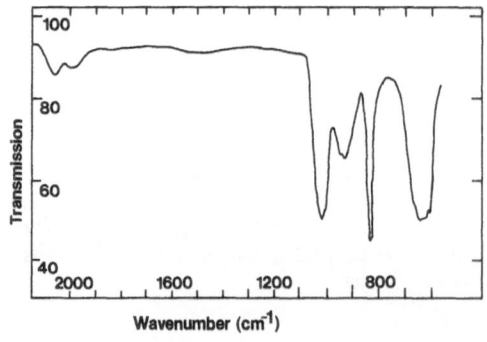

Fig. 7.5. A typical infrared spectrum of a thick sample (7.5 µm) deposited at a nominal temperature of 400 °C and using a gas ratio of $r = 10$ [7.4]

SiF$_4$ gas of *Shimada* et al. [7.22] indicate that the peak could be attributed to a SiF$_2$ vibrational bond.

In the above we have shown that device quality material can be produced from mixtures of SiF$_4$ and H$_2$. Evidence that F is playing an active role in removing the gap states has been provided by the work of *Matsamura* et al. [7.26] which involves the reactive sputtering of Si in Ar and SiF$_4$. We consider this next.

7.3 Sputtered Si with F or H

It has been shown [7.12] that by sputtering Si in a reactive H environment, good quality a-Si : H films can be made. These results suggest that, upon hydrogenation, its dark conductivity is reduced by orders of magnitude and is thermally activated. The hydrogenation is interpreted as causing changes in the density-of-states distribution both in the gap region and in the extremities of the valence and the conduction band.

Matsumura [7.26] has recently reported on a-Si : F alloy which was fabricated by sputtering Si in a reactive environment of SiF$_4$ in mixtures of Ar. From Rutherford back-scattering experiments, the F concentration was determined to be about 20%. In Fig. 7.6 a plot of σ_D vs $10^3/T$ is given for a-Si : F and a-Si : D for various annealing temperatures. The dark conductivity curve for a-Si : F shows typical semiconducting behavior which is possibly an indication of a low density of states in the gap. Upon annealing σ_D does not change, which is not the case for an a-Si : D film. When $T_A \geq 350$ °C,

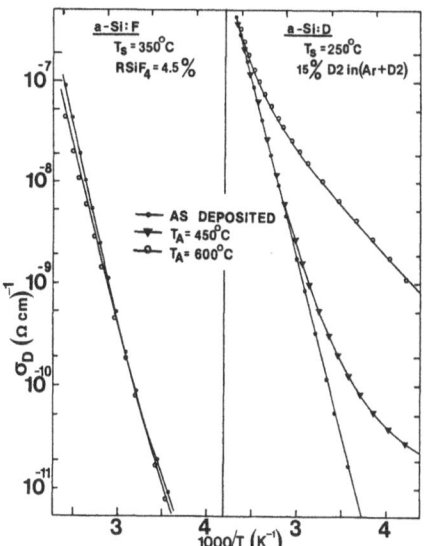

Fig. **7.6.** σ_D versus $10^3/T$ plots for a-Si : F and a-Si : D [7.23]

effusion of deuterium takes place which leads to a high density of localized states indicated by a hopping type conductivity, as shown in Fig. 7.6. This is in substantial agreement with the earlier work of *Fritzsche* and *Tsai* [7.11].

Further, they indicated that a-Si : F was only slightly less photoconductive than a-Si : D film and possessed a value of 10^{-8}–10^{-7} $(\Omega \text{ cm})^{-1}$ with an incident illumination power of 50 mW cm^{-2}. One indicator for the removal of gap states is the efficiency with which substitutional doping can be achieved. With the addition of 4500 ppm PF$_5$ to the reactive gas in the sputtering process, the dark conductivity of the film was changed from 10^{-10} to 10^{-4} $(\Omega \text{ cm})^{-1}$ which is remarkably similar to what has been achieved for the sputtered Si : H alloy [7.41] with the added advantage that the fluorinated films are heat resistant to much higher annealing temperatures.

7.4 The Role of F

Delal et al. [7.24] have suggested that a-Si films grown from SiF$_4$ and H$_2$ mixtures are subjected to strong reactive ion etching with the net growth rate being determined by a balance between deposition and etching. They considered the reactive ion etching to be the major distinguishing feature between the growth of a-Si : F : H alloys in comparison with the growth of a-Si : H alloys. Their model implies that (i) the growth rate is dependent on electrical bias, (ii) under certain growth conditions the films may be subjected to strong ion bombardment, thus promoting microcrystallinity, (iii) if F is to provide reactive ions, then it will still be necessary to satisfy dangling bonds, (iv) it is possible to etch the weakly-bonded Si atoms and hence produce fewer dangling bonds for H to satisfy and (v) it is possible that F could also etch the weakest Si–Si bond and hence promote an increase in the band gap. It is indicated that high conductivity *n*-type doped films can only be produced when the power level on the rf target is very high ($\sim$ several W cm^{-2}). From their Auger data there was no evidence of F incorporation indicating that F concentration is less than 1%. This is in contrast to our experience that high conductivity *n*-type doped films can be achieved with low power levels (typically ~ 0.1 W cm^{-2}) with a F content of the film in the range of 2 to 4%. It has also been suggested by some authors that F is included in the films via the inclusion of SiF$_4$ [7.23] and hence, under these conditions, it is not surprising that F remains inactive in tieing up the dangling bonds in a-Si. It is, therefore, quite possible that the negligible F content observed in the work of *Delal* et al. [7.24] could be via the inclusion of SiF$_4$ rather than by active incorporation.

An attempt to explain some of the divergent results obtained at least for the rf glow discharge in gas mixtures of SiF$_4$/H$_2$ and SiF$_4$/SiH$_4$/Ar has been attempted by *Potts* et al. [7.25]. They emphasized the importance of a Paschen-like curve for characterising the reactor for the particular gas being used. This is obtained by measuring the minimum potential needed to sustain the rf discharge as a function of the *pd* product, where *p* is the reactor pressure and *d* is the interelectrode distance. It is argued that for a given interelectrode spacing, the curve consists of two branches: a low pressure branch where the energy loss in the plasma is dominated by collisions of the electrons with the wall and a high pressure branch where energy is lost through collision with the plasma. The curves for SiF_4 and H_2 exhibit a minima at 0.2 and 0.7 torr, respectively. Since the curve for $SiF_4/H_2 = 8/1$ is quite similar to SiF_4 alone, it is quite likely that for $p < 0.4$ torr, the only charged fragments present in the plasma are those that result from the dissociation and ionisation of SiF_4 alone. For $p > 0.4$ torr, they reason that the plasma chemistry should change significantly as the reactive H species become available. The influence of these changes became clear when they plotted infrared transmission spectra for films grown at different pressures and noted that the intensity of the fluorine related modes (830, 940, and 1010 cm^{-1}) decreased as the pressure increased. (The H_2 concentration of all the films was measured to be in the range 7–10%). However, no results as regards the electronic properties was given. Their conclusion was that the amount of F concentration and presumably its activity in the film is controlled by plasma conditions. This can perhaps explain some of the divergent results which have been reported.

Using Raman scattering, *Tsu* et al. [7.35] have investigated sputtered Si, glow discharge a-Si : H and a-Si : F : H alloys and observed that the line width is 110, 80 and 70 cm^{-1}, respectively. They concluded that the decrease in line width for a-Si : F : H in comparison with a-Si : H alloys is due to a decrease in the internal strain of the film and have attributed this to F. Further, by using Rayleigh scattering and comparing the fluctuation of the dielectric constant of a-Si : H with a-Si : F : H, *Tsu* and *Hernandez* [7.42] noted a decrease in this constant for the fluorinated case. They attributed this to the large electronegativity of F and the ionic nature of the Si–F bond and concluded that F occupies active sites.

In unrelated work involved with the interfacial properties of a SiO_2/Si system for the fabrication of MOSFETS, *Ho* and *Sugano* [7.43] have reduced the interfacial state density to less than 10^{11} states $(\text{eV cm})^{-2}$ near the midgap by annealing in H_2 ambient at 450 °C. However, there still remained an interfacial state density of 10^{12} states $(\text{eV cm})^{-2}$ near the conduction and valence band edges. They reasoned that by replacing hydrogen with a halogen such as Cl or F and introducing this into the oxide to form stable Si–Cl or Si–F bonds at the interface, interfacial states could be removed. There is support for this from the theoretical work of *Sakurai* and *Sugano* [7.44]. Using Cl, the interfacial state densities were reduced to lower than 10^{10} states $(\text{eV cm})^{-2}$ near midgap. Further, the magnitude of the drift mobility of the MOSFET prepared in Cl annealed plasma anodized oxide was 225 $\text{cm}^2(\text{s V})^{-1}$, while that of the MOSFET prepared with a thermally grown oxide was 192 $\text{cm}^2(\text{s V})^{-1}$.

Our work with F, however, shows that F, which is smaller than Cl and has a higher electronegativity, acts more efficiently than Cl in terminating dangling bonds. Indeed, despite some attempts using Cl, high quality a-Si based materials have not been produced [7.45, 46].

Chemical vapor deposition (CVD) of SiF_2 resulting in a-Si-based alloys has been performed by *Janai* et al. [7.27]. They used a transport reaction such as

$$2 \; SiF_2 \; \xrightarrow[1200\,°C]{600\,°C} \; Si(solid) + SiF_4 \; .$$

Films fabricated in this way gave a low F concentration of 1.5% and the spin density, from electron spin resonance (ESR), was $3 \times 10^{19} \; cm^{-3}$. By altering the deposition temperature they found that with increasing deposition temperature, the ESR signal remains constant while the F content in the film decreases and thus concluded that F in their CVD a-Si does not remove dangling bonds. However, it is not clear how F is incorporated into the material. Further, a-Si produced from CVD of SiH_4 gas also has similar problems, as regards the density of dangling bonds. For as-deposited films, the H concentration is only about 0.2% and the quality of the film is only improved by post-hydrogenation in a plasma [7.13].

7.5 Photovoltaic Devices

We first consider one of the major differences that exists between a-Si based materials and crystalline Si-type solar cells. In crystalline Si, the indirect band gap necessitates relatively thick devices ($\sim 100 \; \mu m$) in which the space charge region is of the order of 1 μm in thickness [7.47]. Because of the large mobilities of electrons [$\mu_e \sim 1500 \; cm^2(s \; V)^{-1}$] and holes [$\mu_h \sim 500 \; cm^2(s \; V)^{-1}$], the photogenerated carriers are able to diffuse to the junction where they are separated and hence generate a photocurrent. However, in a-Si-based alloys, which are direct band gap materials, electron mobility μ_e [$\sim 10^{-1} \; cm^2(s \; V)^{-1}$] [7.48, 49] and especially the hole mobility μ_h [$\sim 10^{-3} \; cm^2(s \; V)^{-1}$] [7.45] are low. The low value for μ_h leads to a low diffusion length and hence the contribution made by the diffusion component of the photogenerated current is considered to be insignificant. Therefore, most of the observable current is due to the drift, i.e., the carrier generation that occurs within the space-charge region (SCR). The width of the SCR is inextricably linked to the density of localized states, i.e., the lower the $N(E)$, the larger the depletion width and hence the larger the collection region.

7.5.1 Geminate Recombination

Since the most effective carrier generation is limited to the absorption that occurs within the depletion width, it is important to consider the concept of geminate recombination [7.50] which is usually a limiting factor in amorphous semiconductors such as a-Se [7.51].

Geminate recombination is important when the photoexcited electron and hole do not escape their mutual Coulombic attraction. There are three important criteria associated with geminate recombination:

(i) a rapidly increasing photoconductive yield as the frequency of the incident light increases;
(ii) a decreasing activation energy for photoconduction as the frequency increases (this is associated with an increase in the thermalization separation of the hot photoexcited carriers with increasing excitation frequency);
(iii) an increase in the photoconductive yield at applied fields in excess of 10^4 V cm^{-1} at room temperature.

In Fig. 7.7 we plot for an a-Si : F : H alloy [7.52] the normalized photoconductivity for different wavelengths as a function of the electric field and temperature. We note that none of the necessary criteria required for geminate recombination are applicable for this type of alloy. It should be added that this type of recombination mechanism has been proposed as a limitation to the device performance using a-Si : H [7.53], although conflicting results have been reported by *Wronski* et al. [7.54] and *Yamaguchi* et al. [7.55].

Mort et al. [7.56] using a delayed collection field technique presented evidence for some geminate recombination using 30 μm thick a-Si : H alloy films. It is conceded that the growth kinetics involved in the fabrication of such a thick film could be entirely different in comparison with the 0.5 μm thick samples normally used in devices. Further, it is difficult to understand how geminate recombination takes place since reports of photogenerated collection efficiencies (in solar cells) approaching 90% have appeared in the literature [7.57]. However, if geminate recombination does exist, it probably plays an insignificant role.

7.5.2 Theoretical Conversion Efficiency in Schottky Barrier-Type Cells

Since geminate recombination, at least in a-Si : F : H, does not seem to be important and since most of the current that is to be expected is due to drift, it is, therefore, a simple matter to determine the short-circuit current density (J_{sc}) as a function of the depletion width, which is almost the total collection width.

To determine this [7.58] let us first consider the simplest type of device structure possible, i.e., the Schottky barrier (SB). If a high work function metal such as Pt is contacted with a-Si based alloys, then a space-charge region (SCR) of width W is formed in the a-Si based material. Because of the

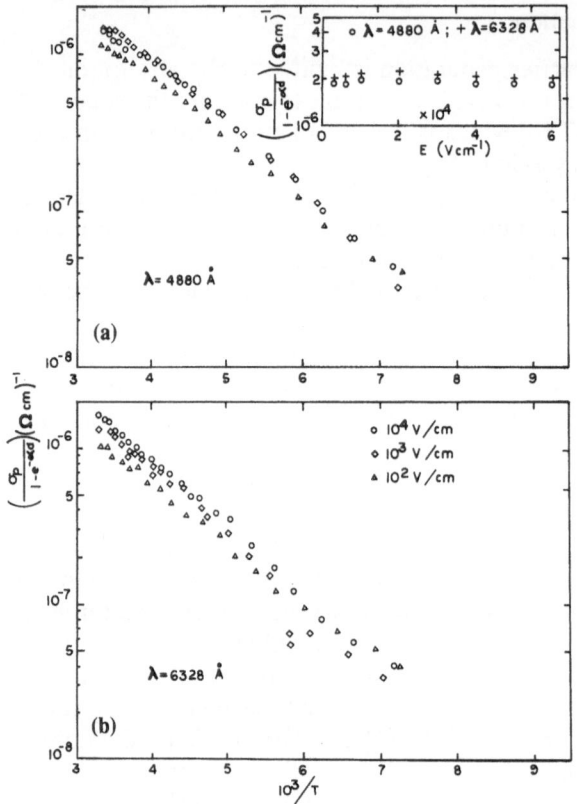

Fig. 7.7. Normalized photoconductivity as a function of reciprocal temperature is plotted for fields in the range 10^2 V cm^{-1} to 10^4 V cm^{-1} for incident illumination of (**a**) $\lambda = 4880$ Å and (**b**) $\lambda = 6328$ Å. The inset shows the normalized photoconductivity of room temperature for fields up to 6×10^4 V cm^{-1} [7.48]

low mobilities of the carriers, transport through the structure is most probably limited by drift and diffusion in the SCR rather than by thermionic emission at the metal-semiconductor (MS) interface. Hence, it is more appropriate to use diffusion theory (DT) which predicts that the quasi-Fermi level varies in the SCR and joins smoothly onto the Fermi level of the metal.

Using DT, the current density can be written as [7.58]

$$J = q\mu N(E_c)kT \exp[-q(E_c - E_{Fn})/kT]dE_{Fn}/dx , \qquad (7.3)$$

where μ_e is the electron mobility, $N(E_c)$ is the density of states at the conduction band edge and E_{Fn} is the quasi-Fermi level. This can be rewritten as

$$[J/kT\mu N(E_c)] \int_0^x \exp(qE_c/kT)dx = \exp[qE_{Fn}(x)/kT] . \qquad (7.4)$$

The integrand on the left-hand side is a rapidly decreasing function of x and becomes nearly independent of x when $x > kT/qE_{max}$, where E_{max} is the maximum field at the semiconductor/metal interface. For a-Si, $kT/qE_{max} < 50$ Å and thus, with the exception of a region adjacent to the metal, E_{Fn} can be regarded as constant within the depletion layer.

Using the simplifying assumption that E_{Fn} is virtually constant in the SCR, we then determine the shape and width of the SCR. This can be analyzed by solving the Poisson equation

$$d^2E_c/dx^2 = \varrho(E_c, E_{Fn})/\varepsilon_r\varepsilon_0 . \tag{7.5}$$

The measured density of localized states can be reasonably approximated [7.14] as

$$N(E) = N_{min}[\exp(E/E_{ch}) + \exp(-E/E_{ch})] , \tag{7.6}$$

where E_{ch} is a characteristic energy and is equal to 0.094 eV and 0.066 eV for a-Si:H and a-Si:F:H alloys. N_{min} is the minimum density of states and is equal to 1.5×10^{17} cm^{-3} eV^{-1} and 2×10^{16} cm^{-3} eV^{-1} for a-Si:H and a-Si:F:H alloys, respectively. Using (7.6), the charge density ϱ can be calculated and using (7.5), the shape of the SCR can be calculated as a function of bias voltage and for various assumed values of E_{Fn} within the semiconductor. Once this is known, using the measured values for the absorption coefficient, the light generated current can be calculated via

$$J_L = q \int_0^\infty F_\lambda Q_\lambda [1 - \exp(\alpha_\lambda W)](1 - R_\lambda)d\lambda , \tag{7.7}$$

where F_λ is the photon flux, Q_λ is the quantum efficiency, α_λ is the absorption coefficient, R_λ is the reflection coefficient, W is the depletion width and λ is the wavelength. In Fig. 7.8 a plot is shown of the light generated current J_{sc} as a function of depletion width, with and without surface recombination [7.59]. The effect of surface recombination was taken into account by assuming that the quantum efficiency $Q_\lambda = 0$ for $\lambda < 400$ nm. It should be noted that for depletion widths of about 0.5 µm, currents of the order of 15 mA cm^{-2} can be generated.

It is known that a-Si based materials possess a nonuniform density of gap states and hence the width of the depletion region $W(V)$ depends upon the position of E_{Fn}, as shown in Fig. 7.9. Once W as a function of bias V has been determined, as discussed above, internal light J-V characteristics can be generated, and by superimposing the dark J-V characteristics to these curves, device parameters such as the open circuit voltage V_{oc}, short circuit current J_{sc}, fill factor (FF) and conversion efficiency η can be determined. In Fig. 7.10 a plot is shown of maximum conversion efficiency as a function of the barrier height for zero back reflection and assuming a back reflection of 80%. Note that conversion efficiencies exceeding 9% are possible using these types of alloys [7.58].

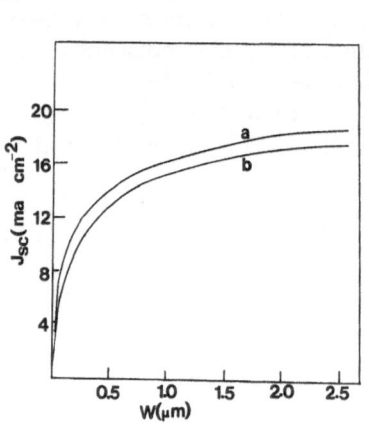

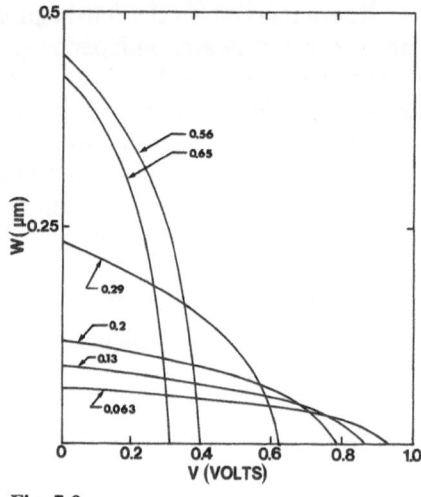

Fig. 7.8.   **Fig. 7.9.**

Fig. 7.8. J_{sc} plotted as a function of the depletion width W for (*a*) no surface recombination and (*b*) with surface recombination [7.56]

Fig. 7.9. The depletion width as a function of the forward bias voltage V plotted for various values of $E_c - E_F$ [7.56]

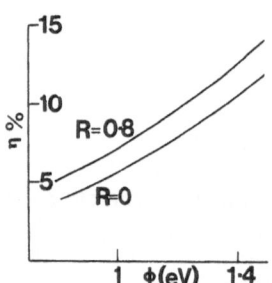

Fig. 7.10. Conversion efficiency versus barrier height for back reflection $R = 0$ and $R = 0.8$ [7.54]

The above efficiency estimates are based on small diffusion lengths. Using a surface photovoltage technique, *Dresner* et al. [7.60] indicated that the diffusion length in a-Si:H samples is about 0.4–0.5 μm. This, coupled with a depletion width of the order of 0.2 μm [7.15], would make the collection width in a-Si:H about 0.6–0.7 μm. However, it should be mentioned that the previous work at RCA had indicated that the diffusion length was 300 Å [7.61] and subsequently, using PEM techniques, it was measured to be 900 Å [7.62]. *Dresner* et al. [7.60] have attributed the increase in the diffusion length to a possible improvement in the material. However, it would be interesting to see if their device efficiencies show a corresponding increase. In contrast, *Snell* et al. [7.63], using a junction recovery method, indicated a diffusion length of about 400 Å using a-Si:H alloys in a *p-i-n-* type device structure. The inclusion of a diffusion length of about 4000 Å in a-Si : F : H would lead to an increase in J_{sc} and a considerable improvement in the (FF) and thus lead to an overall conversion efficiency estimate exceeding 10% for MIS type cells.

7.6 Metal-Insulator-Semiconductor (MIS) Type Devices

The maximum reported conversion efficiency for an MIS-type device is 6.6% [7.5] over an active area of 0.73 cm^2. The device utilized the following: a thin, highly conductive n^+ layer (~ 500 Å in thickness) was deposited onto a conducting substrate such as Mo to provide an ohmic back contact. Next, ~ 5000 Å of active photoconductive a-Si : F : H was deposited using a volume gas ratio of $SiF_4/H_2 \approx 5/1$. Then a 20–30 Å thick oxide was thermally evaporated and contact was made to the device using a 70 Å thick layer of a high work function metal such as an AuPd alloy or Pt. Finally, a 350 Å thick layer of ZnS served as an antireflection coating. Oxides such as TiO_2, Nb_2O_5, SiO_2 and Ta_2O_5 were attempted, but the most satisfactory and consistent results were obtained using Nb_2O_5.

Figure 7.11 shows a microprobe analysis of a typical device, as fabricated above, but deposited on to a crystalline Si substrate. The hydrogen concentration is uniform at about 5%, F concentration is about 2% and the oxygen

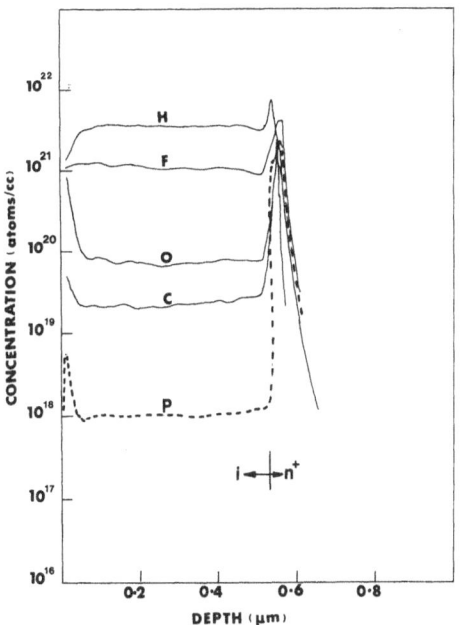

Fig. 7.11. A composition SIMS profile of an MIS-type structure [7.64]

and carbon contamination is less than 0.5%. The figure also shows the P concentration in the heavily doped (n^+) and photoactive (i) layers. It should be noted that there is an abrupt transition of P concentration from the n^+ to i layers with a slight inadvertent P contaminant within the intrinsic layer. Further, the P concentration in the n^+ layer is in the percentage range similar to that reported for the a-Si:H alloys [7.34].

7.6.1 Forward Bias Dark J-V Characteristics

Figure 7.12 shows typical room temperature current density-voltage (J-V) characteristics for a MIS-type device using a-Si:F:H as the active component and utilizing an insulator of thickness δ = 30 Å [7.16]. The curvature that is apparent in the $\log(J$-$V)$ characteristic for forward bias voltage exceeding 0.4 V is due to the resistive nature of the a-Si. This would be an inhibiting factor for a device were it not for the highly photoconductive nature of the material. The slope, from the forward bias characteristics, yields the ideality factor n = 1.2. Figure 7.13 shows the temperature dependence of n for a typical device with and without the oxide [7.64]. It should be noted that at

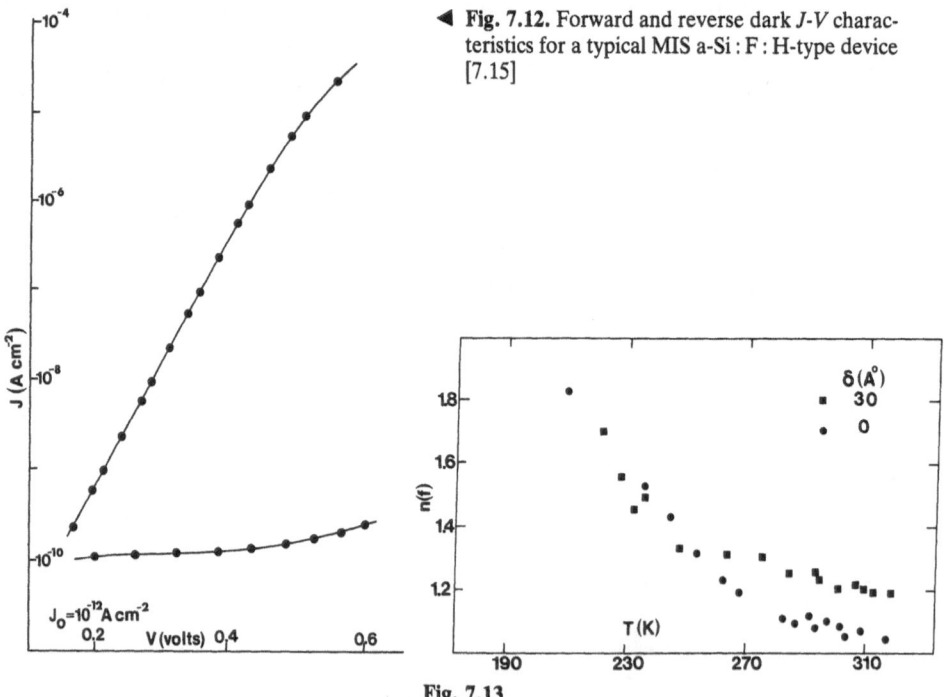

◀ **Fig. 7.12.** Forward and reverse dark J-V characteristics for a typical MIS a-Si:F:H-type device [7.15]

Fig. 7.13

Fig. 7.13. The diode quality factor $n(f)$ determined from the forward bias condition is plotted as a function of temperature for a typical a-Si:F:H device with and without the insertion of an oxide layer of thickness δ [7.62]

high temperatures, n saturates at a value that depends upon the oxide layer thickness, i.e., $n = 1.05$ and 1.20 for $\delta = 0$ Å (corresponding to native oxide on a-Si : F : H alloy) and 30 Å, respectively. At lower temperatures, the value of n for both devices coalesces to a value exceeding 1.5. This we attribute to the recombination current within the space-charge region.

An increase in n with δ has been observed for crystalline Si-MIS devices. *Card* and *Rhoderick* [7.65] have obtained the following expression to describe the behavior of n on δ, i.e.,

$$n = 1 + \delta \varepsilon_s / W(\varepsilon_i + \delta q^2 D_s) , \tag{7.8}$$

where ε_s and ε_i are the permittivities of the semiconductor and insulator, D_s is the surface state density and W is the width of the depletion region.

For a-Si-based MIS-type devices, the above expression (7.8) does not seem to be applicable since if we assume reasonable values for the various parameters such as $\varepsilon_s/\varepsilon_i = 4$, $\delta = 30$ Å, $D_s = 5 \times 10^{12}$ cm^{-2} eV^{-1} and $W = 4000$ Å, we obtain $n \doteq 1.02$ [7.64].

The inadequacy of (7.8) to explain the experimental result of $n = 1.2$ for $\delta = 30$ Å is possibly due to the fact that despite the relatively low density of states, a-Si : F : H still possesses a significant number of trapping levels as shown in Fig. 7.2; hence, in the derivation of (7.8), we must consider the effect of these trapping levels on E_{max}, the field at the semiconductor/ insulator interface.

The effective barrier Φ_{eb} depends upon δ, D_s and the forward bias voltage via the term E_{max}. Φ_{eb} can be written as

$$\Phi_{eb} = \Phi_{ob} - \theta E_{max} , \tag{7.9}$$

where $\theta = \delta \varepsilon_s / (\varepsilon_i + \delta q^2 D_s)$ and Φ_{ob} is the barrier height under flat-band conditions. E_{max} is a complicated function which depends upon the doping density, N_D, δ, D_s and the density and distribution of the localized states within the mobility gap.

Using diffusion theory, our calculations of E_{max} based on the density-of-states spectrum of a-Si : F : H alloys leads to an expression [7.64]

$$E_{max} = (\Phi_{eb} - \xi - V) f(\Phi_{eb})^{1/2} , \tag{7.10}$$

where V is the applied bias, $\xi = E_c - E_F$, (E_c and E_F are the positions of the conduction band edge and Fermi level, respectively), and

$$f(\Phi_{eb}) = (2/\varepsilon_0\varepsilon)\langle qN_D + (qN_{min}E_{ch}^2)/(\Phi_{eb} - \zeta)$$
$$\times \{\cosh[(\Phi_{eb} - E_{i0})/E_{ch}] - \cosh[(\zeta - E_{i0})/E_{ch}]\}\rangle , \tag{7.11}$$

where N_D is the donor doping level density, $N_{min} = 1.5 \times 10^{16}$ cm^{-3} eV^{-1} is the value at the minimum in the density of states spectrum, E_{ch} is a characteristic energy describing the exponential change in the localized state spectrum [7.14] and E_{i0} is the intrinsic Fermi-level position.

The ideality factor n can be defined by

$$n = 1/(1 - \partial\Phi_{eb}/\partial V) .\qquad(7.12)$$

Using (7.9–12), we obtain $n = 1.05$ and 1.14 for $\delta = 5$ and 40 Å, respectively, assuming a zero surface state density. However, if it is assumed that $D_s = 5 \times 10^{12}$ cm^{-2} eV^{-1}, the calculations yield $n = 1.03$ and 1.04 for $\delta = 5$ and 40 Å, respectively. This decrease in the value of n with increasing surface state density D_s is primarily due to the screening effect of the electric field by the interface states.

Comparing these calculations with the experimental values of $n = 1.05$ for $\delta = 0$ Å and $n = 1.2$ for $\delta = 30$ Å as shown in Fig. 7.13, we conclude that the density of surface states in these alloys is relatively small ($< 10^{12}$ cm^{-2} eV^{-1}). This conclusion is also consistent with the interpretation used to explain the variation of the open circuit voltage with the work function of the metal as discussed later.

7.6.2 Reverse Bias Dark J-V Characteristics

The diode equation can generally be written as

$$J = J_0[\exp(eV/nkT) - 1] ,\qquad(7.13)$$

where J_0 is the reverse saturation current density. In MIS or MS-type structures it is appropriate to consider majority carrier current flow, which in this case is due to electrons. Using diffusion theory, the reverse saturation current can be written as

$$J_0 = q\mu N(E_c)kT E_{max}\exp(-\chi^{1/2}\delta)\exp(-q\Phi_{eb}/kT) ,\qquad(7.14)$$

where μ is the mobility of majority carriers (electrons), $N(E_c)$ is the density of states at the conduction band edge and $\exp(-\chi^{1/2}\delta)$ is the transmission coefficient of the electrons through the insulating barrier height of the insulator [eV] and the electron energy.

Figure 7.14 shows the current-voltage $I_r(V_r)$ characteristics in reverse bias for a typical device [7.66]. Note that there is a lack of saturation, in contrast to what is implied by (7.14). The lack of saturation in reverse bias can generally be attributed to one or more of the following causes: (i) image force lowering, (ii) generation current, and (iii) barrier height lowering due to the presence of an insulating layer. The first two give the functional dependencies of $J_r \propto \exp[-q/kT(C-V)^{1/4}]$, where C is a constant and $J_r \propto V_r^{1/2}$ respec-

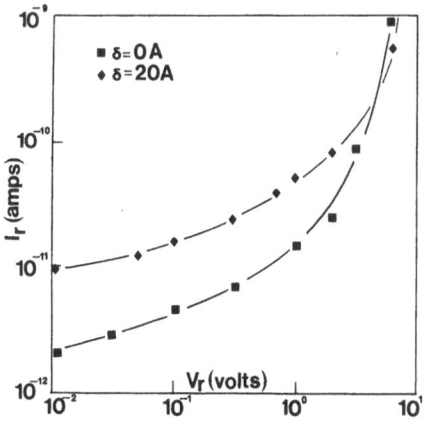

Fig. 7.14. Dark reverse $I_r(V_r)$ characteristics for a device with and without an oxide [7.62]

tively; these cannot account for the observed $J_r(V_r)$ characteristics. The last mechanism (c) can, since under reverse bias the field at the insulator/semi-conductor interface increases and causes Φ_{eb} to decrease. Using the analysis as outlined in Sect. 7.6.1, we can simulate the $J_r(V_r)$ characteristics for the reverse bias case. Assuming that a potential V_i is dropped across the oxide in an MIS-type structure, it can be readily shown that using (7.11), Φ_{eb} decreases with increasing reverse bias and also for increasing oxide layer thickness.

Figure 7.15 shows a computer simulation of the $J_r(V_r)$ characteristics for varying oxide layer thicknesses and the qualitative similarities between this

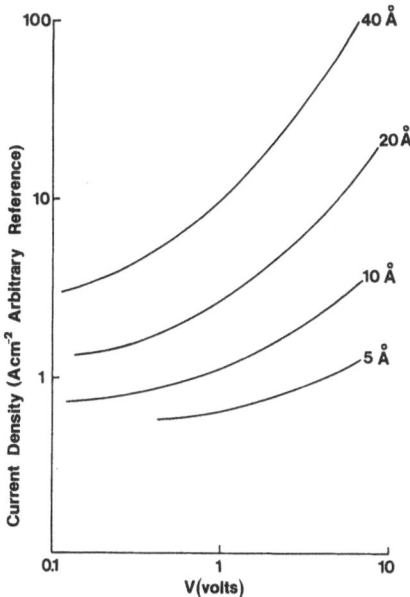

Fig. 7.15. Theoretical reverse $I_r(V_r)$ characteristics for increasing oxide layer thickness [7.62]

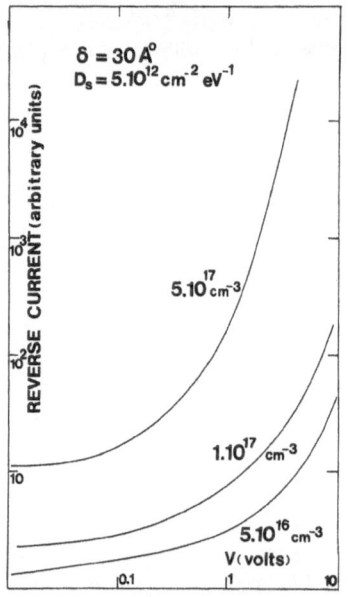

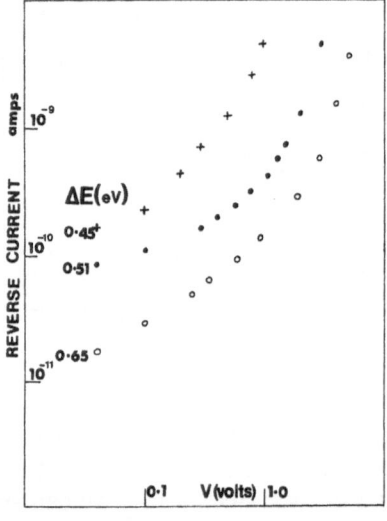

Fig. 7.16. Theoretical reverse $I_r(V_r)$ characteristics for increasing density of states [7.64]

Fig. 7.17. Reverse $I_r(V_r)$ characteristics for various devices in which the Fermi level of the active layer was varied [7.64]

and Fig. 7.14 suggests that the reduction of Φ_{eb} with V_r can account for the lack of saturation in the reverse characteristics rather than a generation current from within the space-charge region.

The above analysis can also provide a means for investigating the behavior of the photoactive material within the device structure. For example, we have observed that the photoluminescence (PL) intensity of the active material as a function of P doping is reduced substantially as the amount of dopant within the alloy is increased [7.66]. The reduction in the intensity suggests that nontetrahedrally coordinated P has introduced non-radioactive recombination centers, in general agreement with what is also observed for a-Si:H alloys [7.67]. In Fig. 7.16, $J_r(V_r)$ characteristics have been derived when N_{min}, the minimum density of localized states in (7.6), is increased. Figure 7.17 shows experimentally obtained $I_r(V_r)$ curves; once again we obtain a qualitative similarity to the theoretically predicted curves. This is in agreement with the observation of *Crandall* [7.68] who used the primary photocurrent technique to deduce an increase in the density of localized states with doping within the gap.

7.6.3 Barrier Height Determination

From the diode equation, the reverse saturation currents J_0 can lead to a measure of the effective barrier height Φ_{eb}. Using thermionic emission theory (TET) which is valid for high mobility solids, J_0 is given by

$$J_0 = A^* T^2 \exp(-q\Phi_{eb}/kT) , \tag{7.15}$$

where A^* is the Richardson-Dushman constant. In the diffusion theory (DT) which is valid for low mobility solids, J_0 is given by (7.14).

If an insulator is inserted between the metal and the semiconductor, Φ_{eb} is altered as discussed earlier and the majority carrier suppression term, $\exp(-\chi^{1/2}\delta)$ [7.65], would have to be included in the above expression (7.15).

The measurement of J_0 as a function of T for the device as shown in Fig. 7.12 and plots of J_0/T^2 and J_0/T versus $10^3/T$ lead to a measure of the effective barrier height Φ_{eb} to be 0.89 eV and 1.0 eV, respectively. It seems that both TET and DT theory can provide an adequate fit to the data [7.69, 70]. However, the difference between the two theories should become more pronounced at much lower temperatures but the highly resistive nature of a-Si introduces experimental difficulties which prevents a test for the applicability of either (7.14) or (7.15) from being made.

Using (7.14) we can estimate the value of Φ_{eb} from the known values of the various parameters. $\mu_e \sim 10^{-1}$ cm^2(s V)$^{-1}$ is the mobility of the majority carriers and is assumed to be the same as in the a-Si:H alloy [7.48, 49]; $N(E_c) \sim 10^{21}$ cm^{-3} eV^{-1}, $E_{max} \sim 10^5$ V cm^{-1} [7.14] and $\exp[-(\chi^{1/2}\delta)] \sim 0.14$ [7.66]. This leads to $\Phi_{eb} \sim 0.91$ eV. However, μ_e may be field dependent and be much larger nearer the top of the barrier where the field has its maximum value. If it is assumed that the μ_e is increased by a factor of say a hundred, then we estimate Φ_{eb} to be ~ 1.02 eV [7.16].

Alternatively, by plotting V_{oc} against T, Φ_{eb} can also be determined since

$$V_{oc} = (n kT/q)[\ln(J_{sc}/J_0) + 1] , \tag{7.16}$$

where J_{sc} is the short circuit current and n' is the diode quality factor under light. n' was measured to be ~ 1.0–1.1 by plotting V_{oc} against J_{sc} for various temperatures and was found to be substantially independent of it. J_{sc} over the temperature range investigated varied by less than a factor of 2. Using $n' \sim 1.0$ gave $\Phi_{eb} \sim 1.12$ eV which is in relatively good agreement with the previous estimates and measurements.

7.6.4 Open Circuit Voltage Dependence on the Metal Work Function

The previous conclusion that a-Si:F:H possesses a low density of surface states also seems to be consistent with the interpretation of the variation of

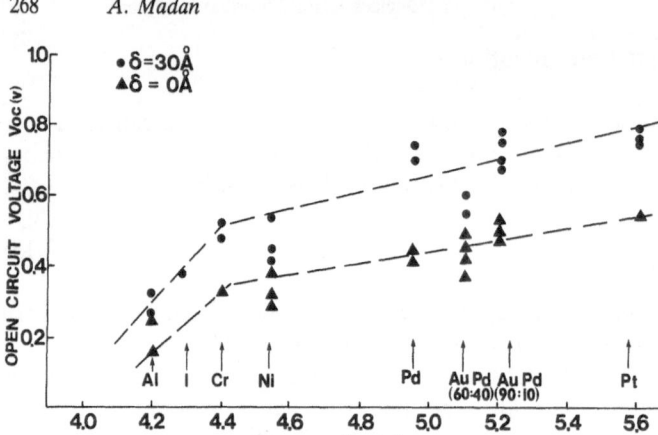

Fig. 7.18. V_{oc} plotted against the work function of the metal [7.62]

V_{oc} (measured under simulator AM 1 illumination intensity) with the work function of the Schottky-barrier contact. Figure 7.18 shows this variation for a variety of top metal contacts for devices fabricated with and without the deliberate insertion of an oxide layer of thickness 30 Å [7.64]. It is apparent that V_{oc} varies linearly for low work function metals ($\Phi_m < 4.4$ eV) and tends to show a sublinearity in V_{oc} for $\Phi_m > 4.4$ eV. For $\Phi_m < 4.4$ eV, the slope of V_{oc} against Φ_m can be described by [7.71]

$$\Psi = \varepsilon_i/(\varepsilon_i + \delta q^2 D_s) . \tag{7.17}$$

For a low density of surface states, $D_s < 10^{12}$ cm^{-2} eV^{-1}, $\Psi \simeq 1.0$ as observed in Fig. 7.18. As the work function increases ($\Phi_m > 4.4$ eV), the Fermi level of the semiconductor within the space-charge region enters a higher density of localized states with the result that the Schottky-barrier profile would exhibit a sharp spike. This would result in the tunneling of majority carriers through the thin barrier and lead to a saturation effect in the effective barrier height Φ_{eb}; this then translates into a weak dependence of V_{oc} with high work function metals such as for Pd and Pt as observed in Fig. 7.18.

7.6.5 Role of the Insulator

In the fabrication of MIS-type devices using a-Si as the base semiconductor, we are necessarily restricted to a deposition of a dielectric rather than thermal growth; this is because at high temperatures necessary for the thermal oxides grown thermally, H$_2$ effusion from the active material would take place [7.11] and would lead to an overall degradation of the device.

MIS structures rely upon ultra-thin interfacial layers such that the wave function of the charge carriers can penetrate the insulating layer and appear on the metallic side and result in current flow within the device. The interfacial layer should be such that [7.72]

$$E_{gi} \geq \chi_s - \chi_i + E_{gs} , \qquad (7.18)$$

where χ_s and χ_i are the electron affinities of the semiconductor and insulator, respectively, and E_{gi} and E_{gs} are the energy band gaps, respectively. The inequality in (7.18) is required in order to guarantee that the device will operate as a minority carrier cell.

This condition implies that an interfacial layer with a moderate band gap but large affinity should serve as an optimum insulating material. It should be stressed that the band structure of such thin layers in all probability differs from its corresponding bulk value. Further, the chemistry and physics of such thin layers is not well understood.

Calculations of the forward J-V characteristics for a MIS (Al–SiO$_2$–Si diode with 2 Ω cm p-type base semiconductor) have been performed for the ideal case without defects at the oxide/semiconductor interface or within the oxide itself [7.73]. In forward bias, a departure from the ideal diode behavior is calculated indicating the changeover from semiconductor-limited operation to tunnel-limited operation. This implies that significant conversion efficiency would not occur for insulator thickness much above 20 Å. The effect of the tunnel-limited region would be equivalent to a series resistance. Hence for $\delta > 20$ Å, one would therefore expect a photocurrent suppression.

This is also expected to be true for MIS-type devices employing a-Si : F : H alloys [7.5]. Figure 7.19 shows a plot of the short circuit current

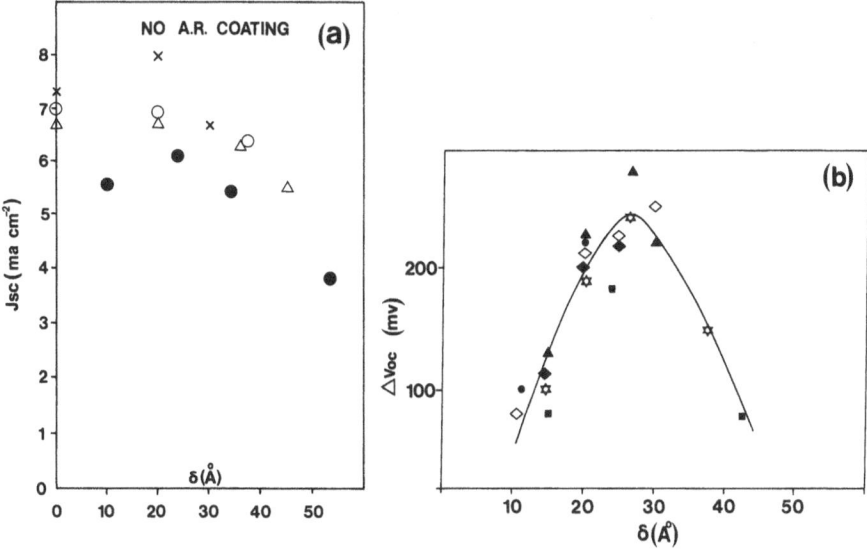

Fig. 7.19. (a) J_{sc} and (b) V_{oc}, plotted against the oxide layer thickness [7.5]

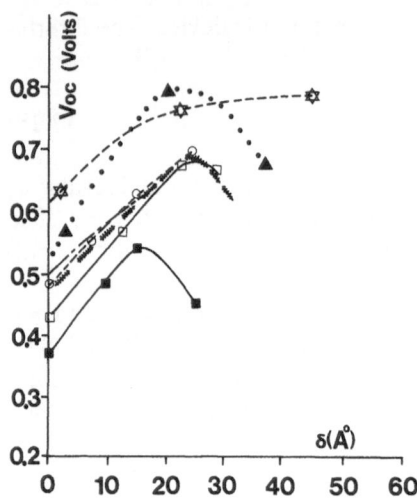

Fig. 7.20. V_{oc} as a function of δ for various devices in which the Fermi level of the active layer was varied [7.64]

density (as measured under simulated AM 1 spectra) and V_{oc} as a function of insulator thickness. The current begins to decrease for $\delta > 30$ Å and the device, according to the above interpretation, is in the tunnel dominated regime.

The reason for using an insulator becomes clear when we consider the change in open circuit voltage with the oxide layer thickness, δ. We note that for $\delta \sim 30$ Å, V_{oc} is enhanced by about 250 mV although in some cases we have observed a change of up to 280 mV. The enhancement occurs without any deterioration in the short circuit current or in the fill factor. This significant change in V_{oc}, according to *Card* and *Rhoderick* [7.65], is primarily due to the majority carrier current suppression.

To confirm that the enhancement of V_{oc} is due to the oxide and is independent of the base semiconductor, a series of devices were fabricated in which the Fermi level of the active layer was deliberately altered by the addition of PH$_3$ in the gas phase [7.64]. Figure 7.20 plots the change in V_{oc} with δ for various devices with the conductivity activation energy $E_c - E_F$ of the active layer varying from 0.8 eV to 0.3 eV. In each case the enhancement in V_{oc} exceeds 200 mV.

Since performing this above work, *Gutkowicz-Krusin* et al. [7.74] have suggested that the introduction of the insulator could also enhance the J_{sc}, primarily at the blue end of the spectrum, by reducing the thermal diffusion of electrons against the electric field. The above data, shown in Fig. 7.19, indicates that J_{sc} may be enhanced. Figure 7.21 shows the spectral response of a cell with and without the insulator [7.5]. It should be noted that the spectral response of some of the cells is improved toward the blue end of the spectrum when an insulator is introduced, in agreement with the above suggestion [7.74].

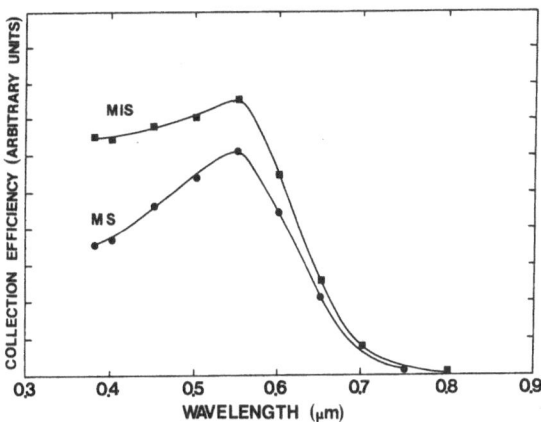

Fig. 7.21. Spectral response of a MS and a MIS cell structure [7.5]

7.7 Reproducibility

In Fig. 7.22 we show the device characteristic data from numerous devices of active area 0.042 cm^2 ostensibly fabricated under identical conditions [7.5]. The fill factor is typically 0.55–0.6, $V_{oc} \sim 0.78$–0.88, $J_{sc} \sim 12$–14 mA cm^{-2} with an antireflection (ar) coating. The device efficiencies are between 3–4% without an ar coating and 5.5%–6.6% with an ar coating.

In Fig. 7.23 we show the overall conversion efficiencies measured for larger area devices. The total areas of the devices were 0.81 cm^2; these utilized a grid pattern (Ag) whose area was subtracted to yield an active area of 0.73 cm^2 which was used in the calculation to obtain the overall conversion efficiency. The efficiencies, with ar coating, for these large area devices is the same as for the small areas of 0.042 cm^2. The best efficiency for a MIS-type structure that we have observed so far is 6.6% with the following characteristics: $V_{oc} = 0.88$ V, $J_{sc} = 13.1$ mA cm^{-2}, (FF) $= 0.57$.

7.7.1 Correlation of the Device Performance with Some Material Parameters

There have been many attempts to determine the quality of a material required for a useful photovoltaic device, using photoconductivity and photo-luminescence spectra as indicators [7.75]. Therefore, to test the validity of these ideas, we have plotted the conversion efficiencies of devices that were fabricated simultaneously with the material under investigation.

In Fig. 7.24 we plot the photoconductivity σ_p under AM 1 incident illumination against the device performance [7.5]. There does not seem to be a clear-cut correlation which is perhaps not too surprising since σ_p measures majority carrier phenomena whereas the devices are minority carrier controlled. However, σ_p must exceed a critical value or the series resistance will inhibit the performance of the device.

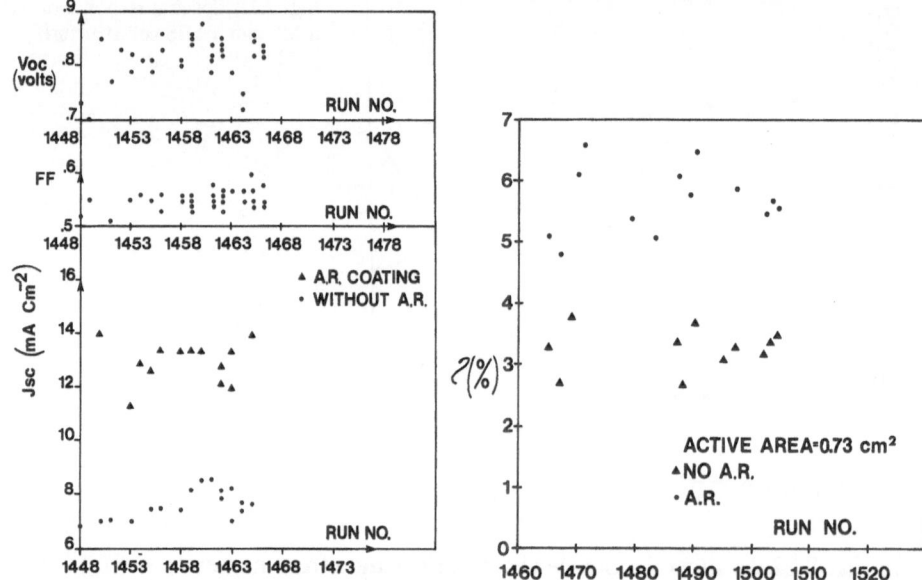

Fig. 7.22. V_{oc}, (FF) and J_{sc} plotted as a function of sequential run number for devices fabricated under identical conditions [7.5]

Fig. 7.23. Device efficiency for large area cells [7.5]

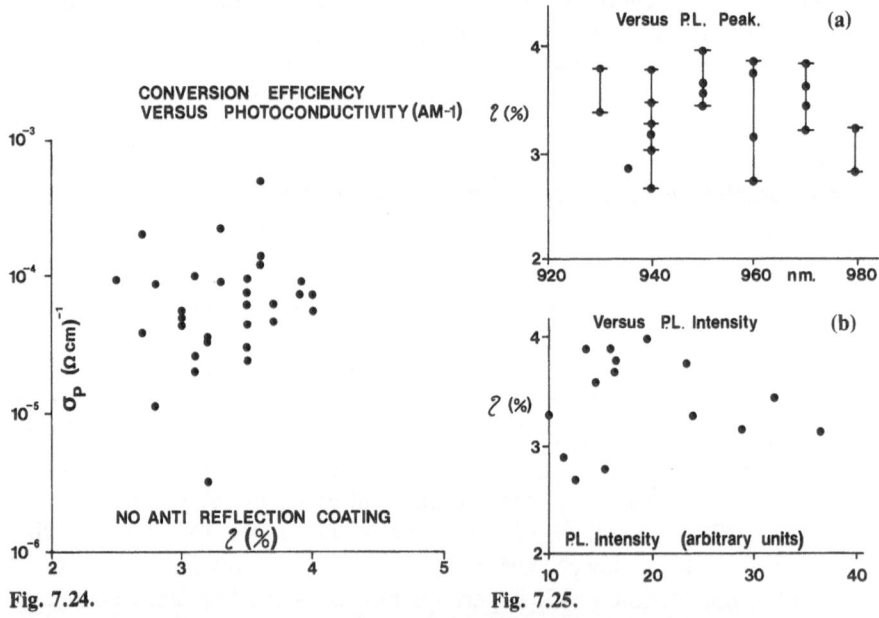

Fig. 7.24.

Fig. 7.25.

Fig. 7.24. Conversion efficiency plotted against σ_p of the active layer [7.5]

Fig. 7.25 a, b. Conversion efficiency plotted against (a) PL peak and (b) PL intensity

In Fig. 7.25 we plot PL intensity and PL peak against device performance. Again, no clear-cut correlation seems to exist.

Device performance is controlled by a host of interdependent parameters such as hole mobility, recombination lifetime, bulk and surface states, etc., and therefore, it is not too surprising that the above-mentioned techniques fail to produce some correlation between device and material parameters.

7.8 *p-n* and *p-i-n* Photovoltaic Devices

In crystalline semiconductors such as Si, *p-n* junctions can be readily formed because of the high quality of the material involved. Attempts to fabricate a *p-n* junction involving a-Si:H alloys have been made [7.76] and have yielded diode quality factors of $n = 2$, which suggest large recombination currents and hence would appear to represent a limitation to the device performance. To understand the reason for the presence of large recombination currents, we consider some aspects of the doping processes.

7.8.1 Influence of B-Doping

As shown in Fig. 7.3, the conductivity of the a-Si based alloys can be altered by orders of magnitude with the addition of B_2H_6 to the gas phase. However, this is accompanied by a decrease in the optical band gap [7.19]. A model based on chemical reasoning for the possible lack of conventional boron doping in a-Si based alloys has been proposed by *Ovshinsky* and *Adler* [7.77]; a model which is also appropriate to the above observations, and for similar results obtained using a-Si:H alloys is due to *Jan* et al. [7.78]. They suggested that a defect level grows in the mobility gap with increasing B incorporation such that the initial matrix element involved with the optical transition from E_v to E_c changes for the most heavily doped material from the defect level to E_c. A reason for this may be that for conventional doping to occur, B has to be tetrahedrally coordinated; since B is trivalent, it is energetically unfavorable for it to over-coordinate, unlike in crystalline Si where it is constrained to enter a fourfold configuration and hence provide an acceptor level. Further, this suggests that conventional *p*-type doping has not occurred in the a-Si based alloys but instead a new *p*-type alloy involving B has been synthesized.

7.8.2 Influence of P-Doping

The situation for *n*-type conductivity is also serious, but not that critical. Figure 7.3 shows that with the addition of P, the conductivity can be changed by orders of magnitude without any discernible change in the optical band gap [7.19]. The doping efficiency is envisioned to be small. The nonsubstitutional P enters sites other than fourfold coordination and seems to create further defect levels, as evidenced by the decrease in the photoluminescence signal [7.67].

The influence of phosphorus (P) doping on the transport properties of hydrogenated amorphous silicon (a-Si:H) has been interpreted in various ways: (i) transport in the conduction band at temperatures T above 200 K [7.79]; (ii) transport in a donor band [7.80] changing to transport in the conduction band above 400 K [7.81]; (iii) parallel path conduction through the donor and conduction bands [7.82]. In a recent study of Schottky-Barrier (SB) heights and photoconductivity by *Viktrovitch* et al. [7.83], strong evidence was presented for the existence of transport through a donor band lying 0.3 eV below' the conduction band minimum for sufficiently large P-doping.

To understand the effect of P-doping in the device configuration [7.84], we fabricated Schottky-barrier diodes by evaporating various metals onto 0.5 μm of a-Si:H, which was deposited onto 500 Å of a heavily doped n^+ film on a Mo coated glass substrate (kept at 300 °C) via the glow discharge in SiH$_4$ gas. The glow discharge deposition process was performed under dc, rf and audio frequency conditions using a premixed ratio of SiH$_4$ and PH$_3$ gases, with the different techniques producing similar device grade films as determined from supplementary field effect, photoconductivity, conductivity-temperature, x-ray and photoluminescence experiments. The Schottky-barrier metal employed was a Au:Pd alloy and the active layer was deposited using audio glow discharge frequencies (30 kHz). The SB structures were fabricated simultaneously with each layer that was independently analyzed. Hence, we have a good level of confidence in the observed correlations between the material and device properties.

The SB structures were studied by monitoring the dark current-voltage $J(V)$ characteristics as a function of temperature and, in particular, by considering the variation of the diode ideality factor n as a function of temperature and bias for different amounts of P contained in the active layer. The dark $J(V)$ characteristics, using an undoped layer, were very similar to those obtained for a-Si:F:H alloys, as shown in Fig. 7.12. n was in the range 1.05–1.1 when avoidable oxides [7.64] were removed and n increased to about 1.2 for an MIS-type structure as discussed earlier. *Tsai* et al. [7.85] have found this to be the case too and have shown that the presence of the insulator raises n to 1.2 or above. The deviation of n from unity is usually ascribed to an insulating layer and to excess currents that flow due to recombination, tunnelling and image force lowering of the barrier [7.71]. For example, in crystalline Si, currents due solely to recombination events that occur near midgap yield $n = 2$. We expect, therefore, that for a SB when $1 < n < 2$, the current can be written as a combination of diffusion/drift and recombination components [7.86]:

$$J = J_{0r} \exp(qV/2kT) + J_{0d} \exp(qV/kT) . \tag{7.19}$$

In the above, the prefactor J_{0d} is assumed to represent a drift/diffusion component expected for a low mobility solid. It is easily shown, using (7.19), that a straight line results over several orders of magnitude of current on a log $J(V)$ plot. Further, the values of n thereby obtained will depend on the relative fraction of the recombination current to the drift/diffusion current. Hence, as T is decreased and the recombination component becomes more important, we expect n to increase as shown in Fig. 7.13. This general trend is found in all our doped samples where recombination currents may be expected to be induced. Indeed, DLTS experiments have shown that the midgap density of states n a-Si:H film tends to increase with increased P-doping [7.87].

Significant changes in the log $J(V)$ characteristics occur as the P concentration within the active layer increases, as shown in Fig. 7.26 where we plot n versus T for different amounts of P introduced in the gas phase. Attempts were made to fit the data shown in Fig. 7.26 by considering a thermionic-field-emission tunnelling component [7.71]. Even when consideration was given to obtaining a realistic value of E_{max}, the field at the metal-semiconductor interface, by including an exponential density of gap states in our calculations, we were unable to fit the data in a quantitative fashion. However, when we included a second channel for the current flow [7.82] via an impurity band located 0.3 eV below E_c [7.83], very good quantitative agreement was obtained. Figure 7.27 shows a comparison between the data and our numerical calculations for one representative case.

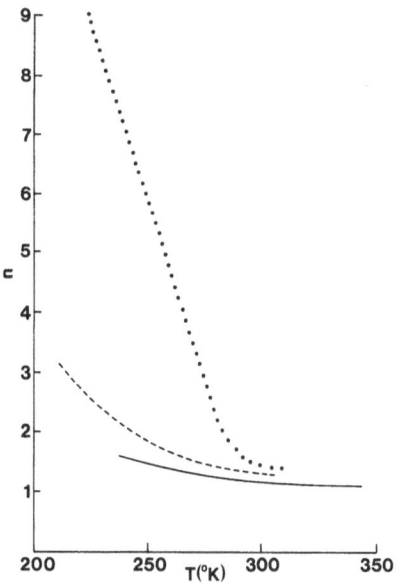

Fig. 7.26. n as a function of temperature for undoped (——), 1000 vppm PH$_3$ (---) and 2000 vppm PH$_3$ (···) introduced into the premix in the fabrication of the active layer for a-Si:H alloys. All data taken when $V_f = 0.2$ V [7.82]

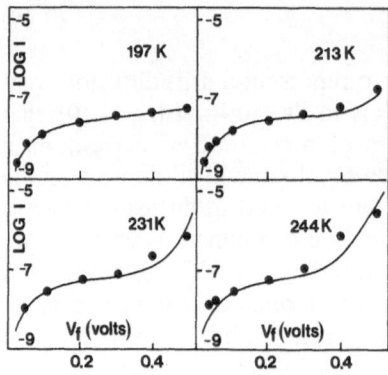

Fig. 7.27. Experimental (···) and calculated (—) results for a SB structure made with 2000 vppm PH₃ in the premix [7.82]

In a two-channel conduction model [7.82], we concur with the view that as the P concentration increases, a donor band forms through which hopping conduction σ_{hop} can compete with extended state conduction σ_{ext}; the conductivity is then given by

$$\sigma = \sigma_{ext} + \sigma_{hop} \, ,$$

i.e.,

$$\sigma = \sigma_{01} \exp[-(E_c - E_F)/kT]$$
$$+ \sigma_{02} \exp[-(E_D - E_F + \zeta)/kT] \, , \qquad (7.20)$$

where E_D is the energy at which the donor band is centered. Here

$$\sigma_{01} = e\mu_0 N(E_C)kTe^{\beta/k} \qquad [> 10^3 (\Omega \, cm)^{-1}]$$

and

$$\sigma_{02} = e\mu_{hop} N(E_D)kTe^{\beta/k} \qquad [< 10 (\Omega \, cm)^{-1}] \, ,$$

μ_0, μ_{hop} are the mobilities of electrons in the extended states and in the donor band, respectively, and $N(E_C)$, $N(E_D)$ are the densities of states at the conduction band edge and in the donor band. β is assumed to be ~2 × 10^{-4} eV K^{-1} [7.28]. ζ is the energy for the hopping process within the donor band. The current is then given by

$$I = I_D + I_R + I_{hop} \, , \qquad (7.21)$$

where the drift/diffusion plus recombination component can be written as

$$I_D + I_R = I_{01}[\exp - (q\Phi_{eb}/kT)] \exp(qV/nkT) \, . \qquad (7.22)$$

We determine the barrier height to be $\Phi_{eb} = 0.79$ eV by plotting J_0/T versus $1/T$. In agreement with the observations of *Viktorovich* et al. [7.83], the barrier heights are reduced from the undoped case by 0.2–0.3 eV for sufficient P doping.

For highly doped samples, we expect the hopping mode to be important at low temperatures and at low values of the forward bias. At low temperatures, few carriers are available to move over the barrier. If a parallel path exists at low forward voltage, such as tunnelling into the impurity band and hopping through it, transport will occur. However, as the forward bias increases, it reduces the field in the space-charge region and drives the bands towards a flat-band condition. Here more carriers are available to move over the barrier and fewer can tunnel into the impurity band. A transition is expected from the hopping to the drift/diffusion-recombination mode; this is clearly seen in the data and in our numerical model. In our model, little current is expected until a specific voltage is reached at which point, the current increases rapidly to a value corresponding to the hopping current. With a further increase in bias, this component saturates and eventually the drift/diffusion-recombination component dominates. We can then write for the hopping contribution

$$I_h = I_{hs}\exp(-q\zeta/kT) , \qquad (7.23)$$

where I_{hs} is the saturation current for the hopping process. By plotting the current at a fixed V_F (0.1 V) versus $1/T$, we obtain $\zeta \simeq 0.09$ eV, which agrees with the estimate in the two-channel model of *Jones* et al. [7.82]. Further, if we plot J_0/T of $1/T$, which corresponds to transport over the barrier, we find $\Phi_{eb} \simeq 1.05$ eV, which is the barrier height obtained for undoped films, where the hopping component is not present [7.69, 70].

7.8.3 The Effect of Doping on Devices

From the above results, the affect of doping on the photovoltaic characteristics becomes clear. The increase in the P-doping of the active layer leads to an increase in the density of states as suggested by *Crandall* [7.68] using a primary photocurrent technique. The effect of this is to increase the excess currents in the dark and hence results in a drop in V_{oc}. With an increase in the density of states, the depletion width is decreased which decreases the J_{sc} and thus the overall efficiency of the device suffers as confirmed by results of *Carlson* and *Wronski* [7.88].

The effect of B is much more serious since residual contamination would affect the material properties severely with the photoconductivity being reduced dramatically [7.34]. This has the effect of introducing a series resistor as well as reducing the depletion width and thus affecting the (FF) and J_{sc}, and hence the efficiency of the device.

With the large increase in the density of localized states in the n and p-type a-Si materials, high quality p-n junction formation becomes difficult as noted by *Spear* et al. [7.76]. To circumvent the above contamination problems, multichamber approach has been suggested for the fabrication of p-i-n cells [7.89].

7.8.4 p-i-n Type Devices

The p-i-n device structure is receiving the most attention at present [7.7–10]. In this type of device, the major carrier separation region should be at the p^+/i interface. Therefore, the optimum design should be for the light to enter from the p^+-side. However, as discussed earlier, the band gap of p-type a-Si:H is narrow and consequently the devices are expected to perform poorly towards the blue end of the AM 1 spectrum unless the thickness of the p-type layer is made sufficiently thin. To circumvent the above problem, the concept [7.90] of a p-type window material has been employed [7.9]. Significant developments have recently taken place: *Tawada* et al. [7.9] announced that using gas mixtures of SiH_4, CH_4 and B_2H_6, an amorphous alloy Si:C:B:H can be fabricated which is p-type and has a band gap between 1.76 eV and 2.2 eV, depending upon the ratio of the gas mixtures involved. Utilizing this as a heterojunction window in a cell configuration consisting of glass/SnO_2/p^+-a-Si:C:H/i-n^+-a-Si:H/Al, conversion efficiencies of 7.55% with J_{sc} = 13.45 mA cm^{-2}, V_{oc} = 0.909 V and (FF) = 0.617 have been reported [7.9]. The area of this device was only 0.033 cm^2. It would be interesting to see if this efficiency is retained for a larger area device.

Instead of the light entering the cell through the p-type layer, an alternative configuration has been chosen by some groups, the inverted p-i-n cell in which the light enters through the n^+ region. Device efficiencies of 6.9% using a-Si:H have been achieved in this way [7.10].

As indicated earlier, several groups have fabricated fluorinated silicon from different gas ratios and mixtures. *Carlson* and *Smith* [7.91] have attempted, from a device point of view, to investigate the effect of the addition of F_2 and SiF_4 to the SiH_4 gas in the rf glow discharge. They note a substantial drop in the conversion efficiency of their p-i-n cells with the addition of F, irrespective of its source, to the SiH_4 gas. From the SIMS profile, it is also evident that the level of oxygen contamination far exceeds the F concentration in the intrinsic layer. Therefore, it is not clear whether the decrease in conversion efficiency, which they attribute to the fluorine component, is in fact due to F or the increased contamination accompanying it.

Using F containing gases in the rf glow discharge, we have recently fabricated p-i-n junctions with conversion efficiencies as high as 7.5% at AM 2 incident illumination, which drops to 7.1% at AM 1 incident illumination. The decrease in efficiency is attributed to the resistive nature of one of the contacts. Full details of this work will appear elsewhere [7.6].

7.9 Stability

The stability of a-Si photovoltaic devices is a major concern since the viability of these types of devices depends on the oft-quoted 20 year lifetime.

MIS devices have generally been considered to be inherently unstable. One possible mechanism for the degradation in MIS devices is that oxygen can penetrate the thin top metallic contact [7.92] and, over a period of time, the oxide layer would widen with the consequence that V_{oc} and J_{sc} drop, as indicated in Fig. 7.19. This is indeed true in unencapsulated devices. In Fig. 7.28 we show the results [7.5] of a-Si : F : H MIS-type devices which were encapsulated by using a dip-coated Tedlar encapsulant. The device under load was subjected to continuous AM 1 illumination and over the time period investigated, no discernible change in the characteristics of the device were evident.

Stability data on a-Si : H *p-i-n* junctions has been presented by RCA [7.93] and the SANYO [7.89] groups indicating a long lifetime. Further, in some devices, reversible annealing effects have been observed [7.93]. From stability studies, *Staebler* et al. [7.93] concluded that the undoped photovoltaic layer is not changed by the action of the light but that there could be changes in the internal field and recombination rate due to the tapping of free carriers which could lead to a degradation.

The reversible effects noted in some a-Si type devices is somewhat reminiscent of the Ag contamination effects that have been observed in

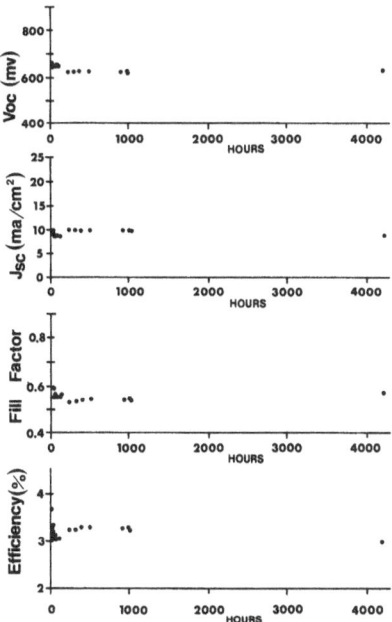

Fig. 7.28. Device parameters as a function of continuous illumination [7.5]

crystalline Si [7.94, 95]. The cell degradation under illumination has been attributed to a decrease in the minority carrier diffusion length. Upon annealing, recovery takes place. This metastable to stable change was attributed to Ag contamination.

7.10 Future Improvement in Device Efficiency

As stated above, one of the limitations to the device efficiency is the localized state density. A further decrease in the density of localized states would lead to a larger depletion width and an increase in J_{sc}. As shown in Fig. 7.8, any significant increase in J_{sc} would have to occur for a much greater depletion width.

Another important limiting factor is the low mobility of the holes, whose collection affects the short circuit current. A significant increase in the hole mobility would therefore lead to contributions made by the diffusion currents which would enhance the efficiency via an increase in J_{sc} and (FF).

Another aspect worth considering is the narrowing of the band gap of the active layer to a more optimum value of 1.5 eV. Attempts to do this with Ge [7.96, 97] have been accompanied by an increase in the density of states; this has the counteracting effect of decreasing the depletion width. Therefore, the gain made by the enhanced absorption is usually lost by the decrease in the depletion width.

Further, as pointed out by *Jackson* [7.98], tandem structures offer a tantalizing prospect for high efficiency devices. However, for this to be possible, problems regarding defect state densities in amorphous semiconductors, such as a-Ge, have to be reduced by orders of magnitude before this concept could be realized. Results using a-Si : H and a-Si : Ge : H in a series-connected tandem cell have given 5.9% conversion efficiency [7.99]. The problem regarding Ge still remains since the efficiency reported for the tandem cell is still lower than the 7.55% device reported for a single device [7.9].

7.11 Conclusion

There has been considerable progress in the last couple of years in the improvement of efficiency and in the increase in size of these types of devices. With a sustained effort over the next few years, we should see a-Si alloys emerge as a useful mechanism for power generation.

Acknowledgements. I would like to thank our group at Energy Conversion Devices, Inc.: L. Christian, W. Czubatyj, M. Hack, A. Hafez, R. Himmler, J. McGill, M. Shaw, R. Singh, R. Stiers, L. Taylor, J. Yang and D. Wickerham for their splendid efforts over the past few years, and ARCO for their support. Special thanks are due to Prof. M. Shur for allowing me to include some of his unpublished theoretical work; I also appreciate the stimulating discussions I have had with him over the course of this work. Finally, I would like to thank S. R. Ovshinsky for his involvement and enthusiasm for this project.

References

7.1   R. C. Chittick, J. H. Alexander, H. F. Sterling: J. Electrochem. Soc. **116**, 77 (1969)
7.2   See, for instance, W. E. Spear: Adv. in Phys. **26**, 811 (1977)
7.3   D. E. Carlson, C. R. Wronski: J. Elect. Mat. **6**, 95 (1977)
7.4   A. Madan, S. R. Ovshinsky, E. Benn: Phil. Mag. **40**, 259 (1978)
7.5   A. Madan, W. Czybatyj, J. Yang, J. McGill, S. R. Ovshinsky: J. Physique **42**, C4–463 (1981)
7.6   J. Yang, W. Czybatyj, A. Madan: To be published
7.7   D. E. Carlson: Solar Energy Mat. **3**, 503 (1980)
7.8   Y. Uchida, H. Sakai, M. Nishiura, H. Haruki: Proc. 15th IEEE Photovoltaic Conference, Florida (1981) p. 922
7.9   Y. Tawada, M. Kondo, H. Okamoto, Y. Hamakawa: J. Physique **42**, C4–471 (1981)
7.10 Sanyo Electric Co., Solid State Technology (July, 1981) p. 22
7.11 H. Fritzsche, C. C. Tsai: Solid State Tech. **21**, 55 (1978)
7.12 W. Paul, A. Lewis, G. A. N. Connell, T. Moustakas: Solid State Commun. **20**, 969 (1976)
7.13 N. Sol, D. Kaplan, D. Dieumegard, D. Durbreuil: J. Non-Cryst. Solids **35 & 36**, 291 (1980)
7.14 M. Shur, W. Czubatyj, A. Madan: Solar Energy Mat. **2**, 349 (1980)
7.15 W. E. Spear, P. G. LeComber, A. J. Snell: Phil. Mag. B **38**, 303 (1978)
7.16 A. Madan, J. McGill, W. Czubatyj, J. Yang, S. R. Ovshinsky: Appl. Phys. Lett. **37**, 826 (1980).
7.17 M. Konogai, K. Nishihata, K. Komori, K. Takahashi: Proc. 15th IEEE Photovoltaic Conf., Florida (1981) p. 922
7.18 S. R. Ovshinsky, A. Madan: Nature **276**, 482 (1978)
7.19 A. Madan, S. R. Ovshinsky: J. Non-Cryst. Solids **35 & 36**, 731 (1980)
7.20 M. Konagai, K. Takahashi: Appl. Phys. Lett. **36**, 599 (1980)
7.21 Y. Nakagome, H. Matsumura, S. Furukawa: Japan J. Appl. Phys. **19**, 87 (1980)
7.22 T. Shimada, Y. Katayama, S. Horigome: Japan J. Appl. Phys. **19**, 265 (1980)
7.23 C. J. Fang, L. Ley, H. R. Shanks, K. J. Gruntz, M. Cardona: Phys. Rev. B **22**, 6140 (1980)
7.24 V. L. Delal, C. M. Fortmann, E. Eser: *Tetrahedrally Bonded Amorphous Semiconductors*, ed. by R. A. Street, D. K. Bieglesen, J. C. Knights (AIP, New York 1981) p. 15
7.25 J. E. Potts, E. M. Peterson, J. A. McMillan: To be published
7.26 H. Matsumura, Y. Nakagome, S. Furukawa: Appl. Phys. Lett. **36**, 439 (1980)
7.27 M. Janai, R. Weil, K. H. Levin, B. Pratt, R. Kalish, G. Braunstein, M. Teicher: J. Appl. Phys. **52**, 3622 (1981)
7.28 N. F. Mott, E. A. Davis: *Electronic Processes in Non-Crystalline Materials* (Claredon Press, London 1971)
7.29 N. F. Mott: Phil. Mag. **19**, 835 (1969)
7.30 W. E. Spear, P. G. LeComber: J. Non-Cryst. Solids **8–10**, 727 (1972)
7.31 A. Madan, P. G. LeComber, W. E. Spear: J. Non-Cryst. Solids **20**, 239 (1976)
7.32 A. Madan, P. G. LeComber: *Amorphous and Liquid Semiconductors*, ed. by W. E. Spear (CICL, Univ. of Edinburgh, 1977) p. 377
7.33 A. Madan: Solar Cells **2**, 277 (1980)
7.34 W. E. Spear, P. G. LeComber: Phil. Mag. **33**, 935 (1976)
7.35 R. Tsu, S. S. Chao, M. Izu, S. R. Ovshinsky: J. Physique **42**, C4–269 (1981)

7.36 A. Matsuda, S. Yamasaki, K. Nakagawa, H. Okushi, K. Tanaka, S. Iizima, M. Matsumura, H. Yamamoto: Japan J. Appl. Phys. **19**, L305 (1980)

7.37 S. Usui, M. Kukuchi: J. Non-Cryst. Solids **34**, 1 (1979)

7.38 T. Hamasaki, H. Kurato, M. Hirose, Y. Osaka: Appl. Phys. Lett. **37**, 1084 (1980)

7.39 G. Lucovsky: Solid State Commun. **29**, 571 (1979)

7.40 G. Lucovsky, R. J. Nemanich, J. C. Knights: Phys. Rev. B**19**, 2064 (1979)

7.41 T. D. Moustakas: J. Elect. Mat. **8**, 391 (1979)

7.42 R. Tsu, J. Hernandez: To be published

7.43 V. Q. Ho, T. Sugano: IEEE Trans. ED-**28**, 1061 (1981)

7.44 T. Sakurai, T. Sugano: Proc. of the Intern. Conf. on Physics of MOS Insulators (Perga-
mon, New York 1980) p. 241

7.45 W. W. Kruelher, R. D. Plaettner, M. Moeller, B. Rauscher, W. Stetter: J. Non-Cryst.
Solids **35 & 36**, 333 (1980)

7.46 R. D. Plaettner, W. W. Kruelher, B. Rauscher, W. Stetter, J. G. Grobmeir: 2nd E. C.
Photovoltaic Solar Energy Conf., Berlin (1979) p. 860

7.47 A. S. Grove: *Physics and Technology of Semiconductor Devices* (Wiley, New York 1967)

7.48 P. G. LeComber, A. Madan, W. E. Spear: J. Non-Cryst. Solids **11**, 219 (1972)

7.49 A. R. Moore: Appl. Phys. Lett. **31**, 762 (1977)

7.50 L. Onsager: Phys. Rev. **54**, 558 (1938)

7.51 D. M. Pai, R. C. Enck: Phys. Rev. B**11**, 5163 (1975)

7.52 A. Madan, W. Czubatyj, D. Adler, M. Silver: Phil. Mag. **42**, 257 (1980)

7.53 R. S. Crandall, R. Williams, B. F. Tompkins: J. Appl. Phys. **50**, 5506 (1979)

7.54 C. R. Wronski, B. Abeles, G. D. Cody, D. L. Morel, T. Tiedje: 14th IEEE Photovoltaic
Specialist Conf., San Diego (1980) p. 1057

7.55 H. Okamoto, T. Yamaguchi, Y. Hamakawa: J. Phys. Soc. Japan **49**, 1213 (1980)

7.56 J. Mort, I. Chen, S. Grammatica, M. Morgan: *Tetrahedrally Bonded Amorphous Semiconductors*, ed. by R. A. Street, D. K. Biegksen, J. C. Knights (AIP, New York 1981) p. 283

7.57 Y. Hamakawa, H. Okamoto, Y. Nitta: 14th IEEE Photovoltaic Specialist Conf., San
Diego (1980) p. 1074

7.58 M. Shur, W. Czubatyj, J. McGill, J. Yang, A. Madan: Proc. of Intern. Electron Device
Meeting, Washington (1980) p. 545

7.59 W. Czubatyj, M. Shur, K. Ng, A. Madan: 14th IEEE Photovoltaic Specialist Conf., San
Diego (1980) p. 1214

7.60 J. Dresner, D. J. Szostak, B. Goldstein: Appl. Phys. Lett. **38**, 998 (1981)

7.61 D. L. Staebler: J. Non-Cryst. Solids **35 & 36**, 387 (1980)

7.62 A. R. Moore: Appl. Phys. Lett. **37**, 327 (1980)

7.63 A. J. Snell, W. E. Spear, P. G. LeComber: Proc. 12th Conf. Solid State Devices, Tokyo
(1980); to be published

7.64 A. Madan, J. McGill, S. R. Ovshinsky, W. Czubatyj, J. Yang, M. Shur: SPIE Trans. **284**,
26 (1980)

7.65 H. C. Card, E. H. Rhoderick: J. Phys. D. **4**, 1589 (1971)

7.66 A. Madan: Trieste Semiconductor Symposium, "Amorphous Silicon and Its Applications",
Trieste (1980); unpublished

7.67 W. Rehm, D. Engemann, R. Fischer, J. Stuke: Proc. 13th Intern. Conf. on the Physics of
Semiconductors, ed. by F. G. Fumi (Rome, 1976) p. 525

7.68 R. S. Crandall: Phys. Rev. Lett. **44**, 749 (1980)

7.69 C. R. Wronski, D. E. Carlson, R. E. Daniel: Appl. Phys. Lett. **29**, 602 (1976)

7.70 A. DeNeufville, N. H. Brodsky: J. Appl. Phys. **50**, 1414 (1979)

7.71 See, e.g., E. H. Rhoderick: *Metal-Semiconductor Contacts* (Clarendon Press, Oxford
1979)

7.72 R. Singh, K. Rajkanan: J. Vac. Sci. Tech. **17**, 376 (1980)

7.73 M. A. Green, F. D. King, J. Shewchun: Solid State Electr. **17**, 551 (1974)

7.74 D. Gutkowicz-Krusin, C. R. Wronski, T. Tiedje: Appl. Phys. Lett. **38**, 87 (1981)

7.75 J. I. Pankove: Solar Cells; to be published

7.76 W. E. Spear, P. G. LeComber, S. Kinmond, M. H. Brodsky: Appl. Phys. Lett. **28**, 105 (1976)
7.77 S. R. Ovshinsky, D. Adler: Contemp. Phys. **19**, 109 (1978)
7.78 Z. H. Jan, R. H. Bube, J. C. Knights: J. Appl. Phys. **51**, 3278 (1980)
7.79 See, e.g., H. Overhof, W. Beyer: J. Non-Cryst. Solids **35** & **36**, 375 (1980)
7.80 G. H. Dohler: Phys. Rev. B **19**, 2083 (1979)
7.81 See, e.g., D. G. Ast, M. H. Brodsky: Phil. Mag. B **41**, 273 (1980)
7.82 D. I. Jones, P. G. LeComber, W. E. Spear: Phil. Mag. **36**, 541 (1977)
7.83 P. Viktvovitch, G. Moddel, W. Paul: *Tetrahedrally Bonded Amorphous Semiconductors*, ed. by R. A. Street, D. K. Bieglesen, J. C. Knights (AIP, New York 1981) p. 186
7.84 A. Madan, W. Czubatyj, J. Yang, M. Shur, M. Shaw: To be published
7.85 C. C. Tsai, R. J. Nemanich, M. J. Thompson: *Tetrahedrally Bonded Amorphous Semiconductors*, ed. by R. A. Street, D. K. Bieglesen, J. C. Knights (AIP, New York 1981) p. 312
7.86 A. Y. C. Yu, E. H. Snow: J. Appl. Phys. **39**, 3008 (1968)
7.87 J. D. Cohen: Bull. Amer. Phys. Soc. **26**, 330 (1981)
7.88 D. E. Carlson, C. R. Wronski: *Amorphous Semiconductors*, ed. by M. H. Brodsky, Topics Appl. Phys., Vol. 36 (Springer, Berlin, Heidelberg, New York 1979) p. 287
7.89 Y. Kuwano, M. Ohnishi: J. Physique **42**, C4–1155 (1981)
7.90 D. E. Carlson: J. Non-Cryst. Solids **35** & **36**, 707 (1980)
7.91 D. E. Carlson, R. W. Smith: Proc. 15th IEEE Photovoltaic Conference, Florida (1981) p. 694
7.92 J. P. Ponpon, P. Siffert: J. Appl. Phys. **49**, 6004 (1978)
7.93 D. L. Staebler, R. S. Crandall, R. Williams: Proc. 15th IEEE Photovoltaic Conf., Florida (1981) p. 249
7.94 V. W. Weizer, H. W. Brandhorst, J. D. Broder, R. E. Hart, T. H. Lamneck: J. Appl. Phys. **50**, 4443 (1979)
7.95 L. J. Cheng, G. B. Turner, R. G. Downing, E. Y. Wang, C. H. Seaman: Proc. of the IEEE Specialist Conf. on Photovoltaics, Washington (1978) p. 1333
7.96 Y. Marfaing: Proc. of the 2nd E.C. Photovoltaic Conf., Berlin (1979) p. 287
7.97 J. Chevalier, H. Wieder, A. Onton, C. R. Guaranreri: Solid State Commun. **24**, 867 (1977)
7.98 E. D. Jackson: Proc. Trans. of the Use of Solar Energy, Arizona (1955) p. 122
7.99 G. Nakamura, K. Sato, H. Kondo, Y. Yukimoto, K. Shirahata: J. Physique **42**, C4–483 (1981)

Subject Index

Amorphous Semiconductors

Editor: **M.H.Brodsky**

1979. 181 figures, 5 tables. XVI, 337 pages
(Topics in Applied Physics, Volume 36)
ISBN 3-540-09496-2

Contents: *M.H.Brodsky:* Introduction. –
B.Kramer, D.Weaire: Theory of Electronic
States in Amorphous Semiconductors. –
E.A.Davis: States in the Gap and Defects
in Amorphous Semiconductors. –
G.A.N.Connell: Optical Properties of
Amorphous Semiconductors. – *P.Nagels:*
Electronic Transport in Amorphous Semicon-
ductors. – *R.Fischer:* Luminescence in
Amorphous Semiconductors. – *I.Solomon:*
Spin Effects in Amorphous Semiconductors. –
G.Lucovsky, T.M.Hayes: Short-Range Order
in Amorphous Semiconductors. –
P.G.LeComber, W.E.Spear: Doped
Amorphous Semiconductors. – *D.E.Carlson,
C.R.Wronski:* Amorphous Silicon Solar Cells.

Amorphous Solids

Low-Temperature Properties

Editor: **W.A.Phillips**

1981. 72 figures. X, 167 pages
(Topics in Current Physics, Volume 24)
ISBN 3-540-10330-9

Contents: *W.A.Phillips:* Introduction. –
D.L.Weaire: The Vibrational Density of States
of Amorphous Semiconductors. – *R.O.Pohl:*
Low Temperature Specific Heat of Glasses. –
W.A.Phillips: The Thermal Expansion of
Glasses. – *A.C.Anderson:* Thermal Conduc-
tivity. – *S.Hunklinger, M.v.Schickfus:* Acoustic
and Dielectric Properties of Glasses at Low
Temperatures. – *B.Golding, J.E.Graebner:*
Relaxation Times of Tunneling Systems in
Glasses. – *J.Jäckle:* Low Frequency Raman
Scattering in Glasses.

Springer-Verlag
Belin
Heidelberg
New York
Tokyo

Fundamental Physics of Amorphous Semiconductors

Proceedings of the Kyoto Summer Institute,
Kyoto, Japan, September 8–11, 1980
Editor: **F.Yonezawa**

1981. 91 figures. VIII, 181 pages
ISBN 3-540-10634-0

Contents: What are Non-Crystalline Semiconductors. – Defects in Covalent Amorphous
Semiconductors. – Surface Effects and Transport Properties in Thin Films of Hydrogenated Silicon. – The Past, Present and
Future of Amorphous Silicon. – Doping and
the Density of States of Amorphous Silicon. –
The Effect of Hydrogen and Other Additives
on the Electronic Properties of Amorphous
Silicon. – New Insights on Amorphous Semi-
conductors from Studies of Hydrogenated
aGe, a-Si, a-Si$_{1-x}$Ge$_x$ and a-GaAs. – Chemical
Bonding of Alloy Atoms in Amorphous Sili-
con. – Photo-Induced Phenomena in
Amorphous Semiconductors. – Theory of
Electronic Properties of Amorphous Semicon-
ductors. – Some Problems of the Electron
Theory of Disordered Semiconductors. – The
Anderson Localisation Problem. – Summary
Talk. – Seminars Given During the KSI '80. –
Photograph of the Participants of the KSI '80.
– List of Participants.

Topological Disorder in Condensed Matter

Proceedings of the Fifth Taniguchi
International Symposium, Shimoda, Japan,
November 2–5, 1982
Editors: **F.Yonezawa, T.Ninomiya**

1983. 158 figures. XII, 253 pages
(Springer Series in Solid-State Sciences,
Volume 46)
ISBN 3-540-12663-5

Contents: Introduction. – Structural Aspects
of Topological Disorder. – Computer Simu-
lations and Analyses. – Some Aspects of
Elementary Excitations. – Statistical Properties in Two Dimensions. – Related Problems.
– Summary of the Conference. – Index of
Contributors.

Applied Physics A
Solids and Surfaces

Applied Physics A "Solids and Surfaces" is devoted to concise accounts of experimental and theoretical investigations that contribute new knowledge or understanding of phenomena, principles or methods of applied research.

Emphasis is placed on the following fields:

Solid-State Physics
Semiconductor Physics: **H. J. Queisser,** MPI Stuttgart
Amorphous Semiconductors: **M. H. Brodsky,** IBM Yorktown Heights
Magnetism (Materials, Phenomena): **H. P. J. Wijn,** Philips Eindhoven
Metals and Alloys, Solid-State Electron Microscopy: **S. Amelinckx,** Mol
Positron Annihilation: **P. Hautojärvi,** Espoo
Solid-State Ionics **W. Weppner,** MPI Stuttgart

Surface Science
Surface Analysis: **H. Ibach,** KFA Jülich
Surface Physics: **D. Mills,** UC Irvine
Chemisorption: **R. Gomer,** U. Chicago

Surface Engineering
Ion Implantation and Sputtering: **H. H. Andersen,** U. Aarhus
Laser Annealing: **G. Eckhardt,** Hughes Malibu
Integrated Optics, Fiber Optics, Acoustic Surface Waves: **R. Ulrich,** TU Hamburg

Coordinating Editor: **H. K. V. Lotsch,** Heidelberg

Special Features:
- Rapid publication (3–4 months)
- No page charges for concise reports
- 50 complimentary offprints
- Microform edition available

Subscription information and/or **sample copies** are available from your bookseller or directly from Springer-Verlag, Journal Promotion Dept., P.O.Box 105280, D-6900 Heidelberg, FRG

Springer-Verlag
Berlin
Heidelberg
New York
Tokyo